D0234376

Environmental Planning

Environmental Planning: The Conservation and Development of Biophysical Resources

2nd Edition

PAUL SELMAN

London · Thousand Oaks · New Delhi

First published 2000

SAGE Publications Ltd
6 Bonhill Street
London EC2A 4PU

SAGE Publications Inc.
2455 Teller Road
Thousand Oaks, California 91320

SAGE Publications India Pvt Ltd
32, M-Block Market
Greater Kailash-I
New Delhi 110 048

British Library Cataloguing in Publication Data
A catalogue record for this book is available from the British Library

ISBN 0-7619-6459-2
ISBN 0-7619-6460-6 (pbk)

Library of Congress catalog card number available from the publisher

Typeset by Dorwyn Ltd, Rowlands Castle, Hants
Printed and bound in Great Britain by Athenaeum Press, Gateshead

Contents

Acknowledgements

The author and publishers would like to thank the following publishers for kindly giving permission for the use of copyright material:

Academic Press Ltd for Box 3.5
Edward Arnold for Box 6.5
Basic Books for Table 1.1
CAB International for Table 2.7
Cambridge University Press for Figures 1.2 and 2.1, and Table 7.1
Carfax Publishing Ltd for Figure 2.6
Chapman & Hall Ltd for Table 7.1
The Commission of the European Communities for Boxes 1.2 and 4.14
The Countryside Commission for Figure 5.5 and Boxes 3.11, 4.12, 7.5
The Countryside Commission and English Nature for Boxes 5.4 and 5.5
The Countryside Commission for Scotland for Table 5.2
Croner for Box 2.2
The Department of the Environment for Figure 3.2, Table 1.4, and Boxes 2.8, 2.9, 3.6, 3.10, 6.3, 7.14 and 7.15
DETR for Figure 5.7 and Box 1.3
Elsevier Science Ltd for Figure 6.2, and Boxes 5.2, 5.3, 7.1 and 7.7
The Environment Agency for Box 5.6
Environment Canada for Box 7.8
The Farming and Wildlife Advisory Group for Table 5.1
The Forestry Commission for Figure 5.1 and Box 4.6
Geobooks for Box 4.15
The Geographical Association for Box 4.1
The Highlands and Islands Development Board for Box 5.8
HMSO for Tables 4.3 and 6.1, and Boxes 3.8 and 4.2
Hodder and Stoughton Ltd for Figure 5.6 and Table 5.1
The Institute for Terrestrial Ecology for Figure 5.4
The International Union for the Conservation of Nature and Natural Resources for Box 3.9
Longman Group for Figures 2.8 and 2.9, and Table 5.3
The Macaulay Land Use Research Institute for Box 7.2
Macmillan for Box 3.3
McGraw-Hill Ltd. for Figures 1.1 and 2.7, and Box 2.4
The Ministry of Agriculture, Fisheries and Food for Box 5.1
Routledge Ltd for Figure 5.2
RSA for Box 6.4
The Royal Society for the Protection of Birds for Box 5.6
UCL Press for Box 7.4
The United Nations Conference on Environment and Development for Box 1.1
The United Nations Environment Program for Table 7.5
University of Strathclyde for Table 1.2
John Wiley and Sons Ltd for Figure 5.3
The author for Box 4.4, from Munton (1990)

Every effort has been made to trace the copyright holders but if any have been inadvertently overlooked the publishers will be pleased to make the necessary arrangement at the first opportunity.

List of Abbreviations

AAPS   Arable Areas Payments Scheme
AEAM   adaptive environmental assessment and management
ALARA   as low as reasonably achievable
ALARP   as low as reasonably practicable
ALC   Agricultural Land Classification
AQMA   air quality management area
ASNW   Ancient Semi-Natural Woodland
AsONB   Areas of Outstanding Natural Beauty

BAP   biodiversity action plan
BAT   best available technology
BATNEEC   best available technology not entailing excessive cost
BGS   British Geological Survey
BPEO   best practicable environmental option
BPM   best practicable means

CAP   Common Agricultural Policy
CBA   cost benefit analysis
CC   Countryside Commission
CCW   Countryside Council for Wales
CEC   Commission of the European Communities
CFCs   chlorofluorocarbons
CGT   Capital Gains Tax
CIS   Countryside Information System
CMP   catchment management plan
CoAg   Countryside Agency
CPRE   Council for the Protection of Rural England
CPS   Country Premium Scheme
CS   Countryside Stewardship
CSEC   Centre for the Study of Environmental Change
CV   contingent valuation
CZM   coastal zone management

DBH   diameter breast height
DETR   Department of the Environment, Transport and the
            Regions
DoE   Department of the Environment
DoT   Department of Transport

DSM	demand side management
DTI	Department of Trade and Industry
EA	Environment Agency/environmental assessment
EAL	environmental assessment level
EC	European Community
EECONET	European Ecological Network
EFILWC	European Foundation for the Improvement of Living and Working Conditions
EH	English Heritage
EIA	environmental impact assessment
EIS	environmental impact statement
ELMS	environmental land management schemes
EMAS	European Eco-Management and Audit Scheme
EMS	environmental management programme
EN	English Nature
ENGO	environmental nongovernmental organisation
EPA	Environmental Protection Act 1990
EQO	Environmental Quality Objective
EQS	environmental quality standard
ES	environmental statement
ESA	Environmentally Sensitive Area
ESCTC	European Sustainable Cities and Towns Campaign
EU	European Union
EUEGUE	European Union Expert Group on the Urban Environment
FAO	Food and Agriculture Organisation
FCGS	Farm Conservation Grant Scheme
FDC	Flood Defence Committee
FNR	Forest Nature Reserve
FoE	Friends of the Earth
FWAG	Farming and Wildlife Advisory Group
FWPS	Farm Woodland Premium Scheme
GATT	General Agreement on Tariffs and Trade
GCM	general circulation model
GDO	General Development Order
GDPO	General Development Procedures Order
GEMS	Global Environment Monitoring System
GIS	geographic information system
GM	gross margin
GPDO	General Permitted Development Order
GPV	gross present value
GRID	Global Resource Information Database

HIE	Highlands and Islands Enterprise (formerly HIDB, Highlands and Islands Development Board)
HOV	high occupancy vehicle
HTML	Hypertext Markup Language
IACS	Integrated Administration and Control System
ICIDI	Independent Commission on International Development Issues
ICLEI	International Council for Local Environmental Initiatives
ICZM	integrated coastal zone management
IDA	Improvement and Development Agency (formerly LGMB – Local Government Management Board)
IDO	Interim Development Order
IEEM	Institute of Ecology and Environmental Management
IEI	integrated environmental index
IEM	integrated environmental management
IFS	Indicative Forestry Strategy
IHT	Inheritance Tax
INCA	Industrial Nature Conservation Area
IPC	integrated pollution control
IPCC	Intergovernmental Panel on Climate Change
IRD	integrated rural development
ITE	Institute of Terrestrial Ecology
ITTO	International Tropical Timber Organisation
IUCN(NR)	International Union for the Conservation of Nature and Natural Resources
JNCC	Joint Nature Conservation Committee
LA	Local Agenda/local authority
LAPC	local air pollution control
LBAP	local biodiversity action plan
LEAP	Local Environment Agency Plan
LFA	Less Favoured Area
LGMB	Local Government Management Board
LUC	Land Use Consultants
LUCC	Land Use Capability Classification
LUPU	Land Use Planning Unit
MA	Management Agreement
MAB	Man and the Biosphere programme
MADM	multiple attributes decision-making
MAFF	Ministry of Agriculture, Fisheries and Food
MCA	mineral consultation area/monetary compensation amount
MLP	mineral local plan
MNR	Marine Nature Reserve

MODM	multiple objectives decision-making
MPA	Mineral Planning Authority
MPG	Mineral Planning Guidance
NCC	Nature Conservancy Council
NELUP	Natural Environmental Research Council/Economic and Social Research Council Land Use Programme
NEPA	US National Environmental Policy Act 1970
NFFO	Non-Fossil Fuel Obligation
NFI	net farm income
NGO	nongovernmental organisation
NIMBY	not in my backyard
NNPA	Northumberland National Park Authority
NNR	National Nature Reserve
NPA	National Park Authority
NPC	National Park Committee
NPP	National Park Plan
NPPG	National Planning Policy Guidance
NPV	net present value
NRA	National Rivers Authority
NSA	National Scenic Area/Nitrate Sensitive Area
NSDS	national sustainable development strategies
NVC	National Vegetation Classification
NVZ	Nitrate Vulnerable Zone
OECD	Organisation for Economic Co-operation and Development
PAHs	polyaromatic hydrocarbons
PAN	Planning Advice Note
PCBs	polychlorinated biphenyls
PDO	potentially damaging operation
PEC	predicted environmental concentration
PPC	production possibility curve
PPG(N)	Planning Policy Guidance (Note)
PRA	participatory rural appraisal
PVCu	polyvinyl chloride (unplasticised)
RAC	Regional Advisory Committee
RAWP	Regional Aggregates Working Party
RCEP	Royal Commission on Environmental Pollution
RDA	regional development agency
RRA	rapid rural appraisal
RSPB	Royal Society for the Protection of Birds
RTPI	Royal Town Planning Institute
SAC	Special Area of Conservation

SDD	Scottish Development Department
SEA	strategic environmental assessment
SEPA	Scottish Environmental Protection Agency
SLOSS	single large or several small
SMR	Sites and Monuments Record
SNH	Scottish Natural Heritage
SoE	State of Environment
SofS	Secretary of State
SPA	Special Protection Area
SSSI	Site of Special Scientific Interest
TAN(W)	Technical Advice Note (Wales)
TCP	Town and Country Planning
TDR	transfer of development rights
TNC	transnational corporation
TPO	Tree Preservation Order
UCO	Use Classes Order
UN	United Nations
UNCED	United Nations Conference on Environment and Development
UNEP	United Nations Environment Program
uPVC	(unplasticised) polyvinyl chloride
URAA	Uruguay Round Agricultural Agreement
URL	uniform resource locator
UV-B	ultraviolet-B
UWE	University of West of England
VB	voluntary body
VOCs	volatile organic compounds
WCA	Wildlife and Countryside Act 1981
WCED	World Commission on Environment and Development
WES	Wildlife Enhancement Scheme
WGS	Woodland Grant Scheme
WHO	World Health Organisation
WIG	Woodland Improvement Grant
WLP	waste local plan
WMO	World Meteorological Organisation
WO	Welsh Office
WOAD	Welsh Office Agriculture Department
WSCC	West Sussex County Council
WSECS	West Sussex Environmental Capacity Study
WTA	willingness to accept
WTP	willingness to pay
WWF	World Wide Fund for Nature
WWW	World Wide Web

1 What is Environmental Planning?

Introduction

Environmental planning is concerned with society's collective stewardship of the earth's resources. The 'environment' refers to the physical and biological systems which provide our basic life support, and which contribute to our psychological well-being. 'Planning' is a generic activity of purposeful anticipation of, and provision for, the future. More specifically, though, it is taken to comprise the actions taken by public sector organisations to regulate land and water use on behalf of society, often through town and country planning legislation. Historically, humans, especially those living in industrial societies, have tended to live by plundering and degrading the earth's resources. The emphasis now, however, has shifted to one of 'sustainable development', in which resources are used in ways and at levels which permit their recovery and regeneration. We have still nowhere near reached a state in which we are in equilibrium with the earth's capacity for renewal; in order to attain this state in the future, we must be prepared to cross a 'sustainability transition'. The ultimate role of environmental planning is to help us achieve this transition.

Both the terms 'environment' and 'planning' are capable of expansion almost to the point of meaninglessness. The environment normally refers to the air, water, soil and rock resources at or close to the earth's surface. Some environmental scientists, however, study the deep earth's interior or the upper atmosphere and beyond. Many social scientists argue that studying the environment only becomes meaningful if we also consider the human users of its resources, together with their economic systems and ethical values which drive and restrain usage. Similarly, the activity of planning exists at all levels, from the individual person to the transnational corporation or global institution, though often we focus on land-use planning at the relatively local level, or economic planning at the national or regional scale. It is, therefore, difficult to set boundaries on a text on environmental planning, and its scope is somewhat arbitrary. In this case, the main emphasis is on the planning and management of land and water use in industrial (even postindustrial) society and its relationship to near-surface natural environmental systems. However, this is a difficult and artificial distinction to maintain. The boundaries must be permeable enough to embrace the economic principles underpinning natural resource transfers, social perceptions of the environment, and the political and technological forces which shape environmental planning options.

Three spheres of environmental study are thus of interest. First, is the *biophysical* environment, comprising the earth's life-support system – air, minerals, soil and water – and the plant, insect and animal matter which these sustain. Secondly, is the *socioeconomic* environment, composed of human social groupings, their cultural activities and the economic processes by which they are inter-related. Finally, some professions are concerned with the *built* environment, consisting of the buildings in which we live and work, and their attributes such as safety, stability, light, warmth, sanitation, water supply, waste disposal and architectural aesthetics. This diversity of interest is unavoidable and is, indeed, one of the principal reasons why the environment has emerged as a central issue of political and popular concern. When ecological issues are addressed in isolation, public attention and awareness tend to wax and wane. It has only been with the more mature realisation that natural resources, social justice, economic sustainability and quality of life are inextricably inter-related that the environment has attained an enduring status as a major focus of political and popular concern.

Also, whilst we have noted that the term 'planning' does not imply a particular professional affiliation, nevertheless there is a common supposition that *town planners* will have a pivotal role to play in environmental protection. Indeed, much of the historical inspiration for town planning has drawn upon the notions of ecology (cf. Patrick Geddes) and a balance in the relationship between town and country (e.g. Ebenezer Howard). Moreover, in spite of the gaps in environmental legislation, many powers do exist which enable planners to safeguard endangered resources, improve degraded sites and create new ecological assets. Planners may also usefully bring their skills of mediation and negotiation to bear on the integration of environmental resource use, which typically has been poorly co-ordinated. Yet, notwithstanding these observations, planning remains a generic activity, and the scope of this book extends beyond the pale of the town planner.

Issues of Environmental Concern

The environment is both a scientific entity and a cultural concept. Although the scientific study of the natural environment has been a fascination from around the seventeenth century and the study of cosmologies from time immemorial, the general concern for and familiarity with environmental issues has arisen from a much more recent awareness of human impacts. Local environmental effects have been a matter of concern for many centuries, though the first significant awareness probably coincided with the Industrial Revolution. From the 1960s there has been a growing acknowledgement that human activities do not only alter local environmental quality but may also destabilise global systems.

Alongside this recognition of human impact has been an awareness of natural and induced hazard. Many people live in the path of natural hazards, such as floods, earthquakes, tornadoes and drought. To many people, the environment is perceived as hostile, and our love for the countryside has grown broadly in parallel with our ability to control and regulate nature's capricious side. Various factors have tended to make us more conscious of the need to protect ourselves against hazard and risk. Equally, there has been a growing awareness that many risks have been heightened by unwise development, such as siting settlements too close to flood zones. More widely still, industrial processes have disseminated a range of additional potential risks into the environment which need to be minimised or eliminated.

There is relatively close agreement over the main impacts presently affecting the world's biophysical resources (Table 1.1). These broadly link the growth of human population and its associated settlements and economies with the quality of air, earth and water, and the effects of urban and industrial waste products on various receptors. In many countries, populations have effectively stopped increasing for many reasons, and are said to have crossed the 'demographic transition'. However, even in these places, populations still exert an ever-increasing demand on natural resources due to their continuously rising expectations. Given that we cannot continue to plunder the environment indefinitely, we must find ways of addressing the reasons why even numerically static populations exert increasing pressures. Stabilising or even reversing these demands, once more, involves crossing the 'sustainability transition' (O'Riordan, 1997).

Until the 1990s, the notion of the environment tended to be restricted to abiotic and biotic systems, perhaps together with parts of the urban fabric. Latterly, partly as a result of increased understanding about the causes and consequences of environmental change, and partly to broaden the relevance of environmental issues to people's everyday lives, the subject has tended to incorporate a much wider set of concerns. These broadly relate to the notion of 'quality of life' and suggest that 'green' issues cannot be tackled effectively in isolation from people's lifeworlds. Ultimately, limits must be placed on this expansion of the 'green agenda', or it will become unmanageable and impossibly diffuse. However, judicious inclusion of 'quality of life' variables seems both a desirable and a necessary extension to the conventional scientific framework (Table 1.2).

Environmental planners are constantly working within the context of policy agendas framed at various tiers in the public and private sector hierarchies. The spatial focus of these ranges from the global to the parochial. Environmental planning raises international issues not only because many environmental problems are transboundary and thus require concerted action, but also because most countries require a degree of external

Environmental Planning

Table 1.1   The main 'environmental' agendas

Population and health

- growth of world population to 8.2 billion by 2025
- accelerating population growth in Africa
- wide disparities in health standards worldwide
- water-related diseases in developing countries
- acute and chronic human exposure to pesticides
- life expectancy at birth
- children's health and rights

Human settlements

- rapid growth of urban populations and urban areas
- waste generation from cities
- environmental, social and political pressures engendered by megacities
- inexorable growth in motor vehicle trips and traffic congestion
- pollution – air, water, toxic, nuclear

Food and agriculture

- despite increasing world food production, levels of hunger and malnutrition are growing
- growth in pesticide-resistant pests
- decreased soil fertility
- issues associated with genetically modified organisms

Forests and farmlands

- clearance of tropical moist forests
- monocultural reforestation in temperate zones, often with non-native softwoods
- soil erosion following logging of forests
- degradation of rangelands
- conversion of farmland to other uses

Wildlife and habitats

- species losses with destruction of large areas of wildlife habitat
- illegal trade in endangered and rare species

Energy

- economic dependence on fossil fuels and nuclear power
- production of 'greenhouse' gases and nuclear wastes
- shortage of fuelwood in developing countries

Freshwater

- very uneven distribution of water availability between countries
- salinisation of water supplies in some arid countries
- pollution from point and nonpoint sources
- destruction and drainage of wetlands
- depletion and pollution of groundwater supplies

Oceans and coasts

- overexploitation of, and declining catches in, some fisheries
- pollution from ocean dumping, oil spills, land runoff and atmospheric deposition
- coastal habitat destruction
- rising sea levels and increased flooding of low-lying areas

Atmosphere and climate

- air pollution
- global warming
- ozone depletion
- interactions between pollutants in the atmosphere

Source: Derived from information in World Resources Institute, 1988 et seq.

Table 1.2 'Quality of life' issues in order of
perceived importance by the 'average' person

- Violent crime rates
- local health-care provision
- levels of nonviolent crime
- costs of living
- education provision
- pollution levels
- employment prospects
- housing costs
- wage levels
- shopping facilities
- unemployment levels
- travel-to-work times
- scenic quality of area
- climate
- sports facilities
- leisure opportunities

Source: Rogerson, 1997.

pressure placed on them before they will agree to behave in accordance with international best practice (which they may not perceive immediately as being in their own interest). Equally, most individuals need to be persuaded, and even compelled, to behave in accordance with the principles of sustainability.

At the global scale, a significant departure point was the Stockholm Conference of 1972, which reflected growing concerns about the unsustainability of industrial and urban trends, especially against a backdrop of continuing war in Vietnam and its associated 'ecocide'. The general climate of international concern was reinforced by a publication of the Club of Rome, called *Limits to Growth* (Meadows *et al.*, 1974), which applied the emergent possibilities of computer simulation to alternative trajectories for global population, pollution and resource use. Although its assumptions were oversimplified, and its mathematical basis was quite primitive by current standards, it nevertheless gave a considerable impetus to how the environmental debate was conducted. An especially important conservation effort was instigated in 1976, when the United Nations Educational, Social and Cultural Organisation instituted the Man and the Biosphere (MAB) programme to promote international co-operation in ecological research and establish a global system of biosphere reserves.

This initial formulation of the global debate, however, tended to polarise the interests of conservation and development, and imply that they were mutually exclusive. This resultant image of environmentalism – as being anti-industrial – inhibited its general acceptance, and it was important, therefore, that scientists offered a more positive alternative, setting conservation in a constructive light. Thus, in 1980, the World Conservation Strategy (IUCN *et al.*, 1980) popularised the notion of 'sustainable development', in which the conservation of living resources was presented as being integral to future human well-being. Development which did not respect the earth's finite capacities was deemed to be illusory, exploitative, short term and, ultimately, counterproductive to future human welfare.

Although this line of argument won many supporters, a continuing weakness was its apparent lack of immediacy for the global poor. Thus, it was easy for the conservation movement to be viewed as a rich nations' club trying to enforce stringent environmental controls on 'third world' countries, most of whose citizens experienced mediocre living standards or even struggled to survive. A parallel international concern was the growing disparity in fortunes between developed and developing countries, and the increasingly disastrous levels of indebtedness of the latter to the former. Consequently, the Brandt Commission (ICIDI, 1980) considered the possibility of levering 'massive transfers' of funds between the rich countries of the industrial or postindustrial 'north' and the poor countries of the industrialising or agrarian 'south'. The specific reforms of trading and lending agreements proposed by the Commission had only limited effect, but its rather contrived depiction of the earth's surface divided by a 'north–south' line was enormously influential.

These two strands – of environmental consciousness and concern for greater economic equity – have been drawn together in a more mature concept of sustainable development. The Brundtland Commission (WCED, 1987, p. 43) subsequently elevated this term to a widely accepted and potentially workable principle, by proposing the pursuit of 'development which meets the needs of the present without compromising the ability of future generations to meet their own needs'. Despite the many competing definitions of sustainable development, several of which seek to give greater emphasis to the intrinsic value of nature, the Brundtland wording has gained widespread official acceptance, not least because it is inherently capable of being expressed in terms of politically credible economic and policy principles. Subsequent to the Brundtland Commission, the 'second world conservation strategy' (IUCN *et al.*, 1991) and the Rio 'Earth Summit' (UNCED, 1992) have further consolidated the scope and purpose of sustainable development (Box 1.1). These agendas have also been reflected in the executive agencies of the United Nations, the work of the Organisation for Economic Co-operation and Development, and some nongovernmental (NGO) bodies, such as the International Council for Local Environmental Initiatives. In parallel with these framework documents, a body of international environmental law, based on protocols and treaties, has been emerging, especially in relation to transboundary effects and the global commons.

Cascading down a level, many of these global agendas are also reflected at the continental scale. For example, the EU has agreed a series of Environmental Action Plans (see Box 1.2), which have gradually moved from end-of-pipe approaches to anticipatory, preventative planning (CEC, 1992). This scale of action has been very effective in introducing a

Box 1.1 Agenda 21: key contents

- International co-operation
- combating poverty
- changing consumption patterns
- demographic dynamics and sustainability
- protecting and promoting health
- promoting sustainable human settlement development
- integrating environment and development in decision-making
- protecting the atmosphere
- integrated planning and management of land resources
- combating deforestation
- managing fragile ecosystems (e.g. mountains, deserts)
- promoting sustainable agriculture and rural development
- conservation of biological diversity
- environmentally sound management of biotechnology
- protection of the marine environment
- protection of freshwater resources
- hazard and waste management
- inclusion of women, children, youth and indigenous peoples
- role of NGOs, local authorities, workforces, businesses, science/ technology and farmers
- means of implementation (finance, capacity building, science, education, awareness-raising, legal instruments, institutions, information)

Source: UNCED, 1992.

Box 1.2 Principles of the Fifth EU Environmental Action Programme (modified)

- Sustainable development cannot be achieved by the EU alone, and so should consider international dimensions of policies for trade, economic and social development
- issues of population should be considered both in Europe and internationally
- integrity of natural systems – soil, water, air and biodiversity – should be preserved/restored
- economic and social development should respect physical limits of resource use and regeneration
- benefits and burdens of policies should be shared equitably by all segments of society (unavoidable serious inequalities should be compensated)
- policies should be based on detailed assessments of risk and environmental impact, and a sensible balance of costs and benefits
- policy evaluation should take into account economic and social development, environmental protection and social equity interdependently, not in isolation
- judgements should be based on the precautionary principle
- decisions affecting sustainable development are a shared responsibility (implying diffusion of information and involvement amongst all society)
- a mix of regulatory and market-based instruments should be used, as should a flexible approach to harnessing the private energies and capital to promote sustainable development
- measures should be introduced on a phased basis to minimise inequalities between winners and losers

Source: CEC, 1992.

reasonable degree of comparability and consistency of methods of environmental assessment, pollution standards, nature conservation networks and grant aid conditions throughout the European Union. In a complementary fashion, the Council of Europe has actively promoted environmental initiatives, such as its Pan-European Biological and Landscape Diversity Strategy to stop and reverse the degradation of biological and landscape diversity values in Europe (Council of Europe, 1995).

At the national scale, domestic agendas are increasingly influenced by the priorities identified at international level. For example, the UK responded to the World Conservation Strategy, as it was exhorted to do, by producing *A Conservation and Development Programme for the UK* (WWF–UK *et al.*, 1983). This was, in fact, orchestrated by quangos and voluntary organisations and was never formally adopted by the government, though it was officially commended. Compliance with European environmental directives during the 1980s was also reluctant, but the growing force of the global agenda led to a quickening of tempo and more positive compliance with environmental standards in the 1990s. Thus, an environmental white paper in 1990 set out a comprehensive programme and set of commitments, to be monitored annually (HMSO, 1990). Subsequently, the Earth Summit was fully reflected in a suite of government publications: a national sustainable development strategy mirroring Agenda 21 itself and policy documents on climate change, biodiversity and forestry reflecting the principal accords of the Earth Summit (Grubb *et al.*, 1993). A replacement UK Sustainable Development strategy, based on an extensive consultation exercise, was produced in 1999 and reflects a significantly stronger emphasis on social issues, balancing out the economic and environmental strands of the earlier strategy (Box 1.3). Sustainable development has also become an all-permeating public policy principle, as exemplified by the comprehensive revision of the DETR's *Planning Policy Guidance* (see Chapter 3) to incorporate its implications. Whilst some of the interpretation and uptake of these policies and principles has been lukewarm, the broad alignment with international good practice has been unmistakable.

Sustainability: The Overarching Agenda

Over the past 20 years, the core concept of using the earth's resources in a sustainable fashion has come to dominate thinking about environmental, and even economic, policy (Selman, 1996). Indeed, it is surprising that the terminology has not become more familiar to the lay public. The key term is that of 'sustainable development', although expressions such as sustainability, sustainable yield and sustainable growth are also frequently used. These all have different nuances, and it is often convenient to use 'sustainability' as a shorthand, whilst being aware of its imprecision as a

Box 1.3 Major elements of *A Better Quality of Life* (DETR, 1999)

The Strategy comprises a set of key aims, priorities and guiding principles:

Aims

- social progress which meets the needs of everyone;
- effective protection of the environment;
- prudent use of natural resources; and
- maintenance of high and stable levels of economic growth and employment.

Priorities

- more investment in people and equipment for a competitive economy
- a fairer society
- promoting choice in transport
- improving urban areas to make them better places to live and work
- steering development and agriculture to protect and enhance the countryside and wildlife
- improving energy efficiency and reducing waste.

Guiding Principles

- putting people at the centre
- taking a long term perspective
- taking account of costs and benefits
- creating an open and supportive economic system
- combating poverty and social exclusion
- respecting environmental limits
- the precautionary principle
- using scientific knowledge
- transparency, information, participation and access to justice
- making the polluter pay.

term. Broadly, sustainability arguments tend to stress the need to view environmental protection and continuing economic growth as mutually compatible, rather than conflicting, activities. Sustainability therefore implies that development programmes should be consistent with natural resource limits and biospherical waste assimilation capacities. In practice, it is helpful to distinguish between

- productive sustainability – the use of an area's natural resources, such as soil and water, such that their long-term productivity is not impaired;
- aesthetic sustainability – the maintenance of an area's natural and cultural heritage; and
- socioeconomic sustainability – the establishment of economically viable communities.

Some indicators associated with sustainability planning in practice are summarised in Table 1.3.

Table 1.3 Indicators of sustainable development for the UK

Broad aims	General topics covered by 'indicators'
A healthy economy should be maintained to provide quality of life while at the same time protecting human health and the environment, in the UK and overseas, with all participants in all sectors paying the full social and environmental costs of their decisions	• the economy • transport use • leisure and tourism • overseas trade
Non-renewable resources should be used optimally	• energy • land use
Renewable resources should be used optimally	• water resources • forestry • fish resources
Damage to the carrying capacity of the environment and the risk to human health and biodiversity from the effects of human activity should be minimised	• climate change • ozone layer depletion • acid deposition • air • freshwater quality • marine • wildlife and habitats • land cover and landscape • soil • minerals extraction • waste • radioactivity

One of the interesting features of sustainable development is that it encourages the consideration of various forms of capital, and thus challenges our traditional concepts of wealth. Ekins (1992) has reinterpreted these concepts to reflect their role, not only in fuelling the material growth of national economies, but also in satisfying fundamental human and ecological needs. Thus, it is argued, that ecocapital, human capital, social and organisational capital, and manufactured capital all contribute to the creation of 'wealth' in its broadest sense, and must be maintained and nurtured by society. One essential requirement is that society should adopt a 'full repairing lease' on the environment, to ensure that the condition of the natural resource base remains sound in perpetuity. It is also possible that governments may wish to ensure that environmental losses are made good by equivalent compensations elsewhere in the resource base, so that a constant environmental capital stock is retained.

One of the most practical interpretations of sustainable development has been to represent the earth's assets as 'natural capital', such that they can be thought of in a similar way to invested capital, which can be inherited, used, created and bequeathed to future generations. As with invested capital, some is effectively irreplaceable (such as a medieval cathedral) whilst some can readily be replaced (such as a bungalow). Thus, environmental capital can be divided into

- *critical natural capital* – effectively irreplaceable elements of the natural or built environment, such as nature conservation sites which have developed over thousands of years, fundamental earth processes such as the hydrological cycle, and buildings of outstanding architectural merit;
- *constant natural capital* – assets which are important to retain where possible, but which can be substituted. For example, a nonexceptional woodland could be felled provided it was replaced by a new plantation of similar size and composition (i.e. the woodland 'stock' was kept constant overall). This principle is often used in 'compensation' strategies; and
- *tradable natural capital* – assets which are often locally important, but which should not stand in the way of higher-order objectives (which may themselves contribute to the overall capital stock in different ways). It may be comparatively easy to compensate for losses of tradable natural capital.

An important way of applying these concepts of natural capital is as a means of moving away from seeing the environment as a set of objects or features, to one of understanding it in terms of the functions or services it provides (CAG/LUC, 1997). Indeed, novel notions about the role of the environment as a provider of free services (e.g. air cleansing, genetic reservoir) have helped to clarify its value in policy and planning decisions. Planners may thus be able to pose a series of key questions to be asked about the development of land or water, namely:

1) What are the characteristics or attributes of this place which matter for sustainability?
2) How important is each of these, to whom, and for what reasons or purposes?
3) What (if anything) could replace or substitute for each of these benefits?
4) Do we expect to have enough of each of them (i.e. are they ever likely to be over or under target)?
5) What kinds of management action are needed to protect and/or enhance each of these attributes to the degree justified by its importance, degree of sustainability and scarcity/plenty?

These questions have the potential to be used as a set of rules in environment-led planning, and may facilitate more informed and sensitive decisions about the consumption, modification and full or partial safeguard of environmental capital.

The Nature of Planning

In the UK and many English-speaking countries, perceptions of the nature of planning has been heavily influenced by the principles underlying the Town and Country Planning Act 1947. Broadly speaking, this nationalised the development interest in land. It defined development narrowly, so that it only applied to certain building, engineering and mineral works, and to certain changes in the use of buildings or land. The right to decide whether this type of development was acceptable no longer lay with the individual, but with the community. Thus, individuals or corporate bodies wishing to undertake development had to make an application to their local authority which, in theory at least, had prepared a development plan depicting acceptable land uses in particular areas, against which planning applications could be assessed. Taking the development plan and other material considerations into account, qualified planners could then make a recommendation to a committee of elected members, who would decide whether to accept, modify or reject the application. In this way, control could be exercised over development. Subsequent planning legislation has altered the detail, but not the principle, of this approach.

It may also be noted that planning operates at different spatial scales, and impacts on every geographical grain between the continent and the street. Thus, some planning policies may operate at transnational levels, such as environmental assessment procedures or urban regeneration funds in the European Union, whilst some countries have relatively strong national physical plans; in Britain, plans are prepared for local authority areas and will increasingly be produced for regions, whereas some planning systems give strong planning powers to the neighbourhood or 'commune' level. Typically, plans at the smaller scales or environmental evaluations based on specific project proposals are more detailed and precise in their format.

The method of 'command and control' tends to be perceived as the normal and proper approach to land-use planning. However, it should more correctly be seen as only one option or ingredient and, in some respects, a very partial solution. Even within the 'command' style of planning, there is considerable scope for discretion. For instance, the UK town planning system is based on flexibility of interpretation of individual cases, whereas most comparable systems (outside the UK and its former colonies) are based on a more rigid system of zoning. In the latter, swathes of land are designated in some kind of 'official plan' for particular categories of land use; if applicants subsequently propose a development which is in accordance with these ordinances, approval is virtually automatic. The former approach is relatively expensive to operate, but a less flexible system might well erode the treasured subtleties of the British landscape and townscape, and this seems to be a price which the public are prepared to pay. However, it does impose a considerable resistance to the idea of extending the planning system to areas such as forestry and agriculture (which periodically has been suggested) in view of its expense and bureaucracy.

Beyond the traditional domain of planning, there are now recognised to be many other ways of approaching environmental resource use decisions.

Thus, command and control procedures are acknowledged to represent only a limited, and perhaps unglamorous, option. During the 1980s, there was a strong swing internationally, spurred on by the collapse of east European communism, towards free market approaches whereby the private use of environmental resources could be influenced by public measures based on economic principles. The strongest advocates of these approaches were neoclassical economists who believed that command and control measures merely distorted the market in unhelpful ways. Thus, there was a lot of argument, supported by tentative practice, in favour of deregulating environmental resource controls and reconceptualising the role of government as that of sending economic signals to consumers to steward resources more wisely. A signal example of this approach has been the price incentive given to users of unleaded petrol (albeit it could also be that countries which legislated to require the conversion of vehicle engines and withdrew sales of leaded petrol achieved even greater success, and the UK has subsequently followed suit).

Other approaches range from informing and persuading, through regulation of designs or emissions, to direct state ownership. Many strategies for conservation and protection have been based on the least regulatory end of the spectrum, and have assumed that an educated and intelligent clientele will behave positively and considerately towards valuable environmental resources. Thus, promotional, informative and educational strategies, exhorting but not compelling citizens to behave in environmentally responsible ways, have often been preferred. Equally, given the complexity and multidisciplinarity of many environmental planning situations, much planning practice now emphasises a collaborative approach between partners. Thus, spatial strategies may focus on establishing agreement and common language between different agencies. These agencies may, in turn, have their own plans (for instance, for directing the use of wildlife, water, atmospheric resources), based on varying levels of persuasion and compulsion. In certain situations, purchase of sites for the protection of nature or prime landscapes may be deemed necessary in order that management agencies can exert adequate control over land use. Similar, though not always satisfactory, conditions may arise by leasing rather than purchasing sites. Where these types of control exist, direct site management planning, rather than indirect land-use planning, may occur.

For most of its history, planning has been delivered in a top-down fashion, debated and implemented by trained professionals normally acting within some framework of political accountability. This approach has been based on assumptions about planners' ability to make rational decisions, frequently on the basis of formal optimisation techniques, and associated with a logical cycle of survey, policy formulation, plan-making, implementation and review. Planners have learned, from their mistakes, that this idealised process rarely works in practice. First, it tends to assume that planners start with a clean sheet and so can produce an ideal solution whereas, other than in the cases of new towns or land resettlement, plans are based on an inherited townscape and landscape. This inheritance seriously restricts the scope for retrofitting sustainability measures. Secondly, it depends upon comprehensive, high-quality, up-to-date information, which is rarely available, so that decisions must be based on patchy datasets. Thirdly, it supposes the existence of long-term and altruistic attitudes, whereby people accept solutions which are for the maximum aggregate benefit for the whole of society. However, planning decisions are more typically associated with political short-termism and parochial actions by people concerned to protect their own 'backyard' interests. Fourthly, it has tended to exclude substantive public debate and has made little conscious effort to include the views of marginalised and under-represented quarters of society. Consequently, many critics would suggest that the 'rational' method of planning is, in practice, something of a fiction. Often, it is more useful to adopt a 'realist' perspective which reflects the rather messier condition of actual practice.

Instead of 'top-down' approaches, there has been considerable pressure for environmental plans to be based more on 'bottom-up' solutions, where local interest groups are facilitated to incorporate indigenous knowledge and preferences. In practice, more of a 'mixed mode' approach is normally preferable, so that professional skills are combined with citizen views and political mediation, to produce a plan which has balance and legitimacy. Equally, plans can be seen as a means of dialogue and debate to resolve key quality-of-life issues for a locality, and can thus stimulate and reflect a wider community discourse. This is associated with transactive or communicative planning, and may be a key to increased citizen involvement and interest in environmental planning. The general nature of these various approaches is reflected in a spectrum of planning instruments, summarised in Table 1.4.

Striking a Balance?

One of the most enduring perspectives on environmental planning is to portray its role as that of 'striking a balance' between the resources we extract from the environment and the wastes which we return to it. This view of planning presumes that there are ecological limits imposed on human activities and that the environment cannot be treated as a bottomless sink, so that the demands created by development must be balanced by the free services afforded by the environment. The impacts of resource consumption and waste emissions on the environment are often studied through a 'materials balance' model (Figure 1.1), in which the economy is portrayed as an open system pulling in materials and energy from the environment and eventually releasing an equivalent amount of waste back (Turner, 1995). Too much waste in the wrong place at the wrong time causes pollution and other externalities as the system is not 'in balance'.

Table 1.4 Spectra of planning instruments

1. **The 'Gilg-Selman' spectrum**
1) Public ownership or management of land via long term leases;
2) regulatory controls, mainly negative, for example, planning permission;
3) monetary disincentives to discourage production and/or undesirable uses;
4) financial incentives to encourage production and/or desirable uses; and
5) voluntary methods based on exhortation, advice, and demonstration, but often backed up with the threat or promise of one of the four methods outlined above.

(Gilg, 1996, p. 11)

2. **Integrated approaches to environmental protection**
An example of a hierarchy similar to the Gilg–Selman spectrum is that of the review undertaken by DETR to investigate how the Environment Agency can take a more integrated approach to its environmental protection responsibilities, overcoming barriers to integration. This study entailed a five-point strategy, namely:

1) prosecution when necessary
2) prevention of pollution by taking action, where possible, before incidents occur
3) sound and cost-effective management, working with all relevant sectors of society
4) influencing politicians, the public and fellow regulators to ensure a better legislative base with which to protect the environment
5) education of current and future generations to ensure that the environment is central to everyday decision-making.

(Based on DoE, 1993a)

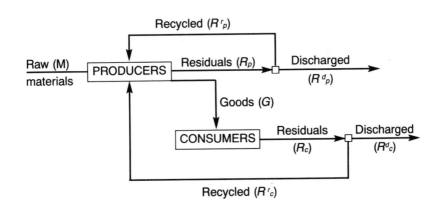

Figure 1.1 The 'materials balance' of environmental resource consumption. According to the first Law of Thermodynamics (Conservation of Matter), flow between extraction and discharges must be equal:

$$M = R^d_p + R^d_c$$

Substituting for M gives

$$R^d_p + R^d_c = M = G + R_p - R^r_p - R^r_c$$

i.e. the quantity of raw materials (M) is equal to the output (G) plus production residuals (R_p) minus the amounts recycled (R^r_p and R^r_c). These are fundamental relationships, and our only room for manoeuvre is to reduce G, R_p or increase R^r_p and R^r_c.
Source: Modified from Field, 1994 and Pickering and Owen, 1997

In trying to strike a balanced solution, however, we encounter a number of challenges. Amongst the most important of these are (Owens and Owens, 1991)

- the intangible nature of many environmental 'goods' (the problem of *quantifying the unquantifiable*);
- enormous uncertainty surrounding complex environmental issues;
- uneven distribution of costs and benefits in time, space and society (the problem of social and intergenerational *equity*); and
- the problem of distinguishing between *need* and *demand*.

Assuming that sufficient progress can be made on these to reach agreement over courses of action, however, there is then a need to regulate users of the environment, which we normally seek to accomplish through intervention in the market by some public agency. This draws us into the realms of policy and planning. For this, Owens and Owens propose an idealised political-administrative sequence (Figure 1.2), reminiscent of the 'rational' planning cycle, which illustrates the fusion of science and policy in the pursuit of environmental goals. We discuss the nature of policy in due course, and we need only note here that this elegant sequence is rarely so straightforward in practice, essentially because of a 'policy–implementation gap'. According to policy analysts, this gap tends to occur for one of three reasons: either the initial policy is inadequately formulated; or there is insufficient skill, time, money or effort devoted to policy implementation; or there is imperfect communication between policy-makers and field staff.

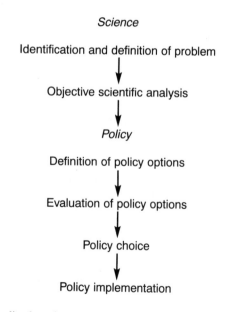

Figure 1.2 An idealised environmental policy process
Source: Owens and Owens, 1991

Various arguments have been advanced as to how our use of the environment might be planned. The following summary makes some gross simplifications about the respective discourses, which, of course, overlap and contain internal disagreements. However, it provides a brief sketch of the principal positions. At one end of the spectrum, free marketeers have argued that there is no evidence of the earth's inability to provide for all our demands. They would claim that no resources have ever run out, that materials thought to be in imminent danger of depletion are now in over-supply, that there is no gross underproduction of food globally (only problems of distribution and politics), and that evidence about global human impact is highly inconclusive. Their approach to environmental management would typically be to ensure strong private user rights of the earth's resources so that these could be used efficiently within an open and competitive market; they would seek to minimise the use of interventionist planning instruments and the use of regulatory measures generally. At the opposite extreme are ecological fundamentalists, who claim that the evidence of our environmental destruction is unambiguous and that we should rapidly develop new types of settlements and economies. Many of them would argue, indeed, that we should renounce materialistic lifestyles, live as simply as possible and have far greater regard for the rights of animals (perhaps even plants). Their approach to environmental planning would be highly interventionist, seeking to enforce strong regulatory measures to block environmentally unfriendly development and to penalise severely any damaging activities. Between these extremes are groups who believe, in varying degrees, that environmental issues are indeed serious, but that a blend of planning and management mechanisms may be used effectively to minimise the adverse effects of, and to encourage optimally efficient resource use within, industrial and urban development.

The regulation of environmental change clearly has political consequences, and much debate has surrounded issues of equal access to resources. Individuals, regions and countries have different abilities to compete for resources and to obtain particular levels of amenities. Traditionally, the explanation of this has been in terms of economics and the redistribution of wealth, but the debate in recent years has tended to shift to issues of gender, ethnicity and culture, which may affect well-being in more subtle and indirect ways. Thus, the popular political debate has swung from issues of 'inequality' to those of 'identity', focusing on the rights of individuals to certain civil liberties and entitlements. This emphasis on the politics of identity has had two major consequences for environmental planning. One is the gendered representation of environments, which is associated especially with arguments that styles of urban design and patterns of resource consumption have unconsciously reflected the needs and values of patriarchal societies. A second is the growing acknowledgement of previously under-represented or unrepresented viewpoints of marginalised groups of people. These may be people in poverty whose basic survival strategies result in environmental deterioration, or indigenous peoples whose traditional environmental management beliefs and practices have been eclipsed by western scientific approaches. More generally, these issues manifest themselves in terms of the notion of the 'other' in society – the relatively 'unheard' person whose views and knowledge systems are at variance with dominant culture and expert knowledge. There has been a growing concern to include 'othered' voices in official plans and surveys. However, the politics of identity should, despite its current popularity, be treated with caution, and it is already beginning to be undermined by accusations of reliance on image and 'spin', and on unrepresentative, anecdotal research. The dominant, positivistic discourse is not necessarily wrong, and the 'othered' voice is not necessarily right: both have significant, but partial, contributions to make.

In terms of principles upon which use of the environment is regulated, perhaps the most fundamental perceptual change has been the shift from trying to 'strike a balance' in our use of resources to 'respecting capacities' in terms of the environment's intrinsic limits. Thus, historically, perhaps the archetypal view of planning has been that of trying to mediate between competing land uses to ensure a 'balanced' outcome which reflected a sage evaluation of the alternative demands on a site. Increasingly, there is an emphasis on the intrinsic 'capacity' of environmental systems, and an acknowledgement of the fact that certain urban or rural areas may be approaching (or have exceeded) their capacity to accommodate new development, and that we must resolve land uses within higher-order constraints. These constraints are often represented as the 'limits of acceptable change'. In some cases, this may also involve reaching agreement by building consensus (to work within collectively agreed capacity constraints) rather than striking a compromise (to accept rationally balanced trade-offs). At present, this is more commonly an influential conception rather than a working principle, and the definition of a capacity is, of course, malleable. However, it does signal a fundamental change in the way we approach land-use options. Once more, this change may reflect fashion and fad rather than genuine advance.

A further key axiom is that of the 'precautionary principle' which, in essence, means that we should not wait until demonstrable proof exists of the environmental damage caused by a particular activity before we seek to curb that activity; rather, once there is a reasonable supposition that hazard may arise, we should adopt a more cautious course of action. Thus, absence of absolute proof of the need for an action should not be used as a reason for indecision. Summarising O'Riordan (1994a), we can observe that the need for precaution in environmental decision-making arose from the high levels of complexity, uncertainty and even chaos in environmental systems, as reflected in

- data shortage – inadequacy of historical records and of monitoring arrangements;
- model deficiencies – environmental models can only be refined, not made representationally accurate; and

- beyond the knowable – many scientists believe that the environment is intrinsically too complex, chaotic and mysterious to comprehend and predict fully.

Thus, a principle of precaution is required in which

- thoughtful action is necessary in advance of scientific proof;
- ecological space is left as room for ignorance;
- careful involvement of the public is necessary in management and planning decisions; and
- the burden of proof should be shifted from the developer to the victim. (After O'Riordan, 1994a).

Thus, the precautionary principle seeks to re-emphasise the responsibility of those who seek to change things to show that they will not cause harm, or to provide a fund for possible compensation to subsequently proven victims. This leads to a 'public trust doctrine', involving a commitment to put back into the earth at least the equivalent of what is being removed in any particular development.

Conclusion

The importance of the environment in land-use planning decisions has been recognised for a long time, and the early twentieth-century pioneers of town planning were alert to the significance of ecological pattern and process. However, our understanding of the centrality of environmental dynamics to planning decisions has greatly increased in recent years. In particular, the new focus on sustainable development rather than on purely 'green' issues has had important consequences. A key role of planning in the twenty-first century appears to be that of helping society cross the 'sustainability transition'.

Sustainable development is now an overarching agenda for planning, providing a meta-narrative within which the main principles of decision-making are couched. This manifests itself in several ways. One is a move away from the time-honoured concept of 'striking a balance' between competing uses to more fundamental assessments of the environment's capacity to accommodate particular types of change. Another is the emphasis on 'quality of life', rather than purely 'green' issues; environmental issues remain important *per se*, but far greater social consensus about the need to cross the sustainability transition will be gained from making the debate more immediately relevant to all spheres of human activity. Perhaps of greatest significance has been the analysis of environmental resources in terms of 'capital' assets which supply a range of essential, and often 'free' (uncosted) services. This opens up possibilities for reflective decisions to be made about changes to different categories of environmental systems and spaces.

In pursuit of these environmental goals, planners have a range of powers. As important as their statutory powers, however, is their potential for engaging society in debates and collective decisions about sustainable futures. In practice, rational and deliberative processes are starting to merge, as decisions are increasingly expected to reflect a broader social consensus. Environmental planning can only flourish and retain its importance within a wider context of concern for and participation in issues which affect the quality of life of present and future generations.

2 Perspectives on Environmental Planning

Introduction

As we noted in the previous chapter, the scope of environmental planning is almost infinitely broad, and it operates in the context of a range of scientific, social scientific and humanistic frameworks. Planners must be aware of all these influences, and of the different perspectives which they offer on policy and practice. An appreciation of the transdisciplinary nature of applied environmental knowledge is essential both pragmatically, in the pursuit of effective decisions, and conceptually, to help reflect on the broader purposes and consequences of professional intervention.

This chapter reviews a range of influences on environmental planning practice. Clearly, environmental planning is the product of a variety of ecological, physical, economic, social and political contingencies. Our awareness of these, and of their relative importance, has shifted over time, broadly reflecting a change from positivistic explanations of resource consumption to humanistic ones reflecting social inclusion in the definition and solution of problems. Consequently, environmental planners must now consider not only the socially constructed nature of knowledge about natural resources and their use but also the factors influencing the regulation of consumption.

Two Integrating Perspectives: Natural Resources and Environmental Systems

An enduring principle of environmental planning has been the wise use of natural resources. (Note that here were are talking about a subject-specific use of the term 'resource', which should not be confused with its use elsewhere to mean finances, personnel or physical facilities.) The nature of 'resources' and their consumption, however, must be addressed from both natural science and social science viewpoints, with the former emphasising the intrinsic limits to our biophysical 'life-support systems' and the latter pointing to human potentials, cultural constraints and economic forces.

An initial unifying framework within which to study environmental planning is thus that of natural resources, particularly in terms of their renewability and nonrenewability (Figure 2.1). Renewable resources are those which can be replenished over time periods of single years to

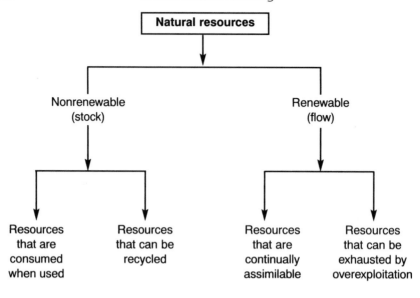

Figure 2.1   A classification of natural resources
Source: Owens and Owens, 1991

centuries. Nonrenewables may only be regenerated over geological time-scales, and effectively their stocks are finite; availability depends on the cost of recovering deposits, potential for recycling (impossible for fossil fuels but technically feasible for most minerals) and development of acceptable substitutes. The main examples of absolute diminution of stocks occur in relation to nonrenewable resources. Equally, however, renewable resources may be seriously degraded by excessive exploitation, pollution or visual intrusion. Many of the more serious pressures on environmental resources are associated with accelerated population growth in developing countries; even so, advanced industrial economies may still experience severe pressure on land and other natural resources as a consequence of factors such as high rates of household formation, demands for extravagant lifestyles and expectations about the appearance and availability of foodstuffs. Thus, resources are both a scientific and social scientific entity: on the one hand, there are absolute quantities of particular chemical deposits in specific geographical locations; on the other, these deposits will only be extracted for human use where economic conditions, human technologies, policy constraints and political-cultural circumstances permit.

The use of resources can be represented in terms of the 'ecological footprint' of a human community, in which a land area is calculated which is arithmetically equivalent to the total quantity of resources consumed (Selman, 1996; Wackernagel and Rees, 1996; Wackernagel, 1998). (For example, food consumption can be represented in terms of the hectarage required directly and indirectly for its production, whilst energy use could be expressed as the area of land which would need to be used for biofuels in order to achieve an equivalent energy output.) This total land area can be represented as a footprint, to give a graphic impression of how heavily a society is 'treading on the earth'. The central idea behind the footprint concept is that, especially in advanced industrial societies, the size of 'equivalent land' consumed is vastly greater than the actual land occupied by the community. Thus, towns and regions may give a false impression of their environment-friendliness, as much of their impact on resources is hidden, and they may be causing serious effects on 'distant elsewheres'. In other words, the population is receiving hidden subsidies, either from other parts of the world, or from other periods of history. Ways in which this can occur are summarised in Table 2.1. A central purpose of environmental planning, therefore, is to seek to reduce our ecological footprint, and thereby minimise levels of destruction and deterioration of the resource stock.

Table 2.1 The ecological footprint

Plundered from the past	Conveyed to the future	Transferred from/to elsewhere
Fossil fuels	Persistent pollutants	Food imports
Minerals	Genetically modified strains	Timber imports
Nonrecharged groundwater	Reduced biodiversity and genetic variety	Energy/mineral imports
Irreplaceable scenery		Wastes for treatment elsewhere
		Imported products associated with polluting manufacturing processes

A second unifying framework in the study of environmental processes is that of the 'system'. A *system* may be thought of as a set of interacting elements which are at least partially enclosed within a boundary. In an isolated system there is no interaction with the surroundings across the boundary; in a closed system, there may be transfers of energy but not matter; whilst open systems experience transfers of both matter and energy across their boundaries. Environmental systems fall into the last of these categories, being characterised by the circulation of both energy and matter. This openness makes them difficult to study, predict and manage.

In natural systems, whereas energy tends to 'flow' unidirectionally from a particular source, gradually becoming diffused, matter is typically 'cycled' and can become concentrated or disseminated in altered forms. Thus, energy is available to physical (abiotic) systems from the direct power of the sun, the earth's rotation, and height and pressure differentials. Biotic systems derive their energy ultimately from the sun, and thus commence with land or water plants which are able to 'fix' solar radiation by photosynthesis.

Whilst all environmental systems are open systems, experiencing continued throughputs of both matter and energy, nevertheless they are characterised by the maintenance of their structure and, barring catastrophes, a
degree of permanence. For instance, a drainage system maintains the organisation of its stream and river channel network and of the contributing
slopes in spite of the continuous throughput of water and transport of
sediment. Thus, the system maintains a more or less stable state defined in
terms of its elements, attributes and relationships through time. This situation may be described as an *equilibrium* (or steady) state which is maintained by *negative-feedback mechanisms*, damping down the effects of
change. Thus, a river may adjust to intense rainfall by spilling temporarily
into its floodplain, before retreating once more to its normal channel.
Sometimes, however, major disturbance (either natural or induced) occurs
to a system and exceeds its thresholds of tolerance, so that its natural
homeostatic mechanisms are overwhelmed. These are termed *positive-
feedback mechanisms*, and long-term directional change occurs. Thus, at
the local scale, 'pulse' events (intense episodes of relatively short duration)
may alter landforms, for example where a slope slumps due to excessive
waterlogging; at a much broader scale, sustained burning of fossil fuels may
alter the gaseous composition of the atmosphere and induce long-term
climatic change. Some processes or events will not occur until the amount
of input has reached a certain level, which is a 'threshold'; once this threshold is breached (in the case above, for example, the bearing capacity of a
waterlogged slope) change may be sudden, dramatic and even irreversible.
Contrary to popular opinion, however, environmental systems are remarkably resistant to positive feedback and catastrophic change.

In practice, natural systems typically involve the linkage of several feedback mechanisms, some positive and some negative, in complex loops. The
phenomenon of feedback implies that a perturbation in one element of a
system can cause adjustments in other elements, and some of these changes
may be undesirable. The more complex the system, the greater the difficulty in predicting responses to perturbation, and the more problematic it
becomes to exert control over one element without causing unforeseen
changes elsewhere. Consequently, many environmental strategies now in-
sist that development should not be permitted unless its long-term and
indirect consequences can be predicted with reasonable accuracy and are
known to be acceptable within a given location (where these cannot be
estimated, the 'precautionary principle' applies). Equally, whilst the most
pressing environmental problems may appear to be global ones (such as
climate change and pollution of the oceans), it is essential to recognise that
many of these have local origins. This may be because the cumulative
impact of individual actions causes positive feedback in environmental
systems, for instance the links between a beefburger in a fast-food takea-
way and the clearance of rainforest for cattle rearing, or the design of local
housing estates which encourage high levels of car dependency and contribute to widespread traffic congestion.

Overall, our level of understanding of natural systems and subsystems is very patchy and, whilst fairly sophisticated models exist for some, so that we may make reasonably reliable predictions about their behaviour in response to external change, our knowledge of others is still very limited. Protection of sensitive natural resources is thus an essential insurance policy for the future. Some prime 'free services' performed by the environment, which need to be managed sustainably, are

- cleansing of air;
- regeneration of crops and wildlife;
- biological productivity of oceans; and
- purification of water.

In addition to these, there are many critical 'free services', as yet poorly understood, which underpin the survival of plants, trees, insects, animals, birds and fish.

Systems concepts are also used, albeit rather more loosely, in relation to the study of social sciences. Indeed, town planning itself went through a phase (in the 1960s and 1970s) of being studied in terms of systems analysis, wherein cycles of policy formulation and implementation were progressively adjusted in response to feedback from monitoring and evaluation studies. Social groups can be studied in terms of their systems of interaction, especially through friendship and kinship networks; economic institutions display systems properties through the flows of money and the adjustment of supply and demand-side factors; political and governmental processes are similarly characterised by the flow of ideas, votes, information and decisions. Perhaps most pertinently to environmental planning, land uses can be seen as interacting systems, driven by the changing stock of land resources, income and expenditure flows, and the personalities and capabilities of land managers. The various systems of concern to environmental planners are summarised in Table 2.2.

The Scientific Perspective

Although the scientific study of geology, botany and physical geography has a long pedigree, it is the relatively recent discovery of our ability to damage the integrity of global systems that has given prominence to environmental science as a principal field of scholarship. The holistic study of the environment can now best be understood in terms of complex interacting systems which drive the physical, chemical and biological properties of the soils, rocks, air and water. The earth's major physical systems centre on climate, oceanic circulation, relative movements of the earth's surface, and the gradual alteration of the earth's soils and crustal rocks by flowing water and other surface processes. Cycling of matter is performed within both *gaseous* and *sedimentary* cycles.

Table 2.2　Key elements of environmental systems

Scientific systems
- atmosphere
- ecosphere
- hydrosphere
- lithosphere

Human activity systems
- urban-industrial
- agriculture and forestry   } productive
- nature conservation
- landscape conservation   } protective
- water catchment
- minerals   } extractive
- military training
- recreation   } user-orientated

Political systems
- elected representatives – central and continental/federal government
- elected representatives – local government
- policy formulation and implementation
- law-making and regulation
- allocation of public monies

Administrative systems
- central government civil servants/local government officers
- nondepartmental government bodies
- nongovernmental organisations

Thus, the key systems close to the earth's surface essentially comprise:

- the geological system (lithosphere), i.e. the outer layers of the solid earth in which rocks and minerals are formed, eroded and re-formed by pressures within the earth and erosive forces from above;
- the ecological system (biosphere/ecosphere), the thinnest of all the layers (generally only up to a few metres thick), comprising organic matter, covering much of the land surface, in which life is sustained by flows of energy, cycles of nutrients (minerals) and availability of shelter;
- the hydrological system (hydrosphere), the layer of water which varies from a maximum depth of more than 11 km in the oceans to shallow and less extensive bodies of water such as lakes and rivers (plus glaciers and ice sheets, and water in soils and rocks), within which water is recirculated between the atmosphere, rocks, plants, lakes, seas, ice-caps and rivers; and
- the atmospheric system (atmosphere) – layers of gases that extend from the earth's surface up to *c*.100 km to the outer boundary of the planet – in which incoming solar radiation and re-emitted radiation from the earth combine with differential air pressures to generate patterns of weather and climate, and where more or less constant volumes of gases serve to support and shield life on earth.

Collectively, these complex, interlocking and highly dynamic systems provide a 'life-support system' (Figure 2.2) for planet earth. As is well known,

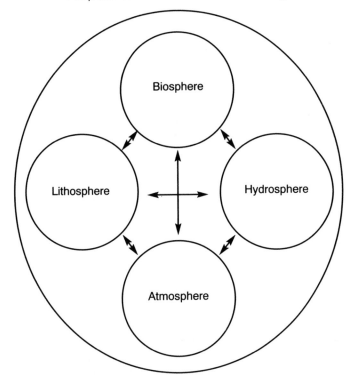

Figure 2.2   The earth's life-support system

it is now within human capability to alter the natural fluxes of these systems, and to risk damaging their integrity.

Of increasing concern is the effects of humans on pathways through these systems. *Biological* pathways occur through food chains and contaminants can travel along these, so that eventual mortality or morbidity may occur in organisms far removed from the original point of consumption. The effect of this is intensified in some circumstances through 'biological amplification' in which persistent fat-soluble pollutants are progressively concentrated in organisms along a food chain. This has been a major cause of unintentional damage to wildlife associated with the application of agricultural pesticides. A celebrated example of *physical* pathway pollution is acid precipitation, in which emissions of sulphur dioxide (SO_2) and oxides of nitrogen (NO_x) undergo complex reactions in the atmosphere before experiencing wet or dry deposition. Their passage can then continue through the vegetation canopy, soil and surface waters, experiencing further reactions and forcing toxic aluminium from base-poor soils into freshwaters. Other common forms of environmental pathway pollution are from waste dumps in which there may be leaching of toxic elements into groundwater supplies, or uptake of contaminants in plants following revegetation of former industrial wasteland.

The Geological-Geomorphological System

Providing the foundation of our life-support system is the physical fabric of the earth itself (the 'lithosphere'). The earth's interior consists of a structure of concentric shells, in which the particles which coalesced at the earth's origin gradually became differentiated under the influence of gravity. Consequently, the denser materials (iron and nickel) now constitute the earth's core, with progressively less dense materials becoming represented towards the earth's crust. The thin outer skin of this system, the *crust*, is around 30 km thick in the continents and some 10 km thick under the oceans; the *upper mantle* extends to about 700 km below the surface, and the *lower mantle* to about 2900 km; whilst the *outer core* (molten) gives way to the *inner core* (solid) at a depth of some 5000 km.

These various layers are characterised by gross similarities in their mineralogical composition (i.e. the geochemistry of their principal compounds). The continental crust is composed of granitic, acidic materials, rich in aluminium and silicon, whereas the oceanic crust mainly comprises basaltic, more basic and generally much younger material. The internal energy driving this system appears to lead to convective circulation of rock, which deforms under intense heat and pressure, and moves as a plastic material within the mantle. This type of motion results in displacements at the crust. Essentially, the earth's surface appears to be divided into a number of 'plates', each of the major plates carrying a continent. Convective currents well up under the earth's surface and result in the gradual, relative movement of these plates – that is, these plates shift against each other at their margins. As the plates are mainly carrying continents, their margins will almost invariably lie under the oceans. If we were able to see these plate boundaries, we would note that in some places there was an upwelling of convected material at the ocean floor, forming a ridge, slowly pushing the continental plates apart; elsewhere, plates would be being forced to slide uneasily alongside each other; or one would be forced under the other to be remolten into the mantle. These situations are termed 'constructive margins', 'transform faults' and 'destructive margins' respectively. The first situation explains why the oceanic crust is of relatively recent origin, as it is constantly being created. The latter two are associated with seismic activity (volcanoes, earthquakes), which is especially noticeable where a plate boundary runs across a continental land mass (e.g. California, Himalayas) or where crustal material is being destroyed (e.g. Pacific 'rim of fire').

These fundamental tectonic processes are responsible for the creation of new rocks at the earth's surface (Table 2.3). Thus, *igneous* rocks are formed by the cooling and crystallisation of molten rock (magma) or mineral fluids derived from the mantle. They may be formed as a result of extrusion, such as volcanic activity, or by intrusion of molten rock into existing crustal material, the latter becoming gradually exposed at the earth's surface as overlying layers are eroded. Extrusive igneous rocks, because of their rapid cooling at the earth's surface, are finely textured, whilst intrusive igneous rocks tend to be coarse grained and contain large crystals. This is also a source of mineral-rich fluids which may crystallise as veins. The second main category of rocks are those of *sedimentary* origin. These may arise from processes of erosion and weathering acting upon existing rocks, which are broken down into their constituent grains and transported (e.g. by wind or rivers) as sediments. The detrital grains which survive this process are typically the most resistant ones, notably quartz (sand). Alternatively, some sediments are of organic origin, especially the shells of marine creatures or the remnants of decomposing vegetation. Over exceedingly long periods, these sediments undergo a process of 'lithification', i.e. being compacted into rock, principally by the weight of overlying layers. Sedimentary rocks characteristically display 'stratification', or layering of visible bands, representing the successive sequences of deposition. Finally, some rocks are formed by alteration of previously existing rocks under conditions of intense heat or pressure (typically by proximity to igneous processes). These *metamorphic* rocks are often described in terms of their 'grade' – that is, the degree of metamorphism which was experienced as a result of how close they were to the heat or pressure source. Occasionally, we may witness spectacular geological change, such as an earthquake or even formation of a new island, but generally it is impossible to imagine the slowness of these processes. Alterations of the earth's surface normally happen imperceptibly, and can only be understood over immense geological timescales (Table 2.4).

The interface of the crust and the atmosphere is a very active zone, and the study of its change is known as geomorphology. Various processes here are of particular interest. One is the release of minerals from surface rocks by *weathering*, caused by chemical and physical attack from air, water and

Table 2.3   Some common rock types

Igneous – intrusive
granite
diorite
dolerite
quartz

Igneous – extrusive
basalt

Sedimentary
sandstone
limestone
chalk
flint

Metamorphic
slate
schist
gneiss
marble

Table 2.4 Geological timescales (simplified)

Era	Period	Ma BP
Cainzoic		
Quaternary	Holocene	
	Pleistocene	2
Tertiary	Pliocene	
	Miocene	
	Oligocene	
	Eocene	
	Palaeocene	65
Mesozoic		
	Cretaceous	140
	Jurassic	195
	Triassic	230
Upper Palaeozoic		
	Permian	280
	Carboniferous	345
	Devonian	400
Lower Palaeozoic		
	Silurian	435
	Ordovician	500
	Cambrian	570
Pre-Cambrian		570 to c. 4000

plant roots. Closely associated with this is the mass *erosion* of surfaces and the transport of eroded material as sediments. Various types of mass movement take place, and they are generally related to processes operating on slopes; the slope is thus an important unifying concept of geomorphology. Of particular interest to planners is the potential instability of some slopes, which can result in landslips. Sliding of a consolidated layer over a layer of unconsolidated material is related both to the 'inherent' factors such as steepness and height of the slope, and the composition of its component materials, and to specific trigger mechanisms such as the excessive loading which occurs after prolonged rainfall (White *et al.*, 1984).

Surface processes also lead to the formation of *soils*. Weathering leads to the release of minerals essential to plant growth and of particles such as quartz which make up the matrix of loose material penetrable by plant roots. Some of these minerals may then be 'leached' down the soil by percolating water, with the most mobile elements being lost from the soil column, but less soluble ones being redeposited some way below the surface. Organic matter, particularly from decomposing leaves and other plant materials, will also be incorporated into the soil, and provide important plant nutrients and chemically active exchange surfaces. Processes such as these lead to the soil column becoming differentiated into recognisable layers (horizons), and to the formation of different soil types. Most soils are thus formed *in situ* by weathering of local rocks, but some are formed by deposition of material from elsewhere (such as alluvial soils, deposited by rivers in flood). Soil development is very much dependent on the type of parent rock. Relatively soft sedimentary rocks, with a high proportion of chemically mobile minerals, such as limestone, will tend to produce very shallow soils. More resistant rocks, such as granite, will weather slowly, to greater depth, and retain the less mobile minerals in the soil column. Climate will also affect soil development, with warmer and wetter conditions tending to produce soils weathered to greater depth. The chemical composition of the host rock will also influence the eventual richness of the soil. In Britain, soils developed on the hard, granitic rocks of the uplands, in cold rainy conditions, will tend to be moderately deep (except on steeper and higher slopes) but relatively infertile. Those on the softer, sandy or alluvial sediments of the lowlands will often be deep and fertile; whilst those derived from chalk or limestone will tend to be moderately fertile, but shallow and stony.

An important influence on the quality of the soil is its *texture*. During the course of weathering, particles of different size are released, namely, sand (quartz grains), the finer silt and the finer still clay (whose particles form chemically active plates). As the predominant size of soil particle decreases, the soil becomes more difficult to work and more prone to waterlogging. However, very light and sandy soils tend to have excessively free drainage and be prone to drought. Consequently, the optimum mix is a 'loamy' texture comprising a blend of all three soil fractions. Soil *wetness* is determined partly by its texture, but also by levels of rainfall and the presence of a permanent or perched water table. Again, the optimum wetness is one in between excessively free drainage and waterlogging. Soilwater also contains free hydrogen ions (especially affected by rock type and precipitation), whose concentration determines levels of acidity. Soil acidity is measured on the pH scale, in which higher concentrations of H^+ are represented by lower pH values (pH 7 is neutral, lower values are acidic, higher ones are alkaline). Extreme pH values can result in essential plant nutrients becoming unavailable; in Britain soil acidity is a common problem, and is often counteracted by applications of lime. Detailed Soil Survey maps showing soil types and their principal characteristics (Table 2.5) are available for most of Britain.

The Ecological System

Ecology is the study of the relationships between organisms, other organisms and their physicochemical surroundings. A common focus of study is the *ecosystem* which is typically defined as a set of organisms interacting with one another and with their abiotic environment. The term may also refer to a specific, stable plant–animal community characteristic of particular conditions, such as a desert, acid grassland, coniferous forest or estuary.

Table 2.5 Main soil types

Type	Characteristics
Podsol (spodosol)	acidic, strongly leached soil, developed on siliceous materials, with a base-deficient litter
Brown earth (alfisol)	rich, fertile, slightly acidic soil, with cation leaching significantly offset by *in situ* enrichment
Calcareous soils	alkaline to neutral soils, reasonably deep and fertile, containing significant quantities of $CaCO_3$
Rendzina	thin calcareous soil found on steep limestone slopes, with rich humus and constantly saturated with $CaCO_3$
Gley soils	occur where there is a high water table (e.g. floodplain), and subject to intermittent waterlogging, which restricts plant growth
Organic soil (histosol)	formed in areas of high rainfall, where partially decayed plant remains are waterlogged and continue to accumulate in the absence of air; raw, acid and infertile

Since ecosystems draw their energy initially from the sun, the most basic biotic component is the primary producer (green plant). Grazing on this are primary consumers (herbivores) upon which predate secondary consumers (carnivores and omnivores). In some ecosystems there is an additional tier of top carnivores, feeding on secondary consumers. This creates a *food chain*, connecting a sequence from green plants to higher mammals, along which energy flows (Figure 2.3). All organisms that share the same general types of food (e.g. primary producers, secondary consumers) within this chain are referred to as a *trophic* (or 'feeding') *level*. Energy flow is inefficient along this chain and, as a broad generalisation, there is about a 90% loss of energy (diffused into the environment as low-quality heat) at each trophic level. Ecosystems also involve the cycling of nutrients, and this is partly the work of a further trophic level – detritivores (detritus feeders and decomposers) – which break down organic remains and return

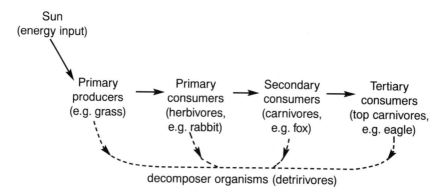

Figure 2.3 The food chain

them to accessible reservoirs. In practice, inter-relationships between organisms are normally highly complex in natural ecosystems and these are more helpfully viewed as a food *web* rather than a simple chain.

The sum total of all biological variation within an area is now often referred to as 'biodiversity', which has been defined as 'all hereditarily based variation at all levels of organisation, from the genes within a single local population or species, to the species composing all or part of a local community, and finally to the communities themselves that compose the living parts of the multifarious ecosystems of the world' (Wilson, E.O., 1997, p. 1). It is more simply thought of as the full range of 'biological diversity' (in terms of species or genestock) within a particular territory. The number of species currently described is about 1.4 million, though it is likely that the total number of species on earth is somewhere between ten million and a hundred million. As well as being of interest for its own inherent qualities, biodiversity is integral to humans' life-support system for a number of reasons, notably:

- its aesthetic values, as humans appear to need natural beauty, variation and inspiration in their surroundings;
- its importance to agriculture, whose versatility of response to changing conditions still depends on genetic variation in wild species;
- bioremediation, or the role of wild species in neutralising pollution and other adverse environmental changes;
- the processes of evolution;
- its value as a source of medicines;
- its provision of 'free services' (fisheries, water filtration, cleansing the atmosphere); and
- its role as an indicator of ecological change and reflector of stress in the biological community.

A fundamental characteristic of ecosystems is the way they develop over space and over time. When starting with a completely unmodified environment – for instance after a volcanic eruption has created a new land area or the retreat of an ice sheet has left bare ground – plants and animals will tend to start to occupy ('colonise') the ground surface in predictable ways. The recolonisation of long-settled land which has been disturbed by recent human activities will be subject to the same laws, although the pace may be quicker and the processes more complex. Essentially, some species will be 'opportunists', expanding their populations rapidly in the new conditions. These species, including many colourful annual plants, are typically short lived with most effort directed at reproduction so as best to exploit disturbed and transient environments. As the processes of predation, disease and competition bring about stabilisation of the community, so other species become dominant. These are better equipped to maintain long-term dominance in favourable and undisturbed conditions. A popular way of classifying plant species' competitive strategies during the process of colonisation is in terms of their response to environmental stress (e.g. Grime *et al.*, 1988; Williams *et al.*, 1998). In situations of high stress, species richness is low because only those plants which are *stress tolerators* are present; where there is minimal stress, only highly *competitive* species are dominant and again species richness is low; whilst in the early stages of colonisation, ground will be dominated by *ruderal* (i.e. wasteland) species. At intermediate levels of stress, species richness is at a maximum because a large number of plants have adapted to so-called 'average' conditions: ruderal species are dominant, but remnants of the stress tolerator and competitor populations are also present. Thus, plants tend to exploit either conditions of low stress and low disturbance ('C' type plants), or conditions of low stress and high disturbance ('R' type plants), or conditions of high stress and low disturbance ('ST' type plants). Hence, the vegetation community which develops in a particular place and at a particular time is the result of an equilibrium established between the intensities of stress (constraints on production), disturbance (physical damage to the plants) and competition (attempts by adjacent plants to capture the same unit of resource).

It should be clear from this principle that the way in which a site is colonised will vary over time, with fairly simple, pioneer communities in the early stages gradually being replaced by more structurally differentiated, mature ecosystems in the later stages. Thus, a series of *seral stages* eventually gives way to a *climax community*, which is stable under prevailing climatic and soil conditions. This process is referred to as *succession* (Figure 2.4). For instance, a site initially colonised by lichen, mosses and annual plants will (as soil develops) be systematically replaced by a more diverse and dense ground cover, perhaps giving way to scrub and eventually to woodland. The woodland – especially in temperate or tropical climates – will itself be structurally complex and contain several layers of trees, shrubs and ground flora supporting their distinctive populations of animal, bird and insect life. Such an ecosystem would be thought of as 'natural' by ecologists. However, millennia of human settlement have led to the widespread modification of climatic climax communities (human agencies thus introduce new conditions of 'stress' and 'disturbance', and deflect the natural climax). In some cases, highly modified agroecosystems (such as intensive cereal cropping) have been established, but in lower-intensity management regimes (such as unsown grazing pastures), the community still comprises mainly local and indigenous species which have established themselves naturally. These latter types of stable, culturally adjusted ecosystems (in which moderate levels of environmental stress are induced by low-impact management regimes, such as light grazing) are referred to as *plagioclimax* communities and represent a semi-natural condition. Often they are of outstanding conservation value as they represent peaks of species diversity.

Finally, it is worth noting that, although ecological texts have tended to emphasise successional processes amongst plant–soil systems, there is a degree of controversy associated with the concept. The stability and real-world existence of climax communities, in particular, has been questioned, and

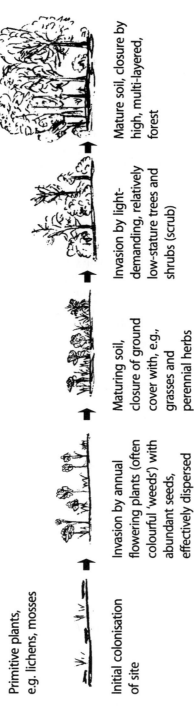

Primitive plants, e.g. lichens, mosses

Initial colonisation of site

Invasion by annual flowering plants (often colourful 'weeds') with abundant seeds, effectively dispersed

Maturing soil, closure of ground cover with, e.g., grasses and perennial herbs

Invasion by light-demanding, relatively low-stature trees and shrubs (scrub)

Mature soil, closure by high, multi-layered, forest

Figure 2.4 Idealised sequence of a vegetation succession

many scientists regard them as idealised notions rather than actual end-points. Nevertheless, theories of succession are practically helpful and are sufficiently robust to be useful guides in conservation and land-use management.

The Hydrological System

The global circulation of water between ocean, atmosphere and land is a closed system with no significant gains or losses. Water initially reaches the land masses by precipitation (principally rain and snow), much of which has originated by evaporation from the oceans. The land loses water by evapotranspiration into the atmosphere (i.e. evaporation from surfaces and transpiration from plants and trees), and by runoff from rivers. An important focus for the study and management of water, and indeed of environmental processes generally, is the 'catchment' – a river system bounded by a watershed and thus reasonably self-contained in terms of matter and energy fluxes (Figure 2.5). A truly self-contained catchment would have a water budget in which the only input was precipitation, whilst evapotranspiration and runoff in streams and rivers would be the main outputs. Water would also be lost by groundwater seepage, at least when the underlying bedrock was permeable. It would also contain storage components – i.e. surface waters, groundwater and soilwater – and transfers may occur between these. In practice, most catchments are not so completely isolated, and exchanges of water do take place to a limited degree. From these basic elements, catchment models may

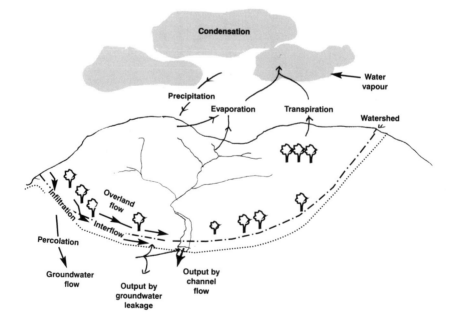

Figure 2.5   The basic elements of a catchment system

be developed, and are frequently used for management purposes. A basic equation underpinning catchment models is:

$$(Pn + R_s + R_g) - (ET + Q_s + Q_g) = \Delta S$$

i.e. (precipitation input + surface and groundwater inputs) – (evapotranspiration + surface and groundwater discharges) = the change in water storage.

Pn and *ET* bring about the interchange of water between the atmosphere and catchment. Runoff increases when *Pn* exceeds *ET*, and decreases when *ET* exceeds *Pn*, so in simple terms a water regime can be characterised by its *meteorological water balance* (*Pn – ET*). Meteorological water balance varies from year to year, and this affects storage components, including river and groundwater levels; longer-term trends, which can be affected by human activities such as water abstraction, can be difficult to detect and can take a long time to recover.

Much of the precipitation falling on a catchment is intercepted by vegetation canopies, and re-evaporates without ever reaching the ground. However, if precipitation reaching the ground exceeds the soil's ability to absorb water (its infiltration capacity), the excess collects in depressions or runs down inclines as overland flow. The nature of the ground surface – various types of vegetation or artificially constructed surfaces – is an especially important influence on the infiltration capacity of the soil. If percolating water reaches an impermeable layer, it accumulates or moves laterally down inclines as 'interflow', and this is often rapid and can contribute (together with overland flow) to 'quickflow', which refers to the *peak flows* in streams and rivers that are generated by rainstorms and can cause flooding. Some rocks are permeable or porous (e.g. many sedimentary rocks) and these may become saturated with water at a fairly stable depth, known as the *water table*. Thus, within the saturated zone, groundwater is held in *aquifers* which may be confined or unconfined. Groundwater flows in the direction of inclines, and may emerge as a spring or spring line, or seep directly into lakes or channels. It is largely responsible for *baseflow* which is fair-weather flow (as opposed to quickflow) in streams and rivers. The potential water storage and flow rates in groundwater bodies depend largely on the nature of the aquifers, especially their *porosity, permeability* (or 'hydraulic conductivity'), *thickness* and *transmissivity* (permeability × thickness). Groundwater is important not only because it is a major water resource, but is also essential for the maintenance of many wetland ecosystems. *Channel flow* (in streams and rivers) is normally the main outflow component from a catchment, but the water in a channel system can also be regarded as a storage component. Indeed, during quickflow periods the normal *bankflow* may be reversed, or the water may spill over the channel banks and cause surface flooding. Flooding of a natural river floodplain is an intrinsic mechanism for accommodating flood waters, but it may become a problem when developments or intensive agricultural lands sited in the floodplain are inundated. Defences constructed to prevent flooding often result in an abnormal surge that causes flooding further downstream.

Water in the environment is never pure, and natural waters vary considerably in the range and concentration of dissolved substances present. They also differ in terms of variables such as suspended particulate material, pH and temperature. Human influence on water quality includes both changes in the concentrations of naturally occurring chemicals such as nitrates, phosphates and metals, as well as the introduction of synthetic substances such as organophosphate pesticides. This distinction is similar to that between biologically active and biologically resistant chemicals, respectively. The former often enrich the nutrient status of water and speed up the rate of plant production. Such 'eutrophication' is a natural feature of the lower, slower-flowing reaches of rivers, and of lowland lakes, but 'cultural eutrophication' caused by sewage and fertiliser runoff can cause excessive growth of algae and other water plants, having adverse consequences for other aquatic life. The latter can cause direct death or damage to creatures. Key aspects of water quality include:

- oxygen levels (differing measures of which are used as general indicators of water quality);
- organic pollution (i.e. oxygen-demanding wastes which absorb dissolved oxygen from the water);
- the heating of natural water bodies by industrial processes (thermal pollution);
- the levels of nutrients (which vary naturally, though 'cultural eutrophication' may occur where excess nutrients are released into natural water bodies by, for example, sewage treatment outfalls and agricultural fertilisers);
- silt and sediment (which may clog watercourses, blanket water plants and reduce sunlight penetration);
- pathogens (such as pesticides leaching from farmland); and
- certain pollutant metals, micro-organics and other chemicals, whose behaviour may be influenced by environmental factors (such as pH) in a complex way.

The Atmospheric System

The air surrounding the earth represents a system which is highly dynamic in physical terms, but whose chemical composition has remained relatively stable over long periods (Figure 2.6). The basic energy input arrives from the sun, and this shortwave solar radiation passes through the atmosphere in a spectrum of wavelengths. Much of this radiation is scattered by materials suspended in the atmosphere and much is absorbed by atmospheric gases. Some reaches the earth's surface, where it is essential to biological processes (especially photosynthesis); some of this is absorbed by the earth, and some is re-radiated into the atmosphere as longwave radiation. One immediate effect of this is to warm the atmosphere, though this warming is not uniform as different layers of the atmosphere have different gaseous

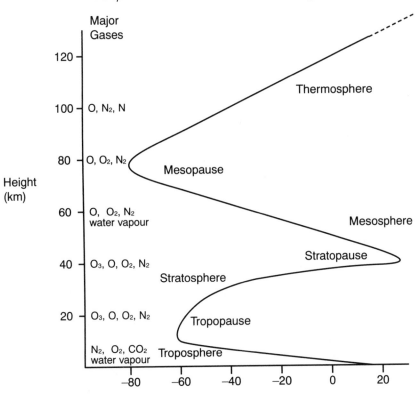

Figure 2.6 The chemical and physical characteristics of the earth's atmosphere: temperature/gaseous zones above the earth's surface

NOTES:

Composition of atmosphere (by % volume):

N	78.09
O	20.95
Ar	0.93
CO_2	0.03 (approx.)

Plus traces of other intert gases and variable amounts of water vapour

Oxygen
- below 60 km, mainly *molecular* O_2
- above 60 km, mainly *atomic* O
- dissociated atoms recombine with molecular oxygen to produce *ozone*
  $O_2 + O \rightarrow O_3$
- maximum production of ozone at 30–40 km
- maximum concentration of ozone at 20–30 km

compositions resulting in their varied capacity to absorb incident radiation. A significant proportion of incident solar radiation is in very short wave-bands, harmful to living creatures, yet this is largely prevented from reaching the earth through absorption by specific gases. For example, ozone (O_3) is very effective in absorbing rays shorter than 0.32 μm, and this gas is relatively abundant in the stratosphere. In practice, beneficial rays reach the earth's surface through 'radiation windows', whilst other, often harmful, rays are filtered out by ozone, oxygen and water vapour. Carbon dioxide (CO_2), water vapour and certain other gases are effective at absorbing longwave (infrared) radiation, and thus trap some of the earth's re-emitted heat. This causes an entirely beneficial 'greenhouse effect' close to the earth's surface, which is essential to the maintenance of equable temperatures: without this insulation, mean global temperatures would be around –17 °C rather than their present levels of about +15 °C.

The balance of radiation varies considerably across the surface of the earth and, combined with the earth's rotation and factors such as slope angles and ground cover, this variably distributed energy causes complex weather patterns. Differential heating of the earth's surface leads to zones of high and low air pressures, which drive air flows and lead to the uplift of water from oceans and lakes, resulting in winds, currents and cloud formation. Characteristic zones of climate occur, and weather patterns are especially influenced by zones where warm/saturated and cool/dry air masses collide. The regular convergence of unstable air masses in this way across the North Atlantic, for instance, results in the changeable and generally damp conditions of the European and North American Atlantic coasts. *Weather* patterns of this nature are relatively volatile on a day-to-day basis, but longer-term *climate* (i.e. the general character of weather patterns in a given region) is comparatively stable.

The earth's climate has changed throughout geological time and is still undergoing change. There have been at least six major cold periods or ice ages; since about 2.4 million years BP, global climate has cooled considerably and the earth has entered the Quaternary period, during which global change has fluctuated between cold (glacial) stages and warm (interglacial) stages, with less intense warm (interstadial) and cold (stadial) periods. Natural causes of global climate change result from internal earth processes such as volcanic activity, processes which are external to the earth (e.g. fluctuations in the earth's orbit and sunspot activity), and catastrophic events such as large meteorite impacts. Past climates have been studied using a variety of approaches, including petrological techniques (which use characteristic sediment and rock types to interpret past climates) and chemical methods (often tracking the changing proportions of stable isotopes of oxygen, carbon and nitrogen). The Quaternary period is most often used in the study of past, and the prediction of future, global climate change, as it is relatively rich in data relative to earlier geological periods.

A great deal of research has been directed into the effects of human activities on the climate and atmosphere. These broadly centre on three issues: the damage to ozone in the stratosphere; the emission of CO_2 through domestic and industrial activities (together with other longwave-absorbing gases such as CH_4), which may enhance the greenhouse effect; and the discharge of a range of other gaseous and particulate pollutants which may affect health, plant life and may acidify rainwater (Table 2.6). Some of the effects of these are readily visible, such as the blackening of buildings by the deposition of soot; however, greater concern is often expressed about the invisible, but much more widespread and potentially irreversible, changes which result from human activity. Estimating the speed, extent and effect of these changes is extremely difficult as the atmosphere is a high-energy system in constant flux, and climates are changing naturally all the time. An account of some of the more serious concerns is summarised in Box 2.1.

A more localised concern is that of the entrapment of atmospheric pollutants close to their source. Air temperature decreases with elevation,

Table 2.6   Key air pollutants and their anthropogenic sources

Pollutant	Anthropogenic sources
Sulphur dioxide	coal and oil-fired power stations, industrial boilers, waste incinerators, domestic heating, diesel vehicles, metal smelters, paper manufacturing
Particulates (dust, smoke, PM_{10})	coal and oil-fired power stations, industrial boilers, waste incinerators, domestic heating, many industrial plants, diesel vehicles, construction, mining, quarrying, cement manufacturing
Nitrogen oxides	coal, oil and gas-fired power stations, industrial boilers, waste incinerators, motor vehicles
Carbon monoxide	motor vehicles, fuel combustion
Volatile organic compounds, e.g. benzene	petrol engine vehicle exhausts, leakage at petrol stations, paint manufacturing
Toxic organic micropollution (e.g. polyaromatic hydrocarbons (PAHs), polychlorinated biphenyls (PCBs), dioxins	waste incinerators, coke combustion
Toxic metals, e.g. lead, cadmium	vehicle exhausts (leaded petrol), metal processing, waste incineration, oil and coal combustion, battery manufacturing, cement and fertiliser production
Toxic chemicals, e.g. chlorine, ammonia, fluoride	chemical plants, metal processing, fertiliser manufacturing
Greenhouse gases	CO_2 – fuel combustion, especially power stations; CH_4 – coal mining, gas leakage, landfill sites
Ozone	secondary pollutant formed from volatile organic compounds (VOCs) and NO_x
Ionising radiation (radionuclides)	nuclear reactors, nuclear waste storage
Odours	sewage treatment works, landfill sites, chemical plants, oil refineries, food processing, paintworks, brickworks, plastics manufacturing

Box 2.1   How might we be altering the climate?

The enhanced greenhouse effect

Carbon is one of the commonest elements in the environment – present in all *organic* substances – and is very mobile, combining with other chemical elements. There is a natural biogeochemical cycle which maintains a balance between the release of carbon compounds from their sources and their absorption in sinks. However, the quantity of carbon released from urban-industrial processes may be combining with the effects of clearance of CO_2-absorbing forests, and this combination may be contributing to an increase in global warming.

Burning of fossil fuels adds 5 billion tonnes of CO_2 each year to the atmosphere, but atmospheric CO_2 level increases only by *c.* 2.5 billion tonnes per year. The remainder sinks mostly to oceans and is taken up by photosynthesis in phytoplankton or in the growth of $CaCO_3$ shells on some organisms. Also, there is some direct diffusion across the air/ocean interface.

Analysis of air bubbles trapped in polar ice provides a historical record of CO_2 concentrations, and this reveals considerable variability. Atmospheric CO_2 is typically 180–285 ppm, but is now around 345 ppm and, towards the end of the twenty-first century, could rise to 500–600 ppm. It is widely thought that a doubling of the CO_2 levels would cause average global warming of 1.3–3.5 °C.

Other important greenhouse gases include:

- methane – CH_4 – of which relatively low levels are present in the atmosphere, but which is three times more effective than CO_2 as a greenhouse gas;
- N_2O (e.g. from denitrification of fertilisers); and
- chlorofluorocarbons (CFCs) – whose production is declining rapidly as a result of international agreements, but which has a long residence time.

Acid rain

Human activities result in the emission of various pollutants (principally SO_2, NOx, NH_3, hydrocarbons and particulate matter) to the atmosphere, which lead to a number of environmental problems such as poor air quality and acidic deposition or acid rain with very low pHs. Acidic deposition (by no means all is anthropogenic) is a particular problem in industrialised regions and countries where the combustion of fossil fuels releases large quantities of SO_2 into the atmosphere. Some regions and countries are polluted because of acidic deposition caused by industrialised parts of other countries up-wind. Acidic deposition results in acidification of groundwaters, surface water, damage to life (particularly forests and aquatic life) and building decay. Buffering reactions due to the presence of certain clay minerals, and because of cations within water in some soils and lakes, may reduce the immediate effects of the acidic deposition on those environments. The susceptibility of a soil to acidification is quantified as its acid susceptibility. Recovery from acidification of the environment can occur and is dependent on the sensitivity of the ecosystem. Managing acidification involves reducing anthropogenic emissions, mainly from fossil fuel-burning power stations, for example by using appropriate clean technologies such as fluidised bed combustion and more efficient combustion engines in both domestic and commercial vehicles.

The most serious impacts of acidification occur in areas underlain by granitic or quartzitic bedrock. There, soils and water are already acidic and so are unable to neutralise ('buffer') the additional acidity in the precipitation. Moreover, in these cold, upland zones, the 'acid surge' associated with snowmelt may intensify effects: as acid accumulates in layers of snow, the spring melt can damage young fish fry – acid may cause reactions in soils which force out aluminium, and this can be very poisonous to fish.

Stratospheric ozone depletion
The importance of stratospheric ozone is its role in controlling the UV-B reaching the earth's surface. The ozone layer absorbs part of the outgoing longwave radiation and redirects it back to the troposphere below, and to the earth's surface. UV-B is a normal component of sunlight, with up to about 5% of the energy reaching the earth's surface under a clear sky at noon comprising biologically active UV-B radiation, but any significant increase in UV-B radiation above natural levels is potentially harmful to human health and the environment. The absolute intensity of UV-B radiation reaching the earth's surface is influenced by many factors, including the angle of the incident sunlight, principally controlled by the seasons and the time of day. Establishing UV-B intensity at the earth's surface cannot be done from measuring stratospheric O_3 levels alone so it is important to obtain accurate UV-B data to establish long-term trends and causal factors. UV-B which passes through the stratosphere may be absorbed and scattered by air pollution, including O_3 in the lower atmosphere.

Measurements of actual quantities of ozone in the stratosphere are lower than would be predicted. This is explained by *catalysts* speeding up the slow natural rate of ozone decay. There are two agencies of this:

- naturally occurring O_3-destroying catalysts – e.g. hydrogen, nitrogen, chlorine oxides; and
- anthropogenic catalysts – of which a major culprit has been the CFCs or 'freons'.

CFCs are inert at the earth's surface, but photochemical degradation of CFCs releases chlorine, which sets off a continuous cycle of ozone destruction. The severity of this is still debatable. For example, the much vaunted Antarctic 'ozone hole' could be caused by dynamic circulation mechanisms (thinning seems to be offset by thickening at 40–50° S).
In essence, the ozone depletion process comprises:

$$\text{Freon 12} \quad \text{(CF}_2\text{Cl}_2) \qquad F - \overset{\displaystyle Cl}{\underset{\displaystyle F}{C}} - Cl \quad \xrightarrow{\text{UV}} \quad F - \overset{}{\underset{\displaystyle F}{C}} - Cl \ + Cl$$

(Dissociation of Freon 12 by ultraviolet light in the stratosphere produces a highly reactive chlorine free radical.)

$$Cl + O_3 \longrightarrow ClO + O_2$$

(The chlorine free radical reacts with ozone (in the stratospheric ozone layer), reducing ozone concentration.)

$$ClO + O \longrightarrow Cl + O_2$$

(The chlorine oxide can then react with atomic oxygen (abundant in the stratosphere) to create molecular oxygen and a chlorine free radical, thus repeating the process.)

Alternatively, chlorine oxide can react with, and destroy, ozone directly:

$$ClO + O_3 \longrightarrow ClO_2 + O_2$$

producing a natural tendency for air to circulate, as warm air close to the ground rises, cools and sinks back towards the earth. However, under certain conditions colder air may rest close to the ground and be unable to rise until, gradually, it picks up heat (this may be related to diurnal and seasonal variations in ground temperature, or to cool air 'rolling down' slopes making valleys especially prone to such effects). This so-called 'temperature inversion', in which near-ground air masses become very stable and unable to disperse industrial emissions, is a particular cause of pollution blackspots.

Pollution impacts have often been assessed in relation to the concept of 'critical loads'. The most important sulphur and nitrogen compounds that pollute the atmosphere are sulphur dioxide (SO_2), nitrogen oxides (NO_x) and ammonia (NH_3). These substances can affect ecosystems both directly and indirectly, the former through direct contamination of the vegetation and the latter via the eutrophication or acidification of the soil. Soil acidification arises because sulphur and nitrogen compounds in the air are converted into sulphate- and nitrate-forming sulphuric acid and nitric acid. A particularly important consequence of soil acidification is the dissolution of aluminium, which causes damage, *inter alia*, to roots, rendering them less able to absorb essential minerals. Moreover, the rapidly weatherable pool of aluminium in the soil, which is essential for sulphur buffering, can become depleted, so that the soil pH becomes very low. The critical load of nitrogen and acid on ecosystems can be defined as the amount the ecosystem can withstand in the long term without suffering damage from changes in the chemical composition in the soil solution. At a series of gatherings, international conventions and agreements have been signed during the last few decades in order to reduce poor air quality and the emissions of SO_2. These have included, for example, the 1985 Vienna Convention for the Protection of the Ozone Layer, and the 1987 Montreal Protocol on Substances that Deplete the Ozone Layer, as amended by the 1992 Framework Convention on Climate Change (Box 2.2).

To assist our understanding of climate change, general circulation models (GCMs) reflect a concerted effort to understand the causal factors which contribute to global climate change. Climate change – short-term changes in global climate, on a scale from tens of thousands to hundreds of thousands of years – appears to be a result of slight oscillations in the distribution and amount of solar radiation or *solar flux* reaching the earth's surface. These result from variations in the orientation and proximity of the earth to the sun. Some researchers suggest that there is a mutual interaction between global climate and ocean current circulation. Whatever the complexities and uncertainties of climate modelling one outcome is clear: human activity is capable of affecting the chemical composition of the atmosphere, global air temperatures and the buffering capacities of oceans, surface waters and soils. Whether this is serious is a debatable matter, but it is clearly something to which the 'precautionary principle' should apply.

Box 2.2 Some key international agreements on air, timber and wildlife

International Tropical Timber Agreement 1994 – ITT organisations' target that all forest products should come from sustainably managed forests; parallel Pan-European Helsinki Process and Non-European Montreal process for boreal and temperate forests.

CITES 1975 – a binding international treaty regulating trade in wildlife and plants to help protect species threatened with extinctions.

Montreal Protocol 1987 of the Vienna Framework Convention on Ozone-Degrading Substances (phasing out use of CFCs) (modified by 1990 London Ministerial Conference on Ozone; 1992 Conference in Copenhagen). Kyoto Protocol 1997 to achieve reduction in greenhouse gases.

In respect of the Kyoto Protocol, it has been noted that:

The target agreed for the industrialised nations is a 5.2% reduction in greenhouse gas emissions from 1990 levels by 2012. This is less than the EU position and nowhere near the level many scientists argue is necessary to halt climate change. The rest of the world agreed to do what they could in the future with no firm timetable. This agreement also has loopholes as it will permit carbon trading with no agreed rules or limits, and the use of forests as sinks (tree planting to absorb carbon dioxide as an alternative or supplement to cutting emissions). Nor are there any compliance mechanisms. There is a real risk that there could actually be an increase in emissions in the early part of the next century, despite the agreement. Funding and enforcement measures are to be put in place at some future date. It remains to be seen what will happen over the next few years. Developing countries who thus far have contributed little to global emissions should not be condemned to continued relative poverty, whilst the US continues to enjoy cheap fuel and emit 25% of the world's total emissions of greenhouse gases.

(Battersby, 1998, p. 3)

IPCC (1996) 2nd Assessment Synthesis of Scientific-Technical Information Relevant to Interpreting Article 2 of the UN Framework Convention on Climate Change 1995 (WMO/UNEP Geneva) – this reaffirmed estimates of the sensitivity of the global atmosphere to increasing greenhouse gas levels; computer models suggested that global temperatures will rise between 1.5 and 4 °C if the atmospheric concentration of CO_2 doubles from its preindustrial level of 280 ppm. Current level is nearly 360 ppmv. The convention also concurred that global mean surface temperatures have increased between 0.3 and 0.6 °C and global sea level has risen an average of 1.0–2.5 mm/y over the past century. Despite reductions in estimates of particulates and phasing out of CFCs, it is still anticipated that surface temperatures will increase by 2.0 °C and sea levels rise by 0.5 m in the twenty-first century.

The Social Scientific Perspective

A traditional failing of environmental planning has been to seek solutions which are based on physical remedial measures alone, rather than on the capacities of people to drive environmental improvement. Indeed, environmentalists have too often viewed people as the problem rather than the solution. A key consideration is that environmental degradation often arises from differential levels of wealth, power and access to resources; thus, some problems are those associated with affluence and high levels of consumption and waste generation; other problems arise from the survival strategies of the poor, who allegedly cannot afford the apparent 'luxury' of environmental stewardship. Thus, issues of economic status and fair trade are central to environmental sustainability, so that environmentalism is often associated with the politics of 'inequality'. Equally, the use of environmental resources may be influenced by the politics of 'identity'. This comprises issues associated with race, age, gender, ethnicity and cultural background, which influence our lifestyles and life experiences. This experience may include poverty and disadvantage, or particular 'ascribed' roles in society, such as nurturing or wage-earning. An understanding of these is both important to issues of inclusion and exclusion within mainstream society, and within the decision-making processes which influence social change. It is hardly realistic to expect alienated, excluded people ('others') to participate willingly in positive environmental actions, especially where these are proposed by the mainstream and privileged voices in society ('dominant, normative discourses').

Policies and Politics

Environmental politics is increasingly concerned with the power relations amongst groups of people, and between people and the environment. This political movement has often been manifest in organisations outside mainstream party politics so that, rather than developing general-purpose manifestos across the range of social and economic concerns, single-issues groups have become advocates of one specific aspect of the environment (such as countryside access or marine sewage disposal). These pressure groups and protesters typically seek to influence official policy or business actions, sometimes by dialogue and sometimes by confrontation. More generally, environmental politics inquires deeply into the nature of power in society, including the ways in which corporate interests strive to influence public opinion and behaviour.

An increasing priority of planning is to incorporate the views of people in designs for place. Although public participation has been a statutory requirement for almost 30 years, this has rarely been undertaken with imagination and vigour, and more innovative approaches are therefore starting to be used to engage citizens' aspirations. In many respects, the paucity of participation in planning mirrors the alleged failings of contemporary decision-making in society more generally. Some observers claim that political decisions have increasingly been reached through covert agreements between politicians and unelected individuals and groups, such as civil servants, quangos and industrial leaders. Indeed, many nationally important decisions happen almost by default, as they are driven by transnational flows of highly mobile capital and by supranational organisations. The largest environmental organisations are themselves not innocent of high-level manipulation of political decisions. This trend, which results in a relative inability of ordinary people to represent their views in an effective manner, is referred to as a 'democratic deficit', and is a matter of considerable current concern.

Two closely linked topics are of importance in helping to redress this deficit. One is the encouragement of a higher level of active citizenship amongst the population. Free citizens in liberal societies normally enjoy a number of rights (e.g. in relation to voting and standards of justice) and recently some of these have extended to 'charters' of entitlements relative to the quality of public services or environmental standards; participation in these universal benefits may be thought of as 'passive citizenship'. Equally, it may be supposed that individuals have responsibilities or duties towards their communities and may engage in public-spirited activities which have the effect of creating social capital and of cementing communities; this may be thought of as 'active citizenship'. In recent years, the idea of citizen responsibility towards fellow humans has extended to ethical duties towards other species and to the earth more generally, so that a concern for active 'environmental' citizenship has emerged (e.g. Hill, 1994; Eder, 1996; Healey, 1997).

Another way in which democratic deficit can be challenged is through public agencies opening up opportunities for citizens to 'reason together' within influential arenas. This possibility is referred to as 'communicative rationality', and supposes that the local and central states can facilitate the expression of argumentation and debate in ways which will enable ordinary citizens to regain greater control over their lives (and, relative to environmentalism, their local places). These considerations are clearly critical to the ways in which environmental planners seek to manage biophysical resources in an equitable and sustainable fashion. In practice, active environmental citizenship is limited by various influences. Harrison *et al.* (1996) suggest that the key constraints inhibiting further acceptance of responsibility among citizens revolve around: the extent to which individuals are able to make judgements about the validity of environmental claims made by experts, pressure groups and the media; the uncertainties about the severity and causes of environmental change, which is encouraging 'cynicism and doubt among lay publics'; and the balance of trust and mutual respect between citizens and governments.

The depletion of environmental resources has often been represented as a straightforward phenomenon of intensive spatial competition for finite assets: this could range from the exploitation of international resources such as fisheries stocks, to local amenities such as rural beauty spots. Whilst this view is often now dismissed as 'naive', it still possesses a fundamental validity, and pressure on resources (especially land) is an important predictor of planning issues and the style of planning in a particular country. Many analyses, however, tend to emphasise a 'political economy' perspective, in which the relative availability of resources is seen to depend on an amalgam of rules and relationships connecting governmental and industrial interests. One variant of the 'political economy' school seeks to establish the relationships between these interests and to explain them in terms largely of the subsumption of external capital into locally fixed assets, which then constrains the way these assets can be exploited (i.e. the way that powerful and wealthy external interests invest in local land and commercial enterprises enabling them to influence political decisions and residents' opportunities).

This critique is often associated with a structuralist standpoint, whereby instances of environmental transformations are seen merely as surface expressions of deep-seated social and economic structures. General structures within society may then be activated by specific agents of change, so that broad interest groups are represented in a particular case by the actions of one individual or group. This leads to explanations of change based on a fusion of 'structure' and 'agency'. Change in structural elements tends to lead to the restructuring and reorganisation of capital assets and in turn influences the restructuring of urban and rural localities. At other times it tends to focus more on the microscale, such as the behaviour of competing interests within local communities, though these of course may link to political and economic affiliations beyond the host locality. Within the *political economy*, therefore, various corporate interests, individuals and their representative organisations, and the state contest the use of the environment. Legislative and landownership frameworks are employed to regulate and legitimise transfers between land-use activities and sometimes to influence the management of natural and social resources. In the context of the British countryside, there are various types of power broker, such as farmers and landowners (supported by their representative organisations); the middle classes, prominent in local government and politics; and development companies, who exercise considerable control over housing production and enjoy corporatist links with the planning system (cf. Marsden *et al.*, 1993; Murdoch and Marsden, 1994). This type of analytical approach is a powerful one, and can be adapted to a great many types of environmental conflict in both developed and developing countries. An alternative, more liberal interpretation of the 'political economy' uses developments of neoclassical economics to explain or influence individual and collective use of environmental resources (Box 2.3).

Box 2.3 Some political economy perspectives

Winter (1996) has reviewed the countryside policy process in terms of certain political science ideas, which revolve around three main pivots of policy initiation, policy formulation and policy implementation. On the input side are the political demands of society expressed through parties, pressure groups and election results. These political demands, together with the resources available to government, combine to determine policy programmes and outputs, which in turn influence society; social fortunes then affect future political demands and, through the economy, the resources available to future governments. A 'structural' model embodies the official or constitutional version of the policy process, i.e. that Parliament represents and interprets the public will through its representatives, who formulate executive policies which are faithfully implemented by civil servants. Thus, the two key actors in the policy process are Parliament and the Civil Service, and the key concept is parliamentary sovereignty. Rather more plausible models of this process have emphasised the role of political parties which strongly influence the types of policy considered; similarly, the Cabinet and Civil Service prove to have a more active role in actual policy production.

The 'pluralist' model is probably the single most influential account to have been developed by political scientists seeking to explain the complex nature of the modern policy process. Pluralists assign a particular importance to the dispersed and noncumulative character of the distribution of power in society, and within that framework to the role of interests and of pressure groups. Thus, it is seen as a theory of bargaining between autonomous, often competing groups and a fragmented state, in which the emphasis is on the flow of influence from the groups to the state within an inbuilt set of checks and balances which supposedly prevent any one group becoming too powerful (Grant, 1985). However fragmented the political structure, some government agencies are more powerful than others and some groups enjoy closer relations with those agencies than do others. In Britain this is quite clearly the case and has led to the idea of 'policy communities' and 'networks' as means of explaining particular features of pluralism.

The 'élite' model emphasises the importance of a 'unified and all-pervasive ruling class'. There is, however, not necessarily an ultimate unity of purpose amongst the élite; rather, there may be a distinction between the political *élite* (comprising those exercising power) and the political *class*, which encompasses all those involved in the policy process, including the élites of organisations not wielding power as such (e.g. opposition political parties). Thus, most exponents of the élite model suggest that in democratic systems there are competing élites, often based around political parties.

The Marxist model builds upon the concept of an élite, and equates this with a specific class interest (where the class is defined by its economic position within society). Within complex society, this power cannot always be wielded directly, but is accomplished in many ways through the exertion of direct and indirect pressure on government and through the legitimation of certain ideas and actions through control of the media. According to some interpretations, the freedom of action by politicians or civil servants is severely limited by the economic power of capital; this can be described as an 'instrumentalist' position, in which the state is seen as an instrument of the bourgeoisie. This position is critiqued by the 'structuralist' Marxist tradition.

Finally, the 'corporatist' model ascribes particular force to pressure groups, and asserts that government agencies are in no way neutral or impartial recipients of policy pressures but are active in sponsoring and recognising particular pressure groups. Crucial to corporatism is the direct link with regulation, whereby representative groups assume some responsibility for the self-regulation and disciplining of their own constituency in return for the privileges afforded by their relatively close relationship with government. A classic example of this is often cited as the policy proximity of the National Farmers' Union to the Ministry of Agriculture, Fisheries and Food.

Another key aspect of the policy process is that of policy implementation – that is, the way in which planning and policy measures move from the abstract world of ideas and objectives to the real world of results on the ground. A helpful way to consider the issue is to examine a number of important preconditions for perfect policy implementation as set out by Hogwood and Gunn (1984), namely, that:

- the circumstances external to the implementing agency do not impose undue constraints;
- adequate time and sufficient resources are made available to the programme, and that the required combination of resources is actually available;
- the policy to be implemented is based upon a valid theory of cause and effect, with a fairly simple linkage between cause and effect;
- dependency relationships are minimal;
- there is understanding of, and agreement on, objectives;
- tasks are fully specified in correct sequence;
- there is perfect communication and co-ordination; and
- those in authority can demand and obtain perfect compliance.

Thus, implementation is not only reliant on formal organisational hierarchies, communication and control mechanisms, but also involves a multiplicity of actors and agencies, together with their value systems and interests, and the associated methods of bargaining. Not surprisingly, these perfect conditions are never met, and so an 'implementation gap' always exists to a greater or lesser degree. This gap may be represented in terms of 'barriers' to the attainment of environmental goals (e.g. Trudgill, 1990), though we must not be unduly negative, and must also recognise the significance of bridges and gateways to successful implementation.

An important part of the policy process is the evaluation of policy performance. This became an especially important process during the 1980s era of 'new right' politics, and the explicit evaluation of performance against defined criteria became an enduring legacy of this political period. In particular, policy evaluation has become a major concern in the UK, as government has sought to ensure that policies are more cost-effective. Key concepts in policy evaluation are:

- additionality (the additional achievement of outcomes which would not have occurred in the absence of a given policy);
- compliance costs (the costs of complying with a particular policy, for example the economic burden on industry of achieving a given reduction in pollution levels);
- cost-effectiveness analysis (in which policy benefits are maximised relative to minimised public and private implementation costs);
- externalities (positive and negative 'neighbourhood' effects arising from a policy or project);
- gearing (the ratio between public monies invested and private funds levered as a consequence of public pump-priming);
- investment appraisal, based on the ratio of costs to benefits (represented as the return on initial investment); and unit costs (the cost of

providing a specific product or service, which can then often be targeted for reduction, in a drive for efficiency); and

- objectives-based performance (the extent to which specific evidence about a programme indicates that it has achieved its initial policy objectives).

All these denote a strong emphasis upon evaluation as a tool for improving economic performance of government programmes and, indeed, introducing market disciplines to public sector operations (cf. Winter, 1996).

Some would argue that this is a limited view of evaluation, which assumes that the policy objectives are clear and can readily be translated into financial terms. Winter (*ibid.*) suggests that a more general approach would be to

- establish the broad policy aims;
- identify specific objectives or desired outputs;
- identify agencies and individuals responsible for implementation;
- specify key stages in implementation; and
- assess achievement.

Moreover, research by economists over the past 20 years has helped us to convert some of these rather brutally financial measures into ones which reflect similar types of achievement in ecological terms. For example, Colman (1994) produced an example of a criterion-based policy 'evaluation by objectives' in relation to conservation instruments. Criteria of 'policy effectiveness' were based on: protection of wildlife sites; protection of large-scale land-use designations or landscapes; enhancement and creation of new interest; timeliness; targetability; monitorability; payment systems; cost efficiency; political acceptability and transparency; and promotion of conservation-mindedness. A range of conservation instruments were then evaluated in terms of their relative performance in relation to each criterion (Table 2.7).

Finally, it is important to have regard to the ways in which various networks and communities influence the behaviour of environmental actors. At the most senior level, it has frequently been observed that government departments and agencies, together with industrial, trade union and other pressure group interests, behave as a loosely coupled community within which policy options are framed. This applies especially where interest groups gain 'insider status' and can obtain formalised conduits into policy mechanisms in return for agreeing to support the policy positions which are eventually reached through the process of negotiation and consultation. This is one specific possibility, but in more general terms there appear to be policy communities circulating around all major public issues, including those concerned with the environment.

Looser still, but none the less potent in terms of influencing long-term change, is the existence of 'knowledge networks' through which users of the environment gain their skills and information as managers. This may

Table 2.7 Criteria for comparative analysis of conservation instruments

Criteria of effectiveness	Management agreements		Standard payments: ESA type	Public sector land purchase/ownership	Grant-aided land purchase	Grants (e.g. s.39 WCA)	Capital tax relief (e.g. from IHT and CGT)	Cross-compliance	Covenants
	Regulatory (e.g. s.15 and s.41)*	Incentive (e.g. s.39)*							
Protection of wildlife sites	Tailor-made for this purpose. Some instances of wilful damage in notification period. Negotiations almost invariably produce an agreement. Fair degree of long-term protection	May aid conservation indirectly by improving the quality of an area	Cannot guarantee to elicit voluntary agreement on key individual sites. Prescriptions too general to hit specific targets. Ten-year contracts offer medium-term protection	Ideally suited to long-term conservation. Provides very strong guarantee of protection, subject to continued political will	Voluntary bodies play a vital role in conservation of key sites. May be some conflict between VB objectives (e.g. access) and conservation. Quality of management high to fair	Not designed for protection in a continuous manner	Linked to individual holdings that successfully apply for tax exemption. Reactive instrument, therefore haphazard. Entails imposing individual management agreements.	Not suited to individual direct conservation benefits	Provides for range of restrictions and positive requirements on individual sites. Not clear how effective they are
Protection of large-scale land-use designations or landscapes	Is employed on relatively large areas, e.g. Leek Moors/Govt Valley. No obvious reason why it cannot be successful	May aid conservation indirectly by improving the quality of an area	Developed for large relatively homogeneous habitat types within larger boundaries. Could be extended to apply nationally to specific land-use systems	Not ideal for large-scale, rapid applications. Opportunities for land purchase are limited. Supporting or supplementary role	VBs may not be interested in all land supported by ESAs. As for public sector purchase, opportunities are limited	Not applicable	Can apply to special smaller sites and larger areas. Provides long-term protection. Use could be expanded	Is envisaged as tied to support regimes rather than designated areas. There are possibilities for controlling stock levels in upland areas qualifying for payments	Not ideally suited as covenants arise property by property
Enhancement and creation of new interest	Not applicable	These are specifically designed to promote enhancement and access subject to opportunity	Incorporate standard contracts for enhancement, e.g. arable to grass, dyke rebuilding, heather regeneration, etc. Capable of widespread application, e.g. as through CPS†	Public ownership can be used for enhancement	Grant aid can be made conditional upon investment in improvements by the VB	One-off standard grants used specifically to promote enhancement. Where only grant aid is available, encouragement biased to more commercial investments	Could encourage enhancement agreements and conservation investment. But Treasury secrecy has meant public unaware of use values	Not ideal for this purpose. Difficult to devise annual investment programmes of right size. Where set-aside is cross-complying condition, could supplement by grants for special works	Limited impact in the absence of complementary grants
Timeliness	Designating SSSIs has been a slow process, and creation of National Parks even slower	Opportunities may take time to mature. Budgetary constraints may relegate incentive agreements to a lower priority level	Rapid once ESA or special area has been designated	Has to be activated quickly when land comes up for sale	Has to be activated quickly when land comes up for sale	Capacity for rapid response	Not timely because driven by tax planning of landowners	Applies across the board immediately legislation is passed	Can be rapidly imposed by existing owners
Targetability	Can be focused very precisely	Can be focused very precisely	Voluntary take-up, cannot target individual sites as opposed to larger areas	Can be targeted precisely, but lack of opportunities may prevent use	As with public purchase	Are transparent. Targetable on types of work and/or specific areas	Not targetable in so far as triggered by holding owner. Targetable in so far as applications may be rejected	A blunt instrument, difficult to target	Target specific

Criteria of effectiveness	Management agreements			Public sector land purchase/ownership	Grant-aided land purchase	Grants (e.g. s.39 WCA)	Capital tax relief (e.g. from IHT and CGT)	Cross-compliance	Covenants
	Regulatory (e.g. s.15 and s.41)*	Incentive (e.g. s.39)†	Standard payments: ESA type						
Monitorability	SSSIs requiring sensitive management may be costly and difficult to monitor	Ready monitoring of works undertaken	Many negative restrictions are difficult to monitor fully	Makes monitoring easy	Rely on committed VB to undertake monitoring	Monitoring easy with capital works, but costly with small grant-aid works	Responsibility for monitoring lies with various public bodies. Effectiveness may depend upon type of agreement	As for ESAs	Monitoring variable. Depends upon resources and interest of covenantee
Payment system	Annual (individualised or standard) and lump sum. Latter carries fewer restrictions	Lump sum or annual payments. Negotiable	Annual, 10-year period. Subject to periodic revision	Purchase cost plus annual management costs partially offset by farming revenue	Part of purchase cost. But, land may be eligible for ESA-type payments or MAs with payments	Variety of systems, but dominated by one-off payments. Current policy seems to make only limited provision for maintenance	Cost is forgone revenue to Treasury plus administrative set-up cost to CoAg. EN	No additional budgetary cost	Donated free, or covenants may be purchased by buying land and selling subject to covenant
Cost efficiency	On a per ha basis, profit forgone payments for s.15s likely to exceed standard ESA payments, but this more than offset by larger-scale uptake of standard payments. Total cost probably lower than for equivalent ESA	Efficient, as payments individually negotiated in relation to costs incurred	Total budgetary cost relatively high. Stimulates uptake on land not under threat. Useful income support instrument	For individual sites this can be more cost efficient than ESA or s.15 management agreements. Is the one case where budgetary cost buys an asset	Given low proportion of purchase cost typically granted, this is likely to be the cheapest option, provided VB management is of good standard	Efficient in sense that grants are based on cost of work done, or on average cost. Can provide some labour income to small farmers	Not possible to assess. Information is confidential. Treasury method of assessment not publicly available	Switch from old support system to new one which is cross-complied may be very efficient. New transfer costs offset by lower surplus management costs	Little evidence. Should be cost effective. Lower budgetary cost than land purchase
Political acceptability and transparency	Sound principle but engenders element of conflict between NCC and farmers. Not transparent	No obvious barriers to acceptability	Has become highly politically acceptable to farmers and MAFF	Acceptable on low-profile, small-scale basis. Would be political resistance to widespread use. Indeed, government pressure to sell public land	Most studies rank this option at the top. No political resistance	No inherent political resistance. Conservation grants favoured above farm investment grants. Transparent	Useful way of protecting heritage. Capable of expansion. Confidentiality restricts transparency	Makes continuance of direct agricultural subsidies more acceptable. Needs to be transparent	No obvious barriers to acceptability
Promotes conservation mindedness	Held to have negative impact	Positive	Positive or uncertain effects	Possibly neutral for farmers	Possibly neutral for farmers	Positive	Weak – main stimulus is tax reduction	Can be positive	Neutral buyer or tenant accepts covenant

Notes:
* s.15, Section 15 of the Countryside Act 1968; s.39 and s.41, Sections 39 and 41 of the Wildlife and Countryside Act 1981.
† Country Premium Scheme.
Source: Colman, 1994.

include formal knowledge mechanisms such as education or extension services, or less formal 'tacit' knowledge shared through personal contact,
which may be more trusted than 'official' government or industrial sources.
Penetrating these knowledge networks is extremely important for organisations which wish to influence environmental behaviour. A further way of
representing scientific and policy decisions is through the process of 'trans-
lation' whereby 'laboratory' knowledge becomes converted into specific
products or methods of problem-solving. In this model, solutions to a particular problem gradually gain widespread acceptance as the more powerful actors in a network displace others, gain status as spokespersons, and
invoke means of enrolling other people in their solution. When a sufficient
level of consensus has been levered regarding the acceptability of the solution, the network stabilises and becomes very difficult to dislodge as the
standard remedy or prevalent wisdom. This process is explained in terms of
'actor-network' theory, and it may effectively be used to describe a range
of environmental planning and management approaches. Although different labels can be put on these various types of network, in practice they
overlap considerably (Selman, 2000).

Social Constructions of Environments

The previous discussion suggests that there is an objective 'reality' of the
environment, which is intrinsically capable of being documented and evaluated rationally. Most people would, however, accept that reality is at least
partly socially constructed, though not everyone would agree with a 'relat-
ivist' stance which abandons any idea of objective reality and which claims
that all descriptions and analyses of a phenomenon are subjective 'representations'. One of the classic representations of rural reality is that of
Murdoch and Marsden (1994), in which they identify two main sets of
concerns in rural sociology. The first is that of 'agrarian political economy',
concerned, primarily, with the role of agriculture in food commodity systems, and focusing on the long-run tendency for the food industry to reduce the contribution of agriculture to the food production process. The
second is the 'restructuring hypothesis', which uncovers the way rural resources of land and labour are increasingly integrated into circuits of industrial accumulation. It challenges the way in which we perceive rural
localities to be distinctive and specific in their nature (cf. the concept of
'specificity'), arguing that their economic complexion derives from past
'rounds of investment' as capital seeks out profitable forms of production.
In many areas this process has led to the subordination of agriculture, so
that rural areas become increasingly attractive to other industrial sectors.
This new source of investment turns out to be uneven in its geographical
incidence, and thus helps to explain patterns of spatial differentiation,
based on both variability in the distribution of suitable resources and in the
incidence of and consumption opportunities for a range of different income

groups. Inevitably, because of the centrality of land-use change as an obser-
vable phenomenon, land-use can be seen as a reflection of the variety of
economic, political and sociocultural relations in the countryside. The discrete social demands on land, the political rules that surround its transfer
between uses and the tendency for capital to become 'fixed' in land thus
produce a series of segmented land development markets comprising: agriculture, forestry, industry, mining, housing and leisure.

As a result of studying development as a dynamic interaction between
'internal' struggles to promote or resist development, and 'external' pressures to promote development, the environment reflects the uses or values
imposed upon it. Relative to 'the countryside' it can be claimed that four
'ideal types' emerge (Marsden *et al.*, 1993), namely:

1) 'Preserved' countryside, which is characterised by anti-development
   and preservationist attitudes, mainly expressed by new social groups in
   the countryside. These may impose their views through the planning
   system on would-be developers, and demand from these fractions may
   also provide the basis for new development activities associated with
   leisure, industry and residential property. The reconstitution of
   rurality is often highly contested by articulate consumption interests
   who use the local political system to protect their positional goods.
2) 'Contested' countryside, which lies outside the main commuter catchments and may be of no special environmental quality. Farmers (as
   landowners) and development interests may be politically dominant,
   but may increasingly be opposed by incomers adopting the political
   positions of their counterparts in the 'pressured countryside'.
3) 'Paternalistic' countryside, where private estates and large farms still
   dominate, and the development process is decisively shaped by
   established landowners. Many estates may be faced with falling incomes and therefore looking for new sources of income, through
   diversification.
4) 'Clientelist countryside', in the remoter areas, where agriculture and
   its associated political institutions (together with forestry, water catchment, mineral extraction, etc.) still hold sway, but are often only sustained by state subsidy. Processes of rural develpment are dominated
   by farming, landowning, local capital and state agencies, and local
   politics tend to be conditioned by the need to maintain employment
   levels and to retain local facilities and residents. Conservation interests
   in these areas are often represented by outsiders (e.g. environmental
   groups located in affluent parts of the country), and may be resented
   by locals.

Economics of the Environment

The economic perspective is nowadays central to understanding our use of
the environment. In essence, planning is necessary because of market

failures, due especially to the existence of 'externalities', i.e. external effects which distort investment decisions and prevent the 'unseen hand' of the market producing outcomes which are of the greatest benefit to society. Planning is based on the notion of internalising externalities in the land market, by including the social interest within decisions surrounding the allocation and use of land. The broader idea of actual or imaginary ('shadow') markets being associated with any environmental good has become integral to the framing and implementation of environmental policy. Some economic principles underlying environmental decision-making are illustrated in Box 2.4.

Earlier, we noted the principle of 'materials balance': this concept can be expressed in economic terms, whereby the economy is portrayed as an open system pulling in materials and energy from the environment and eventually releasing an equivalent amount of waste back. Too much waste in the wrong place at the wrong time causes pollution and so-called external costs (externalities). A certain amount of recycling of wastes is both possible and practicable, but 100% recycling is physically impossible and economically undesirable. A principal contribution of economists, which has emerged from the preceding discussion, is that of striking the most efficient balance between exploitive use of the environment and the protection of natural resources and environmental quality for future generations. Economists generally like to see this issue tackled by market instruments rather than regulatory legal instruments (they see the former as being more efficient than the latter), and Turner (1995) has identified the basic economic instruments as:

- direct alteration of price or cost levels, e.g. emission charges;
- indirect alteration of prices or costs via financial or fiscal means (e.g. incentives to adopt cleaner technologies); and
- market creation (e.g. tradable emission permits).

It is noted that emission targets are related to the quantity and quality of the pollutant and the damage cost inflicted on the environment; user charges have a revenue-raising function and are related to treatment costs, collection and disposal cost, or the recovery of administrative costs (not directly related to damage costs in the environment). Product charges are levied on products harmful to the environment when used in production processes, or when consumed or discarded. Tradable permits may comprise quotas (e.g. of food production or pollution discharge), allowances or ceilings on pollution levels (Box 2.5).

The second main contribution of economics is in relation to the trade-offs between environmental costs and benefits (Figure 2.7). The basic principle of this is that future streams of value associated with an environmental good (such as gold or timber) can be translated into present-day terms by a process of 'discounting' (see pp. 60, 134). Thus, the costs and benefits of exploiting a natural resource can be compared by reference to the ratio between its streams of long-term costs and benefits (cost-benefit

Box 2.4 Trading off resource consumption and environmental quality

Production possibility curves (PPCs) can be used to depict the relationship between marketed output and environmental quality. A PPC is a way of depicting diagramatically the choice faced by a group of people between two desirable outcomes. Consider the graphs in Figure 2.7, in which the vertical axis has an index of the aggregate economic output of an economy, i.e. the total market value of conventional economic goods traded in the economy in a year. The horizontal axis shows an index of environmental quality. The curved relationship shows the different combinations of these two outcomes – marketed output and environmental quality – that are available to a group of people who have a fixed endowment of resources with which to work.

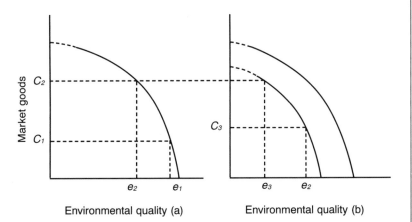

Figure 2.7 Production possibility curves (PPCs) (a) today; (b) at a hypothetical future date
Source: Field, 1994

The PPC is determined by the technical capacities in the economy together with the environmental constraints. Thus, in graph (a) if the current level of economic output is C_1, we can obtain an increase to C_2 only at the cost of a decrease in EQ from e_1 to e_2. But while the PPC is itself a technical constraint, where a society chooses to locate itself on its PPC is a matter of social choice. In the longer term, we may run down the EQ if we continue to deplete our stocks of environmental capital, and a significant drawing down of that may have serious negative effects on the ability of the economic system to sustain itself. So when considering future generations in graph (b), we may find that they have an inner production possibilities curve which has the same level of present marketed output C_2 but a lower level of environmental quality e_3; or they have the same level of environmental quality with a reduced level of marketed output C_3 (Field, 1994).

analysis). This means calculating a demand curve for natural goods, some of which have no market value. Indeed, even if a market value does exist for an environmental commodity, this is probably a serious underestimate of its inherent worth, as it will reflect only the immediate economic needs

Box 2.5   Making markets work for the environment

The former Department of the Environment (1993a) sought to develop economic instruments which could help overcome the problem of externalities by attaching a price to using the environment. They argued that economic instruments could help overcome the problem of externalities by attaching a price to using the environment. The value of environmental assets used would then be allowed for in the production process in the same way as other market costs: producers would be given a choice between investing in pollution abatement or paying to pollute the environment. If there was competition for sales, much of the cost would be passed on to consumers in the form of higher prices (which, in the long run, is an equitable solution).

The advantages of economic instruments prove to be:

- cost-effectiveness (producers know their own possibilities better than regulators);
- innovation (instruments can drive desirable change amongst producers);
- flexibility (producers and consumers are free to take their own decisions);
- information (only individual businesses have the detailed information necessary for improved practices); and
- public revenues (economic instruments usually result in the payment of taxes or permit levies).

However, the fairness of economic instruments must also be borne in mind, as they will tend to make businesses differentially better or worse off.

The key instruments available are:

- *emissions charges* – which put a price on each unit of polluting effluent emitted;
- *tradable permits* – which allocate quotas to producers for particular shares of total emissions, but then allow the quotas to be bought and sold;
- *product charges* – for instance, levies on packaging to reduce the amount of that product used;
- *deposit/refund schemes* – which give users an incentive to return a material or product;
- *legal liability* – a civil legal liability can act like an economic instrument, by encouraging investment in reducing pollution/hazards; and
- *subsidies* – these are not very popular amongst most governments, but involve payment per unit of environmental benefit or pollution abatement achieved.

of the present generation. Its value in perpetuity, or its indirect role in contributing to the maintenance of the earth's life-support systems, will probably be relatively neglected. Consequently, a large part of environmental economics is concerned with calculating realistic costs and benefits of the use of natural resources. Various techniques have been proposed by economists over the last 30 years and an especially favoured one at present is that of 'contingent valuation'. The effectiveness of economic valuation of environmental goods is, however, limited by the key methodological dilemmas, namely:

- it is very hard to measure environmental benefits;
- values are affected by ability to pay (as well as inherent willingness to pay); and

- estimates of willingness to pay depend on respondents' knowledge about the good.

A variety of evaluative methods is available for the estimation of quasi-market values for nonmarket environmental goods (Figure 2.8).

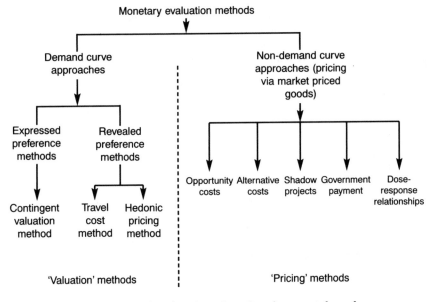

Figure 2.8   Methods for estimating the value of environmental goods
Source: Bateman, 1994

The use of economic principles has also sharpened our understanding of sustainable development to the point at which it can become a significant influence on public policy. This has entailed accounting for biophysical resources in terms of their 'capital' value which, if they have a market value at all (wholesome air or clean seas are 'priceless' for instance), is almost always under-represented by the free market (which takes no account of the broader and longer-term values of a resource, beyond its immediate price as a traded commodity). Thus, it has become important to perceive biophysical resources in terms of natural capital, which may be 'constant' (capable of being replenished, such as buildings or commercial timber plantations) or 'critical' (integral to the earth's life-support system but not easily replaced, either in terms of functionality, quantity or quality, if damaged or destroyed). This allows nature to be costed within, or to act as a constraint upon, development decisions which, of course, assumes that we can place realistic valuations on (unpriced) environmental assets. To achieve this, values may be imputed either by pricing an economic good which is closely related to the asset (e.g. willingness to pay for travel to a scenic viewpoint) or by constructing various types of imaginary ('contingent') market in which interviewees are invited to make carefully controlled bids for unpriced resources (Box 2.6).

Box 2.6   How can we put an economic value on a priceless environmental resource? A contingent valuation example

Willis *et al.* (1996) studied the values placed on ecological improvements in the Pevensey Levels Site of Special Scientific Interest (SSSI), paid for out of public funds under the Wildlife Enhancement Scheme (WES). Their approach entailed many detailed methodological assumptions, but broadly it involved:

- interviewing three categories of respondent (households visiting the SSSI, nonvisiting households living relatively adjacent, and nonvisiting relatively distant households);
- constructing a carefully worded questionnaire based on 'what is the maximum your household would be prepared to pay for the WES in the Pevensey Levels?';
- calculating median bids from each group and estimating the total economic value of the site by aggregating across all the households in the populations from which samples were drawn; and
- calculating the benefit-cost ratio on the basis of the aggregate bid (the ratio of the respondents' 'willingness-to-pay' (i.e. perceived benefits) to the actual costs spent on the WES in the Levels).

Valuing environmental goods and services is essential in order to be able to compare proposed investments with alternative options, including those of leaving the environment in its undeveloped state (Box 2.7). Valuation of the total stream of costs or benefits is achieved through the application of discount rates, which allow comparison of alternative schemes in terms of their ratio of long-term benefits to costs. Discount rates may loosely be thought of as the inverse of interest rates; whereas interest rates permit an initial investment to increase by a specified amount towards a future point, discount rates are used to decrease future income streams by a specified amount so that they can be expressed in present-day worth. In other words, an apparently large sum of money in the future may be worth relatively little in present-day terms, once inflation has been considered, and the 'opportunity cost' of being unable to invest the money in more rewarding ventures has been taken into account. This ability to quantify the expected income stream in each future year, to aggregate these annual streams and to deflate the total sum of income by a suitable discount rate, is crucial to investment decisions regarding environmental resources, as it enables the long-term costs and benefits of one course of action to be compared with another. It also considers distributional issues associated with who gets the benefits and who pays the costs, namely, issues of horizontal 'equity' (treating similarly situated people the same way), and vertical equity (establishing how a policy impinges on people who are in different circumstances).

Box 2.7 Estimating environmental costs and benefits

Environmental cost-benefit analysis requires estimation of the:

Benefits – estimated by

- indirect methods – e.g. travel cost data and materials damage; and
- direct methods – such as contingent valuation.

Costs – which typically comprise

- opportunity costs
- costs of single facilities
- enforcement costs
- regulation costs (enforcement by public sector)
- regulation costs (compliance by private sector)

Risk and Hazard

An important concept in environmental management is that of hazard, which has traditionally often been thought of as a natural or technological issue, but is now seen to be inseparable from human behaviour. The statistical chance of a hazard occurring to a given group of people is its 'risk'. Many political decisions are now organised around minimising the risk of certain events to particular target audiences, such as employees, residents and children. If, however, we were to avoid every conceivable risk, it would result in total inaction and never undertaking any development. In practice, therefore, decisions are based on 'acceptable' levels of risk, one interpretation of which has been defined by the UK government guidance (DoE, 1995c). This states that the precautionary principle should not be used to generate completely hypothetical impacts, but that it is necessary to have some understanding of how a pollutant might cause an effect, even though uncertainties may remain as to its precise nature (Box 2.8).

Pickering and Owen (1997) note a range of natural hazards, which may cause death, injury, and the destruction and damage of agricultural land, buildings and communities. The effects of a natural disaster on a

Box 2.8 DoE – stages in assessing risks

- *Description of intention.* This requires a detailed identification of the nature of the receiving environment.
- *Hazard identification.*
- *Identification of the consequences.*
- *Estimation of the magnitude of the consequences* (broken down into – severe, moderate, mild, negligible).
- *Estimation of probability of consequences.*

Source: Based on DoE (1995c).

community depend upon factors such as the magnitude and extent of the disaster, how prepared the affected population is and its economic resources to mitigate a potential disaster and/or clean up afterwards. Thus, *geological and geomorphological hazards* include building collapse, fires, landsliding (Box 2.9) and subsidence, glacial hazards, tidal waves, flooding, release of poisonous gases, and associated hazards such as contaminated or depleted water sources, disease, famine, injury and death. *Meteorological hazards* include tornadoes, tropical cyclones, flooding by heavy rainfall and storm surges, disease associated with contaminated water sources, thunder and lightning damage, hail storms and droughts. *Biological hazards* include pests and environmental diseases, which may harm crops, animals and people. Increasingly, too, societies have become more vulnerable to *industrial* hazards in which serious and immediate accidents may occur (such as chemical explosions), as well as more insidious, widely disseminated, invisible damage (such as acid rain). *Hazard* thus refers to the nature of the event, whereas *risk* refers to the mathematical probability of its occurrence.

Box 2.9 Landslides: an example of a natural hazard encountered by planners

---

Responses to problems of slope instability:

- **Emergency response and crisis management** *Dealing with the problems only after they arise* – planning for emergency procedures may be needed where development has already taken place without reference to the stability of slopes; this response involves avoidable costs and may be unacceptable where public safety is at risk.
- **Planning for losses** *Spreading the losses by insurance, statutory compensation or fiscal measures* – the risk of losses may need to be accepted where development has already occurred on unstable slopes; this response involves avoidable costs and may be unacceptable where public safety is at risk.
- **Modify the hazard** *By prevention or correction* – slope stabilisation measures, rock-fall protection structures, good practice in property maintenance and restrictions on certain engineering activities may help to prevent or reduce the losses arising from landsliding; this response may be justified where there is significant risk to public safety or buildings and structures; implementation may be constrained by economic landownership considerations.
- **Control the hazard** *By avoidance* – the identification of landslides and the use of hazard assessment in planning control may enable the effects to be avoided altogether by not developing in landslide areas; planning and building controls may enable losses to be avoided or reduced through design and construction measures.

Source: Based on DoE (1996b).

A particularly influential philosopher in recent years has been Ulrich Beck (1992; 1996), who has argued that our former industrial society, which was mainly concerned with the production and distribution of 'goods', is being displaced by an emergent 'risk society', which has to contend with the production and distribution of 'bads'. This may be causing a break from the political concerns of industrial society, in which the main emphasis of party politics was on social inequalities and wealth creation. By contrast, in a risk society, traditional class and party political concerns are largely irrelevant to the ubiquitous nature of environmental risks, which affect everyone more or less equally, regardless of national and class boundaries. Risk society is thus recognisable not only by the problematisation of objective physical-biological danger, but also by a principle of *individualisation* in which agents become ever more free from normative societal expectations and traditional political allegiances. Beck argues that preindustrial cultures were societies of *catastrophe*; in the course of industrialisation they are becoming societies of *calculable risk*.

Society also creates the technology to optimise the benefits of biophysical resource use. Environmentalists have traditionally perceived technology in one of two ways: as a 'technical fix', which can supply ingenious means of reconciling high quality of life with minimum environmental impact; or as a palliative delusion which merely defers moving into a new and less demanding relationship with nature. These two perspectives remain quite polarised yet, in reality, there is potential for a reconciliation. Technology which is driven proactively by a need to optimise resource use, rather than added as reactive bolt-on to fundamentally inefficient devices, is creating new possibilities for sustainable living. Possibilities are even emerging for 'ecologies' to develop between large-scale industrial units so that there are strong functional relationships between waste-producing and waste-consuming plants (see p. 237, 'Industrial ecology').

An Ethical Perspective

Our use of the environment is fundamentally dependent on our ethical position. Biophysical resources can, on the one hand, be viewed as neutral stuff, readily available for humans to exploit in the pursuit of progress and wealth or, on the other, as sacred matter which we only use on a privileged basis and steward with sanctity. The study of ethics, generally, provides us with a set of rules for conducting ourselves morally in relation to other people; a *land ethic* puts limits on the ways in which we are entitled to utilise the earth's resources. It is sometimes argued that western society, with its emphasis on scientific reductionism and trade, possesses a very utilitarian land ethic, and has consequently tended to degrade the environment, whereas many indigenous low-technology peoples treated the land and waters more respectfully and have maintained their productivity. One expression of this is in relation to common resources, where a piece of land or stretch of water is available on trust to a community of people for responsible use in perpetuity: where selfish motives prevail, a few irresponsible people will overexploit the resource and precipitate a scramble for benefits ('tragedy of the commons'), but where a moral economy with reciprocal trust exists, common resources appear to be managed sustainably and responsibly. One line of argument is that the Judaeo-Christian tradition has led to a sense of 'ownership' of the earth by 'man', and that this God-given 'dominion' has led to an exploitative attitude to physical resources and a relative disregard for nonhuman species. Conversely, though, this tradition contains a strong commitment to 'stewardship' which is the very essence of an effective land ethic. Equally, elements of eastern and folk religions are alleged to have a more attuned relationship to the earth, whilst some pundits believe that postmodern attitudes in the next millennium will lead to a more sensitive relationship between society and nature.

Perhaps the only conclusion that can be gleaned from these contrasting viewpoints is that people's attitudes towards the environment have a tendency to arrange themselves along a continuum from reverence to disregard for the environment, but that the majority of attitudes will be somewhere between these extremes and will contain a differential balance between stewardship and human-centred utilitarianism. This spectrum has been most effectively represented by O'Riordan (1994a) whose analysis, despite repeated revisions and challenges, still portrays this fundamental truth. This spectrum may be summarised by suggesting that philosophies of the environment will broadly either be 'ecocentrist' (believing that social actions must always be constrained by ecological limits) or 'technocentrist' (believing that human ingenuity can overcome environmental constraints, and that there are no absolute limits). Characteristic elements of these are the 'technical fixes' of the more enlightened technocentrists, and the 'deep green' attitudes of some ecocentrists (see Figure 2.9). A complementary viewpoint is that of Myerson and Rydin (1996) which separates planners' views of the environment into those which are 'mundane' (e.g. hitting politically sanctioned targets) or 'sublime' (according with higher-order ethical principles). It may be noted (in terms of O'Riordan's taxonomy) that environmental planners and managers are positioned in the technocentric domain and, collectively, are characterised by a 'weak sustainability' stance; similarly, they are probably professionally more attuned to the acceptance of mundane, rather than sublime, sustainability goals.

A final aspect of relevance to the environmental planner is that of professional ethics. At present there is no single environmental profession with a unified set of ethical standards, and Dorney (1989) has considered in some detail the range of professional domains which have some bearing upon this area (e.g. ecologist, agronomist, biogeographer, conservationist, environmental engineer, environmental designer, environmental manager, environmental planner, environmental scientist, human ecologist, wildlife/fishery ecologist and environmental systems analyst). One of the benefits of a more unified profession, Dorney (*ibid.*) argues, would be its ability to integrate a range of philosophical, ethical/moral and knowledge bases. However, although there is no single profession which is representative of all practitioners in this area, all are at least morally constrained by standards which reflect honesty and integrity of practice, especially where

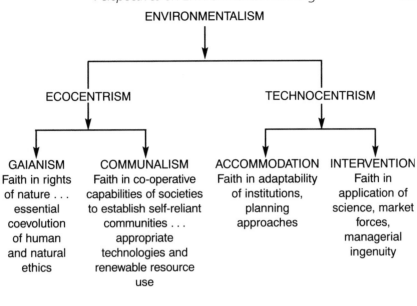

Figure 2.9 'Modes of greenness'
Source: Adapted from O'Riordan, 1994a

working with colleagues, clients and the general public. Indeed, some explicit codes of conduct have been published by professional institutes. These mostly emphasise the establishment, maintenance and promotion of professional qualifications, status and standards. However, it is notable that the Institute of Ecology and Environmental Management (IEEM), for example, emphasises that the responsibility of its members is to promote an 'ethic of environmental care' within the profession, and amongst clients and employers.

Conclusion

All planners need to be aware of the complex systems which sustain the earth's 'free services'. In terms of environmental dynamics, these include fundamental life-support systems associated with the cycling of rocks, minerals and water, as well as the flow of energy through ecosystems. These processes are relatively robust, and the earth has remarkable compensatory mechanisms to dampen the impacts of human activity. However, these systems clearly can be destabilised and even overwhelmed by insensitive use. Whilst planners in isolation cannot prevent nuisance, they can perform a central role in promoting environmentally friendly development.

Traditionally, the environment has been treated as a distinctive and specialised domain of planning, requiring an occasional input from environmental scientists. However, it is now clear that scientific expertise alone is insufficient as a basis for solving environmental dilemmas. Our rationalisation of the use of environmental resources requires an awareness of economic principles and means of analysing policy performance. Thus, social science and natural science methods merge imperceptibly in the search for sustainable ways of living. Equally, the purely rational approach has its limitations, and planners must be alert to the more subtle human beliefs and reactions which condition our perception of natural and industrial hazards.

A purely detached, neutral and rationalistic stance towards environmental decisions, though, may not necessarily be desirable. Professionalism is not simply a matter of making sound and impartial judgements, but also of conducting one's work in ways which uphold a set of ethical standards. In this respect, it is important that planners are aware of the philosophies which underlie our use of the environment. An ethical treatment of the earth's resources is axiomatically a key part of truly professional planning.

3 The Legal and Administrative Framework

Introduction

The complexity of environmental interests means that intervention by public agencies in the use of the environment can only be achieved through a variety of areas of legislation and tiers of administration. This means that planners must be aware of:

- the *policy context* in which environmental actions are framed;
- the *legislation and administration* underpinning environmental action, especially the legal instruments and executive agencies which enable objectives to be implemented; and
- opportunities for *integrating environmental action*.

These are clearly very wide-ranging issues, and the account here deliberately focuses upon those aspects of general environmental policy and legislation which are most pertinent to the spheres in which planners operate.

The Policy Context

Organisations normally operate within a policy framework. Internal policies provide them with a corporate steer on decision-making, whilst external policies (set by other bodies) influence the context in which they operate. There is now an extensive framework of environmental policies at international, national, regional and local levels which bear upon the behaviour of organisations and individuals.

Many people have attempted to define policy, though few have offered a helpful and succinct definition (Ham and Hill, 1993). Generally, policy may be thought of as a coherent and agreed statement of how an organisation proposes to focus on its future mission; it sets out the nature of intended actions or inaction, and sets the boundaries within which these will take place. For public sector organisations, policies will normally express the ways in which legal obligations are efficiently to be discharged, public monies disbursed and permissive powers optimised; voluntary and private sector organisations' policies will reflect the ways in which they aspire to innovate and meet the expectations of consumers, members, trustees and shareholders. Policy statements (Box 3.1) should be sufficiently clear and precise to enable the individual decision-maker to choose a fit course of action, and to allow evaluation of performance against criteria which are related explicitly to the policy.

Box 3.1   Policy statement

As a typical example of a planning policy, one national park authority policy on mineral extraction is that:

> Proposals for mineral working in the National Park will be subject to the most rigorous examination and will only be permitted in the most exceptional circumstances where it can be demonstrated that:
>
> - the proposal is in the public interest and there is an over-riding need for the mineral which cannot be met from an alternative site outside the park or in some other way; and
> - the natural beauty, cultural and natural heritage and quiet enjoyment of the area would not be adversely affected.
>
> (Northumberland National Park Authority, 1995, p. 6)

This, then, relates national planning guidance (on the 'exceptional' nature of development in national parks, and the 'most rigorous examination' to which it will be subjected) to a specific area, whilst guiding mineral extraction companies about their prospects for obtaining future consent. The policy will have been published in draft form to provide an opportunity for people living in the area to comment and, at a democratic level, has given the National Park Committee an opportunity to debate where the balance of interests lies in this particular circumstance. Once adopted, a policy of this nature will provide long-term assurances and consistent treatment of development applications.

Thus, policies are to be found at all levels from world to parish. Frequently, there is a policy hierarchy, in which organisations must ensure that their policy intentions are compatible with those set by 'senior' organisations and, it is hoped, will be accommodated by 'subordinate' agencies and suppliers. In the public sector, it is often the case that EU member states' governments must limit their actions within policies set by the Union, and local governments in turn work within the parameters framed at national level. Equally, though, the principle of subsidiarity ensures that actions should be taken by the lowest appropriate level of government, which ensures that 'junior' agencies have at least some degree of policy independence. In the private sector, environmental responsibility may be cascaded through supply and purchasing 'chains' so that organisations at key points in these, such as major retailers, have a powerful role in driving environmental policy. Private sector enterprises, too, work within frameworks set by national and supranational (e.g. EU) governments.

Some Characteristics of Environmental Legislation and Administration

General Principles

There are numerous ways of seeking to determine the use of environmental resources, and the approaches to this vary according to prevailing political philosophies, traditions and customs of resource usage in a particular cultural context. Broadly, there is a variety of methods available to legislators and planners, ranging from highly interventionist 'command-and-control' approaches to purely advisory styles of persuasion. In Britain, the traditional approach has tended to favour the upper end of the spectrum, based on 'command and control' principles, but during the 1980s the balance shifted down the spectrum towards a more 'enabling' approach. Most people tend, in principle, to favour incentive-based approaches but when confronted with tough environmental dilemmas seem to acknowledge the need for regulation; however, regulatory approaches are invariably complex, imperfect and expensive, and should only be chosen where there is a willingness to devote considerable resources to monitoring performance and to enforcing compliance.

With respect to the environment, legislative powers may be exercised in a number of ways. First, the state may seek to manage land directly by acquiring the freehold or a long-term lease. Outright acquisition of land and its management by the 'highest competent authority' is, for instance, typical for national parks in most countries (but not England and Wales). In Britain, National Nature Reserves are usually owned on a freehold or leasehold basis so as directly to achieve conservation aims. The National Trust, a voluntary organisation, has rendered its land holdings 'inalienable', so that they can only be withdrawn from custodial use by special Act of Parliament. In totalitarian states, acquisition may have entailed minimal compensation or even confiscation, whereas in liberal economies the original owner usually receives full compensation. The latter is evidently a costly course of action, and not available as a general remedy for the pursuit of environmental policies. Thus, for environmental purposes public ownership or long-term lease of land is rarely a viable option on a large scale. It has mainly been restricted to places where direct management by specialist personnel is necessary in order to achieve specific site objectives (usually nature reserves or high-quality amenity land). However, though funds have permitted only a limited amount of land (often relatively cheap land of little agricultural or development value) to be acquired by the state for such purposes, private bodies such as the National Trust and the nongovernmental nature conservation organisations have added substantially to this stock by securing additional and often sizeable areas.

More commonly, the state will seek to modify the rights associated with ownership or the use of resources. For instance, user *rights*, or the development *values* associated with land, may be nationalised, so that the individual's ability to exploit a resource is conditional on state consent, without compensation. This has occurred in Britain, where the right to develop land for building, mineral extraction and various engineering works requires permission from the local planning authority. Permission may be refused, or granted with or without conditions. Compensation is only rarely available, for instance in situations where the authority has 'unreasonably' withheld consent or incurred the applicant in additional costs. Similarly, under the *polluter pays principle*, a user's right to discharge or emit wastes into watercourses, the atmosphere or on to land is strictly curtailed. Such action requires the operator to obtain a consent from the competent authority, and once more this may be refused, granted or granted conditionally. In some countries, discharge consents may be traded, so that each scheduled industry can compete for a share of a quota of pollution allowances and perhaps sell some of its share to other companies should it prefer to invest in additional treatment facilities. The overall quota would probably diminish each year to ensure a progressive improvement in environmental quality. There is typically a right of appeal by an applicant against refusal of, or onerous conditions attached to, planning consents or discharge licences. Other major areas in which state regulation of the environment takes place are in relation to water abstraction, food production and fisheries; in these areas, the use of licences and quotas to limit damage and overexploitation or overproduction is common.

A rather costlier and weaker alternative to planning controls is the concept of 'transfer of development rights' (TDR) (and a cluster of similar approaches) in which selected rights enjoyed by a landowner may be transferred, noncompulsorily, to the local state with full compensation. Thus, the owner may relinquish certain options – such as the right to develop the land for housing – but retain others – such as its use for farming. Such approaches are common in the USA, especially to retain green-belt land, although their use has not always been wholly successful. A somewhat similar arrangement, albeit backed up by compulsory provisions if agreement fails, is the method of protecting wildlife sites in the UK. Here, the Wildlife and Countryside Act 1981 (amended 1985) enables management agreements to be struck for Sites of Special Scientific Interest (SSSIs) in order to maintain them in their existing semi-natural condition. Thus, a farmer is compensated for 'transferring' the right to improve the land, and agrees to farm only at an intensity compatible with conservation objectives. Compensation is set at the net difference between the returns presently received and those which would have accrued had proposals for land-use intensification taken place.

An alternative means of encouraging the adoption of public objectives over privately held resources is to encourage a resource user to pursue certain courses of action. Frequently, this takes the form of grant aid, to assist in the promotion of desirable new investment, such as farm buildings, woodlands or energy conservation. If an action is likely to generate external effects, grant aid may only be paid after satisfactory consultation with affected parties. Under some circumstances, subsidies may be paid to enable a subeconomic system to continue for environmental or social reasons, for instance traditional farming in the Less Favoured Areas (LFAs) of the EU. Less commonly, the state may seek to 'bribe' (though this term is never used officially!) a resource user not to pursue certain deleterious actions. This market-based approach is best suited to pollution control, in which it is theoretically possible to calculate the social benefits of reduced pollution and translate these into a payment which society is prepared to make for industrial enterprises to reduce their emissions.

Monetary disincentives and incentives are starting to become increasingly popular, due to what might be described as the 'rise and rise of neoclassical economics' as an influence on public policy. These include taxation, such as on petrol and landfill; and grants and subsidies, such as planting grant for new woodlands and payments to steward countryside conservation assets on farms. The extraction of 'planning gain' in certain situations is in some respects (though officially it would not be expressed in such terms) a 'tax' on the developer and a 'benefit' to the community. Increasingly, the concept of ecotaxation is gaining prominence as a way of moving towards a more sustainable society, and it is likely that the experience of operating the landfill tax will encourage wider use of similar measures, especially to promote fuel efficiency, reduce pollution and waste, and restrict the demand for car-borne travel and parking. The benefits of ecotaxation are that it provides a clear incentive for people to behave as responsible environmental citizens, and it penalises the production of 'bads' rather than 'goods' in the economy. It may also be directly linked to environmental improvements, such as the option for payers of landfill tax to contribute this directly to environmental trusts. The problems of ecotaxation appear to centre around those of the difficulty of measuring and apportioning tax liabilities accurately, increasing the likelihood of illegal action to avoid paying taxes (the landfill tax, for instance, has resulted in increased fly-tipping), and the political unacceptability of potentially vote-losing measures which charge people to pursue their daily activities.

In addition to these formal mechanisms, encouragement can be provided by simple exhortation, advice, extension (scientific/economic advisory) services and by field/project officers who act as animateurs and 'enablers'. These methods will be relatively futile against organised corporate interests, but may succeed where the main problems hinge upon lack of knowledge or local leadership. Voluntary methods are widely favoured, especially as a supplement to other approaches, as it is accepted that real improvements in our use of the environment will depend on changes in people's attitudes and beliefs. This enlightenment will, it is argued, arise through education in its broadest sense. There is doubtless a significant element of truth in this, though it is hard to measure effect, and it is generally an inadequate remedy where really crucial issues are at stake or rapid change of behaviour is required. Overall, two main types of educational strategy tend to be used: one aimed at a broad public, for example in our treatment of domestic wastes or our use of recreation sites; and one targeted at specific user groups, where an approach based on demonstration of innovative and proven techniques is frequently most effective.

These different measures naturally vary in their relative effectiveness and appropriateness to particular situations. Environmental planners often have to combine the available mechanisms imaginatively, as there is often no obvious off-the-peg legal remedy or managerial solution. The various approaches overlap and should not be seen as mutually exclusive.

Tiers of Policy Specificity

We noted previously that, in terms of public sector action, a major purpose of policy is to set the boundaries within which, and the purposes for which, legislation is to be enacted and incentives are to be paid. Whilst we typically tend to think of this taking place pre-eminently at the national level, it is important to be aware of the increasingly restricted scope for government action as the world becomes more and more of a 'global village'. Thus domestic approaches have become ever more influenced and constrained by the actions of bodies beyond the traditional 'nation-state'. Two powerful forces, simultaneously complementary and contradictory, are acting to 'hollow out' the central state, these forces being *globalisation* and *localisation*.

Globalisation is a process in which strong international trends of trading, communications and cultural homogenisation tend to make places more similar and capital more mobile, so that standards, attitudes and expectations become increasingly universal whilst industrial enterprises operate on a transnational basis. Sometimes a formal framework for this is provided by supranational government, such as in the EU and North American Free Trade Association; elsewhere it is a notional but none the less powerful concept, such as the economies surrounding the 'Pacific Rim'. Equally, strong moral shareholder pressures are forcing transnational corporations (TNCs) increasingly to exercise environmental and social responsibilities, as befits their powerful status.

The counterpoise to this is a trend of localisations, reflected in an evident need of people to attach themselves to a local place. It also often involves enhancing the trappings of visual and cultural distinctiveness, and it has underpinned much of the conservation movement in its broadest sense. This is often accompanied by specific initiatives to build community spirit and action, and perhaps mobilise the planning system in defence of local character and needs. The strength of localisation does not match that of globalisation, yet it remains a significant influence on the political economy. Its effect is especially strong on the 'consumption' and 'production' of amenity and leisure assets, rather than of manufactured goods and services.

Arguably, the biggest influences on British environmental practice in the past 20 years have been external ones, most notably in relation to nature conservation, environmental impact assessment, pollution control and sustainable development. These have all been affected generally by international ideas and experiences, but most specifically by the legislative bodies of the EU. The European Action Programmes have framed successive contexts in which curative approaches, based on end-of-pipe controls, have moved towards preventative, anticipatory approaches. These have included the introduction of pollution standards for industrial, municipal and agricultural wastes; environmental assessment; packaging; and networks of key nature conservation sites. The European Spatial Development Perspective may also herald greater influence over town planning and urban management. Already, it is the case that regional development grants and transport proposals do have significant influences on planning options. Other areas of influence have been international treaties on nature conservation, pollution of the seas, CFCs and global warming. For example, the UN Framework Convention on Climate Change involves 'joint implementation' of progress towards a 50% reduction in CO_2 emissions within the next few decades, which in turn requires the limitation of population growth, dramatic reductions in industrialised country per capita emissions, and sharply constrained growth in per capita emissions from the developing world (Wettestad, 1997). The general pressure to adopt principles of sustainable development, especially via the *Brundtland Report* (WCED, 1987) and the Earth Summit (UNCED, 1992), has also influenced environmental practice and filtered through to national legislation.

Global government is not strongly established, although worldwide organisations do seek to exert concerted action in respect of major issues. These include the United Nations Environment Programme (UNEP), the International Union for the Conservation of Nature and Natural Resources (IUCN) and, in the nongovernmental sphere, the World Wide Fund for Nature (WWF) and Greenpeace. Their influence must rest on research, exhortation, negotiation of protocols and direct action. For instance, one set of UNEP's activities concerns information and monitoring services, through its Global Resource Information Database (GRID) and Global Environment Monitoring System (GEMS) programmes. The United Nations similarly organised a Conference on Environment and Development (UNCED) in Brazil in 1992 with the purpose of obtaining multinational agreement on an 'earth charter', specific legal measures and an agenda for environmental action by the international community (Grubb *et al.*, 1993). The IUCN's activities have included its production and promotion of the first and second World Conservation Strategies (1980, 1991), which relied largely on the supportive response of national governments to commit themselves to specific environmentally sensitive programmes.

A more potent expression of supranational government is at the regional continental level where nation-states collaborate formally over economic and other matters of collective interest. The EU, in this context, passes binding legislation on its member states through the Council of Ministers. Although the council is the sole source of legislative authority, issues are more widely debated in the European Parliament. Of the various types of legislation which the EU passes, regulations (which involve expenditure from the central budget) and directives (which require compliance from member states via domestic legislation) are those which most affect the environment. Agricultural measures, which latterly have contained explicit environmental components, are usually passed as regulations, since they draw on the community's agricultural guarantee fund. The five-yearly Environmental Action Programmes rely mainly on directives, especially relating to pollution control, migratory wildlife and environmental assessment. Article 130R of the revised Treaty of Rome now requires that environmental protection requirements are built generally into EU policies.

The most crucial tier in terms of political authority is the national level (the 'central state') in which sovereign powers are retained with respect to legislation and public expenditure. At this tier, environmental action usually centres on land-use planning, pollution control, water management, nature conservation, promotion of agriculture and forestry, and research, monitoring and information provision. Much debate in recent years has surrounded the notion of the role of the state. The state may be thought of as a conjunction of 'place' with a policy community, including the legitimate government. The state, central or local, will normally reflect an area with a common sense of history and identity, to which its citizens display a degree of loyalty and attachment. It will have its own governmental administration, comprising both elected members and paid officials, but this will be held in tension with a wider range of powerful lobbies and interest groups which help frame the formulation and style of implementation of public policy.

Within the local and central states, change may be understood in terms of the structures and agencies of civil society. Thus, for example, administrative structures condition the ways in which actions may take place. Agents may then choose courses of action within these frameworks. Various people and organisations have an interest in the outworking of public policy. These interests may be termed 'stakeholders', who may organise themselves in various ways to achieve their ends. Most environmental planning situations are characterised by complexity, involving many competing interests and needs, in which solutions need to be bargained between major and minor stakeholders.

The concept of the central and local states is similar, but not identical, to the institutions of local and central government. The 'regional state' is only weakly established in Britain, though the devolutionary and regionalisation programmes instigated in the late 1990s will buttress the relatively feeble and undemocratic tier of regional administration and make it more akin to federal government. Indeed, in many federated countries the provincial/state/regional level is the most potent with regard to environmental and natural resource management, and this may well be reflected in the UK. In addition to the main departments of central government, there are also specialist nondepartmental government bodies, or quangos. However, the study of government is itself being broadened to include the concept of 'governance', which refers to art and science of government, and to the wider (para-governmental and nongovernmental) mechanisms through which public policy is pursued.

Below central and regional tiers of administration is found the 'local state', which may in practice often be responsible for implementing the detail of national legislation or spending the central state's revenue. Consequently the central state may seek to exert close control over this tier of government, which nevertheless may resist any erosion of its autonomy. In local government, elected members serve on planning committees, consider recommendations from officers and set budgets. Some councils have sought ways of ensuring that environmental and sustainability issues are included on the agendas of their departments (for example, all committees may be required to 'balance off' their decisions against sustainability criteria). Officers within local government possess a wide variety of specialisms, several of which are relevant to the environment: they include planners, surveyors, environmental health officers, engineers, landscape architects and environmental managers. In Britain, there is generally little power vested in tiers of government below the local authority, but official bodies such as community or parish councils may still exercise significant responsibility. They may have a particular role in sustaining local pride, commenting on planning applications, protecting rights-of-way and encouraging grassroots environmental action.

The governance of environmental planning varies considerably between countries, and there is no right or wrong place for it to be located. In Britain, there is relatively little impact of European legislation on planning as such, but it does affect the context of planning quite considerably. For instance, it is highly influential on issues of nature conservation and environmental pollution, whilst it has been the subject of important reports regarding the future of European cities and landscapes. The European Spatial Development Perspective has the potential to harmonise planning practice in the longer term, but presently appears unlikely to transform long-established practice. The sustainable development policy agenda has been largely driven by exhortation and moral pressure from the United Nations, though this too increasingly finds itself at the heart of European policy. Thus, Agenda 21 has been followed up by voluntary A21s at the nation-state level ('national sustainable development strategies') which in turn are paralleled by 'local agenda 21s' in local government (Box 3.2). Central government sets broad policy guidance, has a role in approving development plans and operates an inspectorate to mediate in contested planning decisions. On some land-use topics, such as agriculture and forestry, the role of central government is very strong, whilst nondepartmental government bodies such as the Countryside Agency represent an important national voice on specific issues. Regional government has traditionally been weak, and limited largely to framework guidance to set the broad parameters which development plans must take into account, though consolidated government offices in the regions have provided an important administrative focus. However, the creation of assemblies in Scotland, Wales and the English regions is greatly reinforcing this tier. Local government is the tier through which the planning system is principally delivered, and the planning department is often the focus for corporate environmental action. Some consultative powers on plan-making and planning decisions are available to parish councils, and there has even been a modest move in a few areas towards devolving planning decisions to parish level completely.

Box 3.2 National sustainable development strategies (NSDSs)

Dalal-Clayton (1996) has considered the role of national sustainable development strategies (NSDSs) arising from A21. Some of them are dominantly environmental, some are concerned with broader issues of sustainable development, and a number are concerned mainly with federal areas of responsibility; some have clear targets and time horizons; a few are designed around special issue studies. National sustainable development strategies may involve crossgovernment and interdepartmental processes but none have been undertaken exclusively within single ministries/agencies. Extensive stakeholder participation has been a feature of only a few of the strategies, but most have involved some form of consultations with industry, NGOs and the public. Some plans have been directly linked to public budgets.

Basic comparisons between developed and developing country NSDS processes were summarised as:

Developed countries

Approach:
• internally generated
• internally funded
• indigenous expertise
• political action
• brokerage approach

Aims:
• changing production/ consumption patterns
• response to 'brown' issues (pollution)
• environmental focus

Means:
• institutional reorientation/ integration
• production of guidelines and local targets
• cost-saving approaches
• links to LA21 initiatives
• awareness-raising

Developing countries

Approach:
• external impetus
• donor-funded
• expatriate expertise frequently involved
• bureaucratic/technocratic action
• project approach

Aims:
• increase production/ consumption
• response to 'green' issues/rural development
• development focus

Means:
• creation of new institutions
• development of project 'shopping lists'
• aid-generating approaches
• few local links
• awareness-raising

Land-use Planning

The main legal provisions for environmental management are those contained in the town and country planning legislation. Within the UK, this has effectively covered the period after 1947, and the underlying principles have not changed fundamentally since. Town planning is centrally concerned with the operation of 'development', which, as we have seen previously, is rather narrowly defined in legislative terms (essentially, new building, engineering work, mineral extraction and changes in the use or significant appearance of buildings). An agreed future framework for development is enshrined in statutory land-use plans, and applicants wishing to undertake development must submit their proposals to the relevant local administration so that they can be controlled in accordance with the policies and proposals contained within the plan. In order to ensure a reasonable degree of consistency, central government has certain powers in reserve to modify plans and to resolve contested planning decisions. A similar, but generally less flexible, approach adopted in some countries involves the designation of *zones* for particular categories of use. Typically it is assumed that all land uses will conform to the zoning regulation for that area, whereas under the more flexible British system prior nonconforming uses could remain unless very unneighbourly (in which case a discontinuance order may be confirmed, entailing heavy compensation). Elements of this have been introduced for specific purposes in Britain, notably Enterprise Zones and Simplified Planning Zones. In both cases preconditions are set and, provided incomers comply with these, planning regulations are reduced to a minimum.

An appreciation of the nature of town planning cannot be gained from its statutory role alone, however. Planning, perhaps more than any other area of environmental or local government activity, has been subject to critical examination about its nature and purpose. Perhaps this is because of deep-seated tensions in what a redistributive profession can hope to achieve within a capitalist economy, and in its intermediary position between wealth-generating development activity and the quality of life of individuals and communities. It is also perhaps a profession which originated with very ambitious goals and ideals, yet must in practice seek much more pragmatic outcomes.

Very briefly, the original perception of town planning in the 1940s and 1950s was one of 'naive public administration', in which simplified assumptions were made about the connections between development plan production and the subsequent implementation of public policy (for a helpful account, see Rydin, 1998). Then, there was a belief that democratically agreed aims and objectives could be expressed in spatial terms, and that new development would be controlled in accordance with these statements in a fairly uncomplicated manner. In practice, relatively few development plans from this era were ever completed, there were many departures from the expressed provisions within the plan, and implementation of major public expenditure schemes often resulted in poor and unattractive living conditions. The 1960s were characterised by a more 'scientific' approach to planning, in which the influence of engineers and architects tended to wane relative to that of geographers, and the development process was seen as a 'system' to be controlled in a rational manner by professionals who were skilled in forecasting and responding to future conditions. A weakness of this viewpoint was that the typical cycle of planning was, in reality, much less rational than systems theorists supposed. Thus, a process based on a seamless loop of forecasting, plan production, implementation and monitoring proved impossible to sustain, and a much greater awareness of the role of political influences and messily negotiated solutions was necessary. Complementary to this somewhat technocratic view of planning were more radical models derived from American experience of community development, which presented the planner as an 'advocate' on behalf of particular sectors of society (especially deprived groups, who needed professionals to articulate their case in influential quarters).

The 1970s were seen by many as a period of crisis for the planning profession. One particular critique was based on organisational theory, which sought to explain the problems of the planning system in terms of its internal limitations, especially those of policy formulation and implementation. Rather than concentrating on methods for devising the ideal plan, therefore, the key issue became the constraints on achieving policy goals and hence the best way to put policy into practice. An important variant of this was 'procedural planning theory', in which researchers focused on identifying resource availability and enhancing monitoring mechanisms (i.e. means of improving the 'procedures' of practical planning). A second avenue of inquiry emphasised the process of implementation, representing this as continual renegotiation between policy actors. Thus, networking and bargaining were seen as valuable skills to aid realistic achievement of planning aims – planners were seen to negotiate in a relatively impartial manner, loyal to the goals of policy rather than any particular sectional interest group. However, a second strand of 1970s theory contested this supposed neutrality of planners, and examined more closely the relationship between planning and economic processes: this resonates with the 'political economy' approach previously mentioned. A 'liberal' political economy variant drew upon welfare economics, notably theories of market failure, so that the task of planners was seen as one of trying to achieve more efficient/equitable market outcomes. For example, through techniques such as cost-benefit analysis and environmental impact assessment, planners could act as expert assessors of the impact of environmental change. The 'radical' political economy variant was based on a Marxist model of society, and considered ways in which the state influenced the flow, rate of turnover and accumulation of capital. Planning was thus seen as an agent of the capitalist state serving its interests either generally or in relation to specific class fractions (e.g. professional incomers to villages) who were able to lever the workings of the planning system. These various themes developed further during the 1980s, but were accompanied by an ever-deepening threat to the planning profession arising from the dominance of free-market political discourses. By the 1990s, it has been argued that four key theoretical domains had emerged to describe the nature and purpose of planning (Box 3.3).

Box 3.3 The four key theoretical domains to describe the nature and purpose of planning

---

1) *The 'new right'* – with its continuing rationale for reducing the scope of local authority activity, centralising power at the nation-state level, and treating the public primarily as consumers of competitively priced and quality-assured commodities (i.e. they are essentially 'consumer-citizens'). In this view, planners appear as bureaucrats within the 'nanny state', and are seen to deliver few real benefits.

2) *The 'new left'* – which is still struggling with ideological implications of the collapse of east European communism, and is thus trying to reinterpret its analysis of linkage between an economy based on capital accumulation processes and the nature of social and political activity in a liberal representative democracy. This theory raises questions both about the role of state professionals in allocating resources as urban managers, and about the struggles that occur in and around the central and local state, and the ways in which disadvantaged groups can best achieve a greater share of policy benefits.

3) *Liberal 'political economy'* – characterised by the renaissance of neo-classical economics, in which the planner can recoup a role as an objective, expert professional, armed with a range of influential techniques which assist the transition to sustainable development.

4) *Institutional approach* – in which the main focus is on the way in which planners, as state professionals, engage with individuals who represent social groups and economic interests in society, and in which planners perceive these stakes and negotiate with individuals. This viewpoint is strongly associated with transactive methods of planning, which consider both the ways in which groups of professionals and politicians collaborate within differing institutional contexts, and the means of drawing citizens into dialogues over 'reasoned futures'. Thus planning becomes a deliberative activity, concerned with constructive argumentation between stakeholders and overcoming the democratic deficits which threaten the legitimacy of state action.

Source: After Healey, 1997; Rydin, 1998.

Whilst the statutory purview of town and country planning is severely circumscribed by the statutory definition of development, the range of interests addressed by planners is much wider. This reflects the more general ways in which plan policies can affect, or be affected by, a range of activities. Directly or indirectly, planners may engage with numerous environmental interests, including:

- industrial and economic development

- transport
- housing
- design of urban and rural spaces
- safeguard of fine buildings, archaeological sites and historic or architecturally important areas
- pollution
- climate change
- retailing and service provision
- public participation and community development
- alternative energy
- industrial development
- nature conservation
- landscape planning and management
- minerals extraction and recycling
- waste management.

Whilst planning has a significant role in relation to these tasks, though, it must generally work in collaboration with other departments and agencies, and respect their expertise and jurisdictions.

In practice, the principal piece of town planning legislature in England and Wales is the Town and Country Planning Act 1990, supplemented by the Planning and Compensation Act 1991 (parallel provisions exist in Scotland). This system is based on a combination of *development plans*, which set out the preferred pattern of land use, and *development control*, which ensures (in theory at least!) that individual planning consents conform to the forward-planning framework, as well as satisfying detailed local considerations. 'Development' is taken to include building, civil engineering works and mineral extraction, but not agriculture or forestry – an exclusion which has profound implications for the management of change in the biophysical environment. Changes in the use of buildings may also legally constitute development. Since the early 1970s, strategic issues have been addressed by 'structure' plans and more detailed policies and proposals by 'local' plans. (*Policies* are essentially those statements which set a framework against which to appraise new proposals emanating from the private sector and other agencies, whilst *proposals* largely comprise an authority's own measures for land-use change.) In the past, progress on local plan production has been slow and patchy, but this is now gradually being rectified by the requirement for complete territorial coverage under the Planning and Compensation Act 1991. Most unitary authorities produce only a 'unitary' development plan, which is a hybrid of the broad-brush statements of the structure plan and specific measures of the local plan.

Since 'development' may clearly involve a whole range of relatively trivial structures or changes of use, minor cases are exempt from planning control. Modest new structures and engineering works (such as small house extensions, many farm buildings and farm/forestry tracks) receive exemption under the General Development Order (GDO), whilst changes in the use of buildings which do not raise material planning issues (for instance from a bookshop to a newsagent) do so under the Use Classes Order (UCO). However, even small developments may raise legitimate concerns in urban conservation areas or high amenity landscapes, and so a 'direction' may be made under Article IV of the GDO requiring planning permission to be sought for normally exempted developments. (Technically, the GDO is now split into the General Permitted Development Order [GPDO] and the General Development Procedures Order [GDPO].)

An applicant who is aggrieved by a planning authority's decision may appeal and this appeal may be heard in public by an inspector (reporter in Scotland). Similarly, very important planning applications which raise issues of national significance or depart substantially from an approved development plan may be 'called in' by the Secretary of State to be determined directly (perhaps following a public inquiry). Indeed, the Secretary of State exercises ultimate authority, especially through reserve powers over structure plans, determination of appeals (through the planning inspectorate) and (where objections are received) confirmation of Article IV directions.

Pressure for new building is clearly endemic in the countryside, especially within the urban fringe. Given that carefully planned new development can sometimes lead to a rationalisation of land use and the creation of new landscape features, some observers feel that this can be a driving force behind environmental improvement. One option which is increasingly being considered is the scope for negotiating planning agreements (or 'planning obligations') with developers (under section 106 of the Planning Act 1990, amended by section 12 of the Planning and Compensation Act). These have been used widely in a number of contexts to secure miscellaneous benefits which are only indirectly (but more than minimally) related to the planning application or lie outside the development site, and so which cannot be specified as formal conditions attached to a planning permission. Nevertheless, developers often perceive them to be worthwhile concessions, and may thus be willing to let the local authority extract 'planning gain' (as it is informally described) in this way. The extension of this approach to environmental matters is often referred to as 'landscape gain', with new features being created, or sensitive sites being conserved in conjunction with development proposals at the developer's expense. Examples include off-site landscaping, donation of wildlife reserve sites and financial contributions to site maintenance and other operational costs of conservation bodies. However, as with all planning 'gain' (i.e. agreements/obligations) it is better to frame isolated site benefits within more strategic and comprehensive approaches to land use.

Not surprisingly, after more than 50 years of operation, the concepts and practices underlying town planning have come under scrutiny, and it is possible that significant reform of the system will take place in the early years of the twenty-first century. The author's view is that this should be a substantial review, focusing on the capabilities of integrated legislation to deliver 'bottom line' improvements in environmental quality. In Britain, there has been considerable reform of other fields of environmental law and, in administrative terms at least, an impressive rationalisation of statutory agencies, albeit there are still difficulties of achieving genuine integration of functions at the field level. The debate surrounding town planning may well seek to merge it more seamlessly with broader environmental policy and practice.

Various approaches internationally have marked the ways in which integrated approaches to environmental planning can lead to the flexible pursuit of quality targets, rather than the prescribed implementation of specific controls and procedures. For example, in The Netherlands, the National Environmental Policy Plan has shown how targets can be set to move towards sustainability over a range of topics, involving a dialogue between planners, industrialists and others. In Germany, landscape plans are produced which propose integrated solutions at various levels from site to region, including compensatory measures to replenish environmental capital where habitat damage has been caused by infrastructure developments. In New Zealand, the Resource Management Act has rationalised a range of planning and zoning legislation so that solutions can now be approached through flexible use of a range of instruments in order to pursue an 'environmental bottom line'. Although this has achieved significant success, it is salutary to note that doubts have been expressed about its fairness and effectiveness, and about equality of treatment in obtaining planning consents. In Metropolitan Toronto, there has been a merger of seven municipal administrations, such that there now exists an opportunity to achieve integration of official planning on a sustainable development basis. More generically, the sustainable services planning framework promulgated by the International Council for Local Environmental Initiatives has established a medium through which a network of municipal authorities are pursuing sustainable development. They have produced evidence of ways in which this framework is leading to increased interdepartmental integration and effective incorporation of community views. Although the British town planning system still retains an impressive statutory base, and has been streamlined and buttressed, it is likely that truly integrated environmental planning will have to await major legislative reform, perhaps within an EU context and influenced by wider international experience.

Environmental Quality and Resource Management

A major area covered by environmental legislation is that of pollution control, especially the permitted levels of contaminants emitted to air and water. These affect both in-pipe concentrations and end-of-pipe emissions. A strategy of *dispersal* is normally adopted for airborne pollutants by the use of high chimney stacks, above the temperature inversion layer; a *dilution* approach is taken to aqueous discharges, involving the outfall being placed at a point where water volume is sufficient to reduce concentrations and far enough from other outfalls to avoid mixing of pollutants. Of course, when contaminants display a high degree of persistence, dilution and dispersal strategies do not remove pollution, they merely delay or transfer it. 'Acceptable' concentrations of a pollutant are based on a variety of criteria, particularly those which relate to its severity, extent, toxicity and its relative tendency to persist, accumulate or degrade in the environment. Whilst planners have few statutory responsibilities related to pollution control, their role in contributing to pollution management strategies has been affirmed by *Planning Policy Guidance Note 23* (DoE, 1994c), which nevertheless notes the importance of liaison and nonduplication.

The preferred approach to pollution control in many countries is to set maximum end-of-pipe concentrations especially, though not exclusively, for aqueous discharges into confined water bodies. Substantial moves have been made in this direction in the EU, notably in respect of the most dangerous and persistent wastes. Usually, a set of determinands is specified to establish whether the discharge and receiving water body comply with the set standard. Pollution control in Britain, which has one of the longest pedigrees in the world, has, by contrast, traditionally been based on a more flexible principle. This principle was known as 'best practicable means' (BPM) and enabled the pollution inspector to take into account the economic circumstances of the industrial concern and the body into which the pollutant was discharged. Thus, an economically marginal enterprise, which might be bankrupted if it had to install instantly the most sophisticated control technology ('best technical means'), would be set a different standard from a highly profitable multinational corporation. Alternatively, a factory discharging into a confined watercourse would have to achieve a higher standard than one discharging into the open sea, where dispersal mechanisms were much more powerful. This flexible approach had much to commend it, and it was often argued that in practice it achieved a more sustained improvement in air and water quality than would have occurred under fixed standards, and was especially necessary in a country like Britain with its early onset of industrialisation and ageing factories. However, not only did the 'flexible approach' come under pressure from the European Commission, it also proved to be increasingly flawed in the light of nonlocal pollution impacts – for instance, acid precipitation, global warming and cumulative deterioration of the quality of the North Sea.

In practice, Britain has moved towards the EU approach through the medium of Environmental Quality Objectives (EQOs). These are articulated as Water Quality Objectives in which stretches of river are designated in terms of their desirable condition (e.g. suitable for industrial, recreational or domestic use), and Air Quality Objectives related to public health criteria. Discharge consents are then issued by the 'environmental protection agency' (Environment Agency (EA) or Scottish Environmental Protection Agency (SEPA)) only if they are compatible with this quality status. In practice, the levels specified in this consent are normally as demanding as the fixed standards in specific European directives. Similarly, with regard to air pollution, any factory not complying with the 'presumptive standards' set by EA/SEPA would be assumed not to be applying the most appropriate means of control. The public have an opportunity to comment before any authorisation is granted and information on authorised processes resulting in pollution, including breaches of authorisation, are held on a public register, though commercially confidential information and information affecting national security are excluded. There is also a separate set of controls over particularly hazardous materials.

Thus, the notion of best practicable means has effectively been superseded by terms such as BATNEEC (best available technology not entailing excessive cost) or, for pollutants where a safe lower threshold is difficult to quantify, ALARA/ALARP (Figure 3.1) (as low as reasonably achievable/practicable). Although these still sound rather fuzzy concepts and remain couched in the language of the 'British discretionary approach', they are increasingly being interpreted by detailed specifications for a wide range of industrial processes. Furthermore, the onus of proof has very significantly shifted towards the polluter, who must now demonstrate the efforts made to comply with specified levels under the Environmental Protection Act 1990, and must satisfy the inspector that a safety culture is in place, together with effective and intelligible emergency procedures. The operator must also often produce an optimum plan for disposing of all pollutants, ensuring that the cumulative and collective effects of all airborne, aqueous and solid wastes are minimised. Thus, a reduction in one emission cannot simply be

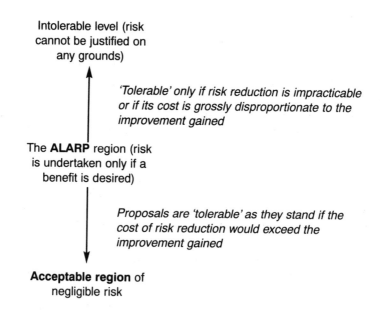

Intolerable level (risk
cannot be justified on
any grounds)

*'Tolerable' only if risk reduction is impracticable
or if its cost is grossly disproportionate to the
improvement gained*

The **ALARP** region (risk
is undertaken only if a
benefit is desired)

*Proposals are 'tolerable' as they stand if the
cost of risk reduction would exceed the
improvement gained*

Acceptable region of
negligible risk

Figure 3.1 Basis for evaluating the acceptability of pollution risk

offset by a corresponding increase in emissions elsewhere. This creates a situation of 'integrated pollution control' (IPC), where a single optimum solution – the best practicable environmental option (BPEO) – replaces individually conceived measures for separate emissions. Thus, BPEO is:

> the outcome of a systematic consultative and decision-making procedure which emphasises the protection and conservation of the environment across land, air and water . . . [it] establishes, for a given set of objectives, the option that provides the most benefit or least damage to the environment as a whole, at acceptable cost, in the long term as well as in the short term.
> (RCEP, 1988, p. 14)

The concept of BPEO has been clarified by the Environment Agency through published guidance. This demonstrates how operators must adopt conditions for achieving certain objectives (Boxes 3.4, 3.5).

Box 3.4   Concepts of pollution control

Pollution control practice must ensure that BATNEEC is used for preventing releases or, where that is not practicable, minimising them and rendering them harmless. In addition, for processes which could involve a release to more than one environmental medium, BATNEEC should be used for minimising the pollution to the environment as a whole, having regard to the BPEO. The requirements concerning the use of BATNEEC are expressed in terms of harm, which is often difficult to predict. In practice, 'harm' is related to statutory environmental quality standards (EQSs) which are used to provide constraints which must be taken into account in determining authorisation conditions. EQSs are thus being used to define the upper bound of the concentration of a substance in the environment which can be considered tolerable. Thus, where there is an EQS for a substance, the environmental assessment level (EAL) is set equal to the EQS. Also, to assist targeting of action, certain releases are considered a priority if the predicted environmental concentration (PEC) resulting from the release is greater than 80% of the EAL. The PEC is the sum of the concentration resulting from the release of the substance from the process and any already existing concentration of the substance in the environment.

A key element in the assessment of the BPEO is the evaluation of the effects of releases on the environment as a whole. This assessment is based on the ratio of the concentration of each released substance (the 'process contribution') in each environmental medium (e.g. air, water) to the EAL for that substance in that medium as a comparative measure of the effect of the release. This ratio is known as the environment quotient for a given substance in a particular medium. Where there are several substances released to more than one environmental medium, the sum of the environment quotients can be used to indicate the overall effects. The sum over all releases to a single medium is termed the environment quotient for that medium and the sum over all substances and all media is termed the integrated environmental index (IEI). Where a process involves the release of substances to more than one environmental medium, an application for an authorisation under IPC needs to show that the proposed operation represents the use of BATNEEC to minimise the effects on the environment as a whole, having regard to the BPEO.

The Integrated Pollution Prevention and Control Directive (1996) sets out to establish a Europe-wide authorisation system that would require most medium-sized and large industrial installations within the EU to obtain an integrated operating permit that lays down limit values for emissions to air, water and land simultaneously. Although the directive was similar to the existing UK method of integrated pollution control, it embraced a wider range of processes (e.g. intensive livestock units, certain food and drink manufacture). A principal challenge will lie in applying the concept of best available techniques (BAT) for pollution control which, after much redrafting, is fairly vague. Although the use of BPEO is assumed to be compatible with the definition, the European term clearly excludes the 'NEEC' part of the acronym, yet it is apparently not the intention that the directive should be applied without regard to financial cost. However, these economic factors appear to refer broadly to a particular industrial sector, rather than to individual operators.

The management of air quality entails a twin approach by local authorities and the EA, under the statutory powers afforded by the Environmental Protection Act (EPA) 1990 and Clean Air Act 1993, and local nuisances are dealt with by local environmental health authorities. In general, this is reactive control, coming into effect after the nuisance is created, though anticipatory action is possible through the use of smoke control orders (which require the use of smokeless fuel in designated areas). The EPA extends the powers of local authorities to deal with the air pollution aspects of a number of industrial processes ('Part B processes'); more complex industrial operations ('Part A processes') must comply with IPC measures agreed with the EA. IPC measures are also paralleled at the local level by local air pollution control (LAPC) and air quality management areas (AQMAs). In respect of long-range transport of air pollutants, countries must now seek to bring down 'critical loads' (i.e. the levels of pollution loading at which unacceptable damage starts to occur) of key pollutants (Wettestad, 1997).

The EU Directive on Ambient Air Quality Assessment and Management is now setting the pace in pressing member states to set air quality objectives, especially for densely populated areas and areas where the air quality is poor or improving. These set targets (substance limit values) which member states should seek to achieve in relation to the principal atmospheric pollutants. In a similar vein, the Environment Act 1995 required the government to produce a national air quality strategy, and this was published in 1997. It contains:

- air quality standards and air quality objectives for eight priority pollutants;
- a timetable for achieving the air quality objectives (2005); and
- methods of achieving the objectives.

The 'air quality standards' represent concentrations of the pollutants in the atmosphere at which there would be either no effect, or a minimal effect on

Box 3.5 BPEO: general principles and specific application

The methodology set out for assessing the BPEO in 12th report (1988) of RCEP involves six steps:

1) defining the objective
2) generating options for achieving the objective
3) assessing the options
4) summarising and presenting the assessment
5) identifying the BPEO: choose BPEO on basis of impacts, risks and costs
6) reviewing the BPEO: search for unforeseen effects, monitor and audit.

Example of BPEO for a proposed transport scheme (based on Bond and Brooks, 1997) following a similar (but not identical) sequence to the RCEP model:

Step 1: Describing the alternatives

- Categories of alternatives: alternative ways of meeting identified transport needs.
- Alternative routes (if selected a category which requires a new route).

information on routes

information on nature of environmental impacts

Step 2: Gathering baseline information
Baseline data on social and environmental components for each location

checklists of potential components

maps

site visits

consultations

Step 3: Scoping components and actions

examination of likely actions to be associated with each alternative

consultations

Step 4: Impact identification matrices

identify likely actions during construction, maintenance and operation phases

Step 5: Impact research

quantitative assessments
qualitative assessments

Step 6: Assessing impact significance

assess relative permanence of changes

identify effects on environmental quality standards

assess importance of components and of actions

Step 7: Comparing alternatives and identifying the BPEO

compare significance matrices for each alternative and location

involve political decision-makers in value judgements

public health, based on scientific and medical evidence. 'Air quality objectives' represent intentions in terms of improvement of air quality and are set with reference to the air quality standards. While the latter are based purely on scientific and medical evidence, the objectives also take into account the technical feasibility, the cost of achieving the air quality improvements and the timescale. The strategy identifies areas of achievement which are to be met by the three principal sectors involved: business and industry, transport and local authorities.

Water management used to be a wholly public sector activity undertaken in England and Wales by regional water authorities (Rydin, 1998), until water privatisation split responsibility between the private water companies and the regulatory body (EA) (a quango arrangement exists in Scotland). Whereas in the past most investment went into new reservoir capacity, attention is increasingly being focused on the loss of supplied water through leaky pipes; demand management is also receiving more attention, and is leading to the more widespread use of water meters. The EA remains responsible for overall water resource management alongside functions with respect to water pollution, flood defence and land drainage, fisheries, navigation and harbours. It is supported in these activities by a number of regional and local advisory committees. Under the Water Resources Act 1991, the execution of these functions must have regard to the conservation and enhancement of natural beauty and amenity, conservation of flora and fauna and recreation. The water management function charges the EA with conserving, redistributing and augmenting water resources in England and Wales and securing the overall proper use of such resources. This is achieved by agreeing water resources management schemes with water companies, determining minimum acceptable flows and minimum acceptable levels of volumes for inland waters and issuing certain licences (for abstracting or impounding water).

The collection and distribution of water have increasingly focused on the use of rivers, and on reservoirs for compensation/regulation rather than direct supply. Much of our water is also abstracted from aquifers, some of which are becoming over-used experiencing damage to their geological structure or suffering from agricultural or industrial pollution, and these have become depleted and their geological structures damaged. Increasingly, planners are being urged to protect aquifer recharge zones, though an equal concern is rising groundwater levels under certain cities where groundwater abstraction for drinking purposes has ceased because of pollution. Sustainable development requires that water resources are used as efficiently as possible, and an important approach in recent years has been recycling systems aimed at maximising the use of 'grey' waters.

The coastal zone, despite being a complex environment of exceptional importance, is planned and managed in a fragmentary fashion. Latterly, there have been renewed attempts to prepare integrated strategies for coasts and estuaries. English Nature, for example, is aiming to achieve the production of management plans for all English estuaries, and Kidd (1995) has described the key objectives employed in the preparation of one such plan for the Mersey, namely:

- estuary dynamics – to allow the estuary to function as naturally as possible and in a self-sustaining way by controlling human interference in intertidal and marine areas having regard to the natural conditions and processes of the estuary;
- water quality and pollution control – to support continuing improvements in water, air, land, noise and light quality and the adoption of environmental good practice within the estuary zone;
- biodiversity – to conserve and where relevant restore the natural biodiversity of the estuary zone; and
- land use and development – to promote careful stewardship of land resources, landscape and townscape within the estuary zone.

Integrated coastal zone management (ICZM) has been widely advocated but, in practice, it relies heavily on advisory guidelines and collaborative administrative arrangements. Topics which should be considered within an integrated framework have been reviewed in various publications (DoE, 1992c; 1995e; Nicholas Pearson Associates, 1996) (Box 3.6). Clayton and O'Riordan (1994) have argued that ICZM needs firm measures if it is to be effective, and have proposed four challenging principles – namely that:

1) natural processes of defence and protection should be encouraged, costed properly and fully incorporated into any plan or management scheme;
2) natural zones essential to this purpose, such as headlands, dunes, salt marshes and wetlands, should be adequately protected by law, and carefully monitored for their continuing role (controversially, it is suggested that they should be cleared of existing settlement, with compensation if necessary);
3) coastal defence works should always be designed sympathetically and encourage the retention of a natural beach, and cost–benefit analyses should recognise the essential linkage between the two; and
4) land-use planning should formally take into account the vulnerable areas of coast subject to sea-level rise and increased storminess, so that no new settlement or economic activity is permitted in such areas (again, it is also advocated that existing buildings are left unprotected, with compensation where necessary).

They note that these are tough requirements, which are made significantly more difficult by the complex ownership and legal situation at the coast.

Coastal erosion, coupled with the possibility of rising sea levels, has led to the acceptance that coastal defences are not appropriate in all situations. In the longer term, it may be necessary to plan for an orderly withdrawal of farming and even settlements away from vulnerable coastlines, and this is referred to as 'coastal retreat'. The evidence about the likelihood of sea-level changes is highly conflictual, though it does appear that some parts of

Box 3.6   Key elements of integrated coastal zone management

- coastal planning and development
- coastal defence
- dredging of marine aggregates
- ports and harbours
- construction and development in the sea
- shipping
- fisheries
- oil and gas
- waste disposal at sea
- water quality
- landscape
- nature conservation
- the historic environment
- leisure, sport and tourism.

Source: DoE, 1992c; 1995e; Nicholas Pearson Associates, 1996.

the coast are at some significant risk. Conservation organisations have argued that prime wetlands (e.g. saltmarshes) are likely to be lost as a result of the 'coastal squeeze', as they lie between rising sea levels and sea walls. They argue that, in some situations, the landward defences should be breached, permitting tidal surges to inundate existing farmland, so that conservation uses can be extended inland.

Although this account has emphasised heavily the command-and-control approach based on regulations, we have also noted the increasing consideration being given to 'economic' instruments. Turner (1994) has identified the key economic instruments for pollution control, namely:

- direct alteration of price or cost levels, e.g. emission charges;
- indirect alteration of prices or costs via financial or fiscal means, e.g. subsidies or fiscal incentives to encourage the adoption of 'cleaner technologies'; and
- market creation, e.g. emission permits (tradable).

Economists argue that charges of various sorts are an efficient and straightforward way to price the use of the environment, and these include: *emission charges* (related to the quantity and quality of the pollutant released and the damage cost inflicted on the environment); *user charges* (with a revenue-raising function, not directly related to damage costs, but reflecting the treatment cost, collection and disposal cost or recovery of administrative charges); *product charges* (related to the environmental damage costs associated with the target product, levied on products that are harmful to the environment when used in production processes, or when consumed or disposed of); and *marketable permits* (quotas, allowances or ceilings on pollution levels). The last of these may be traded subject to a set of prescribed rules so that operators, or even countries, which find it difficult to comply with enhanced control standards may purchase permits from those which find it relatively easy. As well as imposing charges, incentives may provide an additional mechanism: these require payments rather than specific compliance, which should encourage the economically rational polluter to change behaviour. Despite the popularity of these approaches with economists, some critics feel that they can be unfair and possibly even unethical, by, for example, 'bribing' industrialists and favouring producer interests.

In addition to air and water pollution, domestic and industrial sources also generate large quantities of solid wastes, and some 90% of solid waste is disposed of in around 4,000 controlled landfill sites in the UK. Sites for disposal of waste must be licensed (i.e. they require a waste management licence and a valid planning consent) and controlled waste can be disposed of only at a licensed site. Waste regulation authorities also have to draw up waste disposal plans which forecast future waste generation and consider the adequacy of existing disposal facilities; county councils and national park authorities must also prepare a waste local plan. The principal piece of legislation in this context is the Environmental Protection Act 1990.

General government policy towards waste management is based on a hierarchy of:

- reduction
- reuse
- recovery (including recycling, energy recovery and composting), and
- safe disposal (i.e. the last resort, after the previous three steps have been successively exhausted).

Appropriate management options in any particular case will vary according to the waste stream and local considerations, but generally there is an increasing emphasis on sustainable solutions. A further policy requirement is that of the *proximity principle*, under which waste should be disposed of (or otherwise managed) as close to the point at which it is generated. This is viewed as creating an increasingly responsible and therefore sustainable approach to the generation of wastes, as well as limiting the pollution from transport.

The EC Framework Waste Directive requires competent authorities to draw up plans which (in addition to considering special arrangements for particularly demanding wastes, and suitable disposal points) set out:

- the type, quantity and origin of waste to be recovered or disposed of, and
- general technical requirements for waste management.

These are implemented through the Waste Disposal Plans (under section 50 of the EPA); however, sanctioning sites and setting criteria for new facilities is the role of the planning authority. Consequently, the Waste Management Licensing Regulations (1994) place a specific duty on planning authorities to include in their development plans policies in respect of suitable waste disposal sites or installations. The Planning Act 1990 similarly places a duty on county/unitary councils and national park authorities to prepare waste local plans (WLPs) (or combined mineral and waste local plans). General planning considerations in this respect are established in the structure plan (which will identify broad areas of search for appropriate sites for various types of waste disposal, and the accessibility of sites by nonroad transport), whilst the more detailed considerations of the local plan include:

- constraints to development which may arise from compliance with environmental quality standards/objectives;
- identification of land (or establishment of criteria) for the location of potentially polluting activity;
- the scope for separation of potentially polluting and other development on land use;
- the environmental consequences of former land uses (e.g. leading to contaminated land);
- completed landfill sites that would be suitable for development or other use;
- identifying restoration and planning standards sufficient to ensure that land is capable of an acceptable after-use; and
- the identification of possible locations for potentially polluting development, in the light of wider social and economic needs.

WLPs should draw together these various considerations, in a way that reflects the proximity principle (i.e. that regions should take a more or less self-contained approach to waste management) (Figure 3.2).

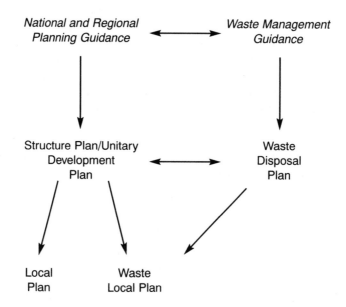

Figure 3.2   The relationship of guidance and plans concerning waste matters
Source: DoE 1994c

It is evident that WLPs can perform quite a wide-ranging function relative to pollution issues, and are not purely confined to solid waste management. This reflects the broader role which land-use planning may play in the reduction of pollution nuisance. In general terms, planners require to be aware of the pollution implications of applications for industrial development, and also of the role of the local authority as a producer of wastes in its own right. More specifically, planners can play a unifying role in respect of the many legislative provisions for land use and pollution through the proper use of environmental impact assessments which, because of their integrative nature, may provide a step on the road to achieving BPEOs. Moreover, land-use planning can help to secure the separation of polluting activities from pollution-sensitive zones, and so take anticipatory pre-emptive measures in relation to environmental nuisances.

Planning for Special Qualities of the Land Resource

Planners have been centrally concerned with the stewardship of the land resource. From the point of view of environmental resources, three particular qualities of land are especially important: the ecological (i.e. biodiversity and nature conservation value), the visual (i.e. scenic and aesthetic quality) and the positional (i.e. the location of land relative to a source of demand for specific uses such as recreation or house building). These qualities can be related closely to the sustainable development notions of critical and constant natural capital, where assets which have taken millennia and longer to develop or which have unique positional attributes can be treated as essentially irreplaceable and thus absolutely worthy of safeguard (critical capital), whilst those which are more recent or commonplace may still be important, but could perhaps be subject to trade-offs and compensation (constant capital). Sustainable development of countryside resources has been related to: the adoption of a long-term strategic vision; giving greater recognition to the character of individual areas; and using new methods to measure and protect what people see as valuable (Countryside Commission, 1998a).

Nature conservation is the primary responsibility of specialist quangos, whereas landscape conservation is more evenly divided between planning authorities (e.g. national parks) and quangos. The bodies dealing with these are English Nature, the Countryside Agency, Scottish Natural Heritage and the Countryside Council for Wales. The latter two reflect the realignment of amenity and scientific conservation interests in Scotland and Wales during the 1990s. As with other aspects of environmental planning, there is a spectrum of possible approaches to the protection of critical natural capital. In this case, the key mechanisms are those of site acquisition, site protection by land-use regulation and by striking an individual agreement with the landowner.

The major powers for designation and protection are to be found in the National Parks and Access to the Countryside Act 1949, the Countryside Act 1968 (1967 and 1981 in Scotland) and the Wildlife and Countryside Act 1981. The first of these provided for the designation and planning of National Parks and Areas of Outstanding Natural Beauty (AsONB), the improvement of access in the countryside including the creation of access agreements, the acquisition of nature reserves, and the formation of a Nature Conservancy and National Parks Commission. These bodies were superseded by a Nature Conservancy Council (NCC) (in 1973) and a Countryside Commission (in 1968; 1967 in Scotland), although they were merged in Scotland and Wales in the early 1990s to create the Countryside Council for Wales (CCW) and Scottish Natural Heritage (SNH). The NCC in England became English Nature (EN) in 1991 although an umbrella GB Joint Nature Conservation Committee (JNCC) continues to harmonise scientific standards between EN, SNH and CCW. The Countryside Commission in England merged with the Rural Development Commission in 1999 to form the Countryside Agency, reflecting the need to create a more powerful body to promote quality of life and standards of living in the countryside. As from 1974, all national parks also had a chief officer and dedicated planning and management staff, and have acquired more comprehensive planning powers during the 1990s.

The principal nature conservation designation in Britain is the Site of Special Scientific Interest (SSSI), a small proportion of which are acquired or leased as National Nature Reserves (NNRs). The fundamental rationale behind nature conservation designations is that they comprise biotopes and formations intended to represent the range of biological and geological variation across Britain. Nationally, there are around 240 NNRs, covering nearly 1,700 km^2, with over 40 in Northern Ireland. Local authorities can enter into similar agreements and create Local Nature Reserves (numerically and spatially, these are equivalent to about one-tenth of the stock of NNRs). There are also Marine Nature Reserves (MNRs) (e.g. Lundy and Skomer). The SSSI network generally is much more extensive, and there are nearly 5,700 in Britain covering just under 18,000 km^2 or 7% of the land. The Wildlife and Countryside Act 1981 covers protection of individual species (Part I), conservation of nature and landscape (Part II) and public rights of way (Part III). Most attention has been focused on Part II, especially the provisions for entering into agreements with landowners to secure site protection. Planning decisions concerning any SSSI automatically involve consultation with EN/SNH/CCW. SSSIs are designated on the basis of scientific criteria, reflecting the nature conservation significance of the land. EN/SNH/CCW are required to gauge each potential site in terms of its regional importance, based on criteria such as rarity and representativeness; hence essentially similar sites may or may not qualify for designation, depending on regional and local circumstances.

Superimposed on the network of nationally recognised sites are various types of international designation endeavouring to encourage the management of features which, because of their linear and continuous structure or their function as stepping stones, are deemed to be essential for migration, dispersal and genetic exchange (Box 3.7).

Box 3.7   International nature conservation obligations and interests

The Bern Convention on the Conservation of European Wildlife and Natural Habitats: refers to wild plants, birds and other animals, with a particular emphasis on endangered and vulnerable species and their habitats.

The Ramsar Convention on Wetlands of International Importance especially as Waterfowl Habitat: seeks the conservation of wetlands, especially sites listed under the convention.

EC Council Directive on the Conservation of Wild Birds (the 'Birds directive'): applies to birds, their eggs, nests and habitats, and requires member states to take measures to preserve a sufficient diversity of habitats for all species of wild birds naturally occurring within their territories, especially regarding certain rare and migratory species. Leads to the designation of 'Special Protection Areas' (SPAs).

EC Council Directive on the Conservation of Natural Habitats and of Wild Fauna and Flora (the 'Habitats directive'): requires the conservation of biodiversity through measures to maintain or restore natural habitats and wild species at a favourable conservation status. Leads to the designation of Special Areas of Conservation (SACs).

The Bonn Convention on the Conservation of Migratory Species of Wild Animals: requires the protection of endangered migratory species listed, and encourages separate international agreements covering particular species (e.g. an agreement on the conservation of bats in Europe was amongst the first to be reached).

Under the 1981 Act, EN/SNH/CCW has to inform landowners and other relevant bodies of the intent to designate an SSSI; and landowners have a reciprocal duty to inform the relevant conservation of any intended operations within such a site, giving four months' notice. If EN/SNH/CCW objects to proposed operations they have a limited period of time within which to negotiate a management agreement for the land, with compensation being payable for operations forgone by the landowner. These agreements normally involve the payment of compensation, and this may vary from a small 'consideration' for inconvenience or management effort, to substantial sums for loss of profits.

Currently 6% of the area of SSSIs is subject to management agreements. At the end of the negotiation period EN/SNH/CCW can apply to the Secretary of State for a section 29 extension order; however, at the end of this extended period (up to 12 months) the only remedy is compulsory purchase (which is extremely rare). Added protection is available for sites of national importance, where written approval from EN (or equivalent) is required for operations. Similar notification conditions for national park authorities and local authorities are also included in sections 41 and 39 of the Act. The Act also provided for the protection of an SSSI to be taken into account by the Minister of Agriculture when giving farm capital grants, and this essentially negative measure has effectively given way to the Wildlife Enhancement Scheme which aims to promote positive management of SSSIs. In addition to these area-based policies there is a list of protected species which encourages safeguard of their habitats and constrains development in the vicinity, as well as preventing killing or capture in specified cases. If these 'command and control' powers appear unduly negative, it is worth noting that the monies and efforts associated with preventive management agreements declined sharply during the 1990s, and were progressively being replaced and supplemented by positive agreements with and payments to farmers with SSSI land to enhance the nature conservation resource (Box 3.8).

Box 3.8   Roles and responsibilities of conservation agencies in Britain

Nature conservation bodies have responsibility for:

- establishing, maintaining and managing NNRs and MNRs
- identifying and notifying SSSIs (and their management agreements)
- providing grants for nature conservation
- providing advice and information about nature conservation
- supporting and conducting research, and
- advising the government on relevant nature conservation matters.

The JNCC, *inter alia*:

- sets standards and criteria by which the national bodies work in notifying SSSIs, and
- advises the government on Britain's international obligations.

Countryside agencies are concerned with:

- conserving and enhancing the natural beauty of the countryside
- encouraging the provision and improvement of facilities for open-air recreation
- designating, subject to ministerial confirmation, national parks and AsONB in England and Wales and equivalent designations in Scotland
- providing advice on the conservation and enjoyment of the countryside to the government and other public bodies whose activities affect it
- providing grant aid to landowning farmers, local authorities and others to take action on conservation and recreation provisions, and
- operating a range of economic and social programmes to assist the rural economy.

Source: HMSO, 1993.

One of the most distinguishing features about the British environmental movement has been the very prominent role of nongovernmental (voluntary) organisations. These were the pioneers of conservation work in the late nineteenth and early twentieth centuries, and have grown to be influential actors. They continue to be important in lobbying nationally for wildlife, clean air, noise abatement, proper use of the seas, protection of the countryside, access to open land and safety from risk and pollution. They not only lobby, but also provide important briefings to ministers on conservation topics. Some of the memberships of NGOs are very large and they represent formidable political forces. Most are well organised in relation to lobbying central government and have been instrumental in changing policy and shaping legislation; many are also well organised at European level. Other NGOs place their emphasis at the local level and seek to protect the amenity of particular areas. Equally, some national and international organisations seek to influence local government: for example, the Council for the Protection of Rural England marshal considerable research in relation to planning issues for use by their county-based offices, whilst Friends of the Earth (FoE) have helped to pioneer environmental auditing and state of environment reporting within local authorities. A further way in which they may act is through campaigns, aimed at raising awareness of particular issues and trying to place them on the policy agenda. For instance, FoE have been especially successful in raising awareness of the need to reduce the amount of traffic on the roads.

Whereas key nature conservation sites have frequently been acquired outright or are subject to stringent management controls, areas valued for their landscape qualities have generally been less strongly safeguarded. National parks in England and Wales, for example, are 'protected' principally through planning mechanisms; planners' restricted scope for intervention on land management matters has only limited effect in the face of external forces for change. Internationally, the concept of a national park is generally more cogent, and state acquisition and management by the 'highest competent authority' is the norm. Various levels of protection to areas of 'critical natural capital' thus prevail internationally and are recognised by the International Union for the Conservation of Nature and Natural Resources (IUCN). A mix of approaches is often desirable, for instance in biosphere reserves and some national parks, where outright acquisition may be appropriate for the core zone, and agreements with landowners in the buffer zones (Box 3.9).

Box 3.9 Categories and management objectives of protected areas

I.	Strict protection (i.e. strict nature reserve/wilderness area).
II.	Ecosystem conservation and recreation (i.e. national park).
III.	Conservation of natural features (i.e. natural monument).
IV.	Conservation through active management (i.e. habitat/species management area).
V.	Landscape/seascape conservation and recreation (i.e. protected landscape/seascape).
VI.	Sustainable use of natural resources (i.e. managed resource protected area).

Source: adapted from International Union for the Conservation of Nature and Natural Resources, various sources.

It is clear that acquisition is an impractical approach for large areas, and so more indirect methods must be sought to protect the broader countryside; yet these are significantly weakened by the exclusion of agriculture and forestry from planning control. Nevertheless, within areas of critical natural capital, planning and other land-use legislation may be used more stringently than usual to restrict undesirable changes. Since this, as the initial step, requires areas to be designated, the establishment of AsONB, SSSIs, etc., is commonly termed 'designation'; alternatively, in some countries it is referred to as 'greenlining'. The more general characteristics of greenlining (Mason, 1994) are that it does not preclude human occupance or economic growth, but helps guard against poorly planned and uncontrolled development, entails a participatory process with various tiers of government, and can involve a menu of potential land management techniques. These techniques are typically drawn from options including full acquisition, purchase and lease-back, conservation easements, purchase or transfer of development rights, tax incentives, regional zoning plans, strict control of new development, public funding for projects and co-operation with private trusts. Problems of greenlining/designation include those of delineation of area boundaries, choice and blend of management options, the problems of gentrification and diminution of the low-cost housing stock, conflicts between long-term residents and incomers, and the 'museumisation' of working landscapes.

It is government policy that sound stewardship of wildlife and natural features should not rest on centralised quangos and ministers alone; it depends also on the decisions made by local planning authorities, landowners and others who influence the development and use of land (*Planning Policy Guidance Note 9* – DoE, 1994b). Thus, making adequate provision for development and economic growth whilst ensuring effective conservation of wildlife and natural features as an important element of a clean and healthy natural environment, is seen to be collective responsibility. Policy guidance emphasises that action for nature conservation extends beyond formally designated sites, as local designations may be important to individual communities, often affording people the only opportunity of direct contact with nature, especially in urban areas. It is suggested that statutory and nonstatutory sites, together with countryside features which provide wildlife corridors, links or stepping stones from one habitat to another, combine to form a network underpinning the maintenance of the current range and diversity of flora, fauna, geological and landform features and the survival of important species. In some rural areas the maintenance of traditional agricultural practices is important for nature conservation objectives. In developed areas, sensitive landscaping and planting, the creation, maintenance and management of landscape features, and the skilled adaptation of derelict areas may also be appropriate.

Clearly, a great deal of change in the countryside is associated with agricultural and forestry operations, both of which are exempt from planning control. Planners are generally at a loss to influence such change, other than by localised management agreements in national parks and various types of conservation advice. 'Development' (i.e. agricultural and forestry buildings, tracks, related mineral extraction) is, in principle, subject to planning control, but in practice is largely excepted. Thus, whilst new construction, alterations or extensions to farm buildings or excavations and engineering operations do constitute development they are usually, if reasonably necessary for the purposes of agriculture, permitted under the GDO. More extensive rights are available to agricultural units of at least 5 ha in extent. Since 1992, a system of prior notification for such development has operated over substantial tracts of agricultural and forestry land in England and Wales, and this allows local authorities to regulate certain aspects of siting, design and appearance.

More generally in the British countryside, a restrictive regime has tended to prevail towards development involving agricultural land and sporadic development. Planning authorities must consult the agriculture ministry/ department if applications (either singly or cumulatively) affect more than 20 ha of grade 1, 2 or 3a farmland which, in the authority's opinion, would conflict with the adopted plan for the area. This arrangement is somewhat less restrictive than it used to be and planners are expected to take a more balanced view of overall 'countryside' need, rather than one based purely on farmland protection. The most extensive guidance on planning attitudes towards rural land and its development, set within the principles of sustainable development, is to be found in *Planning Policy Guidance 7* (DoE, 1996c) (Box 3.10), and this requires planners to

- meet the economic and social needs of people who live and work in rural areas, by promoting the efficiency and competitiveness of rural businesses, and encouraging further economic diversity to provide varied employment opportunities (especially in areas still heavily reliant on agriculture);
- maintain or enhance the character of the countryside and conserve its natural resources, including safeguarding the distinctiveness of its landscapes, its beauty, the diversity of its wildlife, the quality of rural towns and villages, its historic and archaeological interest, and best agricultural land;
- improve the viability of existing villages and market towns, reduce the need for increased car commuting to urban centres, and reverse the general decline in rural services by promoting living communities which have a reasonable mix of age, income and occupation, and which offer a suitable scale of employment, affordable and market housing, community facilities and other opportunities; and
- recognise the interdependence of urban and rural policies.

Overall, planners are asked to respect the processes of change in the countryside. This requirement recognises the major shifts which have occurred in the rural economy and society, as a consequence of substantial

Box 3.10 Types of development plan policy appropriate to the countryside

- Encourage rural enterprise;
- protect landscape, wildlife and historic features;
- safeguard the best and most versatile agricultural land;
- have regard to the quality and versatility of land for use in forestry and other rural enterprises;
- protect other nonrenewable resources;
- strengthen rural communities by encouraging new employment, facilities and adequate supply of affordable and market housing, and underpinning services and community facilities; and
- achieve good-quality development which respects the character of the countryside.

Source: DoE, 1996c.

expansion of settlements (mainly from migration) and the capture of new industry and commerce. The marked changes which have taken place in the manufacturing and service sectors, often as a result of 'new' technologies, now mean that almost a quarter of employees live in the countryside, despite sharp losses of employment from agriculture, forestry and quarrying. This has had major implications for levels of car usage, provision of facilities and the composition of rural communities. More extensive and positive recommendations on planning the countryside have been issued, emphasising the promotion of quality and the need for integrated approaches (Box 3.11).

Many of these policy topics reflect the role of land as a 'positional' good. A recurrent theme associated with this is that of the development potential of land, and thus of safeguarding land to prevent swathes of new building merging continuously across the countryside. Thus, a designation of particular importance, and which has found many parallels throughout the world, is the green belt. In its original conception, this was to be land girdling a city, purchased by the municipality. With the advent of planning legislation, however, it became possible to afford protection largely by the exercise of planning powers (i.e. through 'designation'). In Britain, the major purposes of green belts have been repeatedly confirmed as the restriction of urban sprawl, safeguarding the countryside from further encroachment and preventing neighbouring towns from merging (DoE, 1995b). More recently, their role in assisting urban regeneration by deflecting development pressure back to brownfield sites, and in providing accessible recreation opportunities and conservation resources, has been recognised. The concept of 'green belt' is often misunderstood, and it refers only to limited areas which have specifically been designated in development plans, rather than to undeveloped peri-urban land generally. Only 14% of England is covered in this way, and green belts in Scotland and Wales are even more limited; contrary to popular opinion, most prime developmental land around towns and cities has no explicit protection. Green belts are not always even especially scenic, and it is their

Box 3.11   Ideas about planning for quality in the countryside

---

The Countryside Commission guidance suggests that only development which requires a rural setting should be permitted in the countryside. Any such development should:

- fit its setting in terms of location, scale, design and materials, and respect local character;
- compensate for any net loss of countryside where greenfield land is used, by enhancing the quality of the remaining countryside and its sense of place;
- make appropriate provision, or a contribution to larger-scale provision, for dealing with its wider off-site effects, such as additional travel demands, and the improvement of local services; and
- be sustainable, in terms of both construction and use, in materials, energy and transport requirements.

In this context, it argued that planning documents should

- provide a long term *vision* for the area they cover;
- support this vision by specific *objectives for countryside character, quality and accessibility*;
- as far as possible, *integrate environmental, economic and social objectives*, providing guidance to developers on how new development can best meet each of these sets of objectives;
- take a firm view about the extent to which proposed development – especially on rural or greenfield sites – meets *the public interest*;
- specify clearly the *criteria of scale, design, location and respect for local character* against which the acceptability of development proposals will be assessed;
- ensuring that development makes more *efficient use of resources*, such as water, minerals and energy, in terms of both construction and subsequent operation;
- make greater use of *planning obligations* in protecting and enhancing the countryside as new development takes place. Especially, this should compensate for loss or deterioration of countryside; all residents should have new opportunities to enjoy well managed countryside close to where they live;
- be used as a *public policy context* not just for development control, but also for other programmes and strategies affecting the countryside; and
- be *adequately resourced*.

Source: Countryside Commission, 1998a.

permanence (at least 'as far as can be seen ahead') rather than their greenness, which is critical.

Green-belt policy has never been wholly effective and has in places been undermined by ambiguous (perhaps draft, interim) plan policies or by developers winning their case on appeal, contrary to the desire of the planning authority. In countries without comparable planning legislation, purchase of the land or perhaps 'transfer of development rights' may be necessary. Current guidance in Britain stresses the need for the adoption of very long, stabilising planning horizons to ensure the integrity of green belts, but also to rationalise them so that land not permanently essential to their character can be available to be released for development. Apart from buildings related to agriculture or forestry, outdoor sport, cemeteries and a small number of other cases appropriate to a rural area, new development is normally automatically refused. Conversion of disused farm buildings, however, may receive favourable consideration in view of their potential contribution to economic diversification (although it is known that most conversions are typically to residential use).

Another important type of 'positional good' in the countryside is the outdoor recreation facility, reflecting increases in leisure time and the widespread use of private cars in accessing recreational sites. Some recreation sites may be especially apposite for certain activities – ski pistes and large water bodies, for example – whilst others are relatively ordinary areas, but reasonably attractive and accessible. It is generally felt that day visits to the countryside will be more focused if there is a specific destination in mind, and that 'honeypot' sites are potentially useful means of deflecting visitor pressure from ecologically and visually sensitive, or excessively congested, locations. The need to cater for this demand has led to the planned provision of various types of countryside visitor facility, notably:

- country parks owned and managed by local authorities (accessible areas of at least 10 ha which have some variety of interest and are actively managed to improve the visitor experience);
- country parks owned and managed privately, often grant aided from public funds;
- regional parks (in Scotland);
- smaller sites and trails in the urban fringe;
- national trails (long-distance paths); and
- Millennium Greens (usually in the countryside, but sometimes also in urban settings).

Frequently grant aid is or has been available to support the creation and management (especially through countryside ranger services) of some of these facilities. The recreational role of national parks should also be noted, as should the possibility of integrating visitor access with conservation objectives in nature reserves.

Although the early emphasis, from the 1960s, was on catering for recreational enjoyment, frequently with the interests of the car user uppermost, later concern was directed towards environmental improvement in the more accessible countryside closer to towns. This reflected a recognition of the recreational and amenity potential of such areas, which often lay unrealised because of their neglected or fragmented appearance. If urban fringe locations could be made more attractive and open to walkers, this could satisfy the needs of noncar-borne and less affluent users, as well as improving the backdrop of the principal residential catchments. A major difficulty, from the environmental planner's perspective in achieving this, however, lay in the lack of planning controls over agricultural and forestry land.

However, bringing these uses under full planning control would create problems of its own, as it could result in cumbersome and undesirable bureaucracy; moreover, many urban fringe land-use issues are not appropriate to solution by 'planning' mechanisms as they require practical and localised site improvement. Consequently, an approach based on local project management evolved, recognising that many problems are relatively local and small scale, intractable to resolve, and require responsive and entrepreneurial action. The (former) Countryside Commission has strongly advocated, and financially supported, this 'countryside management' approach since the early 1970s. It has particularly been directed towards issues such as changes in field boundaries and clearance of tree cover; poorly organised access to, or across, private land; incompatible recreation activities, traffic congestion and indiscriminate parking; minor eyesores and fly-tipping; and underutilised and run-down land.

A typical 'countryside management' response to these involves delineating an area for action and the appointment of a project officer, usually with the assistance of grant aid. The project officer then acts as a catalyst and focal point for local initiatives, either seeking to involve local residents on an *ad hoc* basis or trying to foster more of a self-sustaining *community action* approach. The project officer, whilst having to report to a regular committee cycle, will have some delegated financial powers to fund modest capital works and may achieve considerable autonomy by attracting industrial sponsorship. Typical areas in which this approach is taken include the urban fringe, uplands and coast. Countryside management has had a mixed reception and, although it is now widely established, its effectiveness will always be limited by lack of basic statutory powers. The success of project officers will also clearly depend on such factors as the goodwill of landowners, amount of private sponsorship available and the general prosperity of farmers in the area. Three broad purposes have been identified where project management can succeed in achieving practical results, namely: 'area management', or the resolution of small-scale conflicts within a clearly defined stretch of countryside; the implementation of management agreements (either informal or legally binding) for specific sites; and the production of management plans or similar broad policy statements for tracts of countryside.

Integrating Environmental Action

A recurrent theme of environmental planning is the problem posed by sectoral government and reductionist training (both of which, of course, also have great virtues). The complex interlinkages within the environment clearly require the adoption of an interdepartmental, team-based approach. In practice, policy spheres (such as pollution, agriculture and conservation) are generally conceived in isolation, which then restricts the scope for integrated implementation. Moreover, individual landowners or agency personnel frequently have undertaken a specific professional training which, whilst imparting a high level of expertise in a particular field, may leave them poorly versed in the wider implications of their work for other interest groups or environmental subsystems. Furthermore, public administrations have often been fragmented in respect of their delivery of environmental services. This type of organisation (sometimes referred to as 'departmental silos'), whilst suiting most purposes of the public and private sectors, is contrary to the inherently interconnected nature of environmental issues.

Typically, bureaucratic departments have been organised on the basis of a 'functional principle', in which line management is orientated towards the attainment of clearly defined, primary functions (such as defence or education). This leads to a vertical division of work between departments rather than, for example, apportioning responsibility horizontally between departments so that each one deals with the overall needs of a client group. Thus, whilst client groups may have needs which are multiple and complex, an individual department will only address a subset of those needs and must pass on the client elsewhere for further assistance and advice. Although this leads to clarity of objectives and purpose intraorganisationally, it may be confusing to the client and there may be inefficient areas of overlap or omission between departments. In a similar fashion, the western scientific tradition has compartmentalised fields of study on a strongly disciplinary basis. Again, this leads to high levels of expertise, but has not developed laterally directed inquiry into cognate issues or nurtured crossdisciplinary solutions. An approach more geared to environmental issues is that of horizontal management in which tasks cut across departmental responsibilities. However, this leads to unwieldy departments, and may cause separations of activity which are detrimental to other issues. The Department of Environment, Transport and Regions, created in 1997, is probably at the limits of manageability, yet still has no responsibilities for a number of key environmental areas, such as those related to industry, energy, farming, education and forestry. Thus, a fundamental challenge in environmental planning is to resolve the tension between the forces of 'vertical integration' and 'horizontal integration': truly integrated solutions are ones which reconcile the problem-solving capacity of the latter with the clarity and focus of the former.

As an example of disintegrated policy-making, consider the situation which has arisen in British agriculture since 1945. Initially, policy was aimed purely at increasing food production and maintaining viable incomes for farm units, and so its functional isolation within an agricultural ministry was uncontentious. However, from the 1960s onwards, it was clear that modern farming methods were giving rise to significant environmental issues, especially those associated with loss of wildlife and scenery. During the 1980s it also became clear that major water pollution problems were arising from the discharge of nitrates and phosphates, and that these were jeopardising, amongst other things, compliance with European legislation.

During the 1990s, food quality emerged as a key political concern, linked more widely to influences such as public safety issues and farm animal welfare. However, the location of agricultural policy-making within one dominant government department tended to marginalise many of these issues for considerable periods; moreover, when policy responses were eventually devised, these typically involved a reduction of food production and the pursuit of nonagricultural objectives (e.g. ecological features) which were politically difficult to fund through the agricultural budget. It has taken considerable political will and public pressure for even limited 'horizontal integration' to transpire.

At the governmental level, there are a number of imperfect solutions which can be adopted to the broader challenge of horizontal integration – or, as it is known in modern parlance, of 'joined-up government'. One is that of having a 'green minister' in each department with a specific brief to consider environmental implications of departmental activity. Thus, an important outcome of the 'environmental white paper', *This Common Inheritance* (HMSO, 1990), was that each government department had assigned to it a minister with a specific environmental brief, in order to encourage horizontal functional integration. Similar arrangements may also operate at local governmental level, where departmental 'green officers' are usually desirable complements to a specific 'green team' within a particular department or the Chief Executive's Office. Another mechanism is to have interdepartmental working parties to consider issues with crossdepartmental implications, and these may operate both at administrative (civil service/officer) and political (minister/councillor) levels.

Beyond the intragovernmental context, the main way in which horizontal integration can be pursued between private, voluntary and public bodies is that of partnership. Partnership structures may be very formal or very informal, and they may be amongst equals or have major/minor stakeholders. Environmental partnerships are now well established in many contexts, ranging from the international to the parochial. In the field, team-based approaches are often adopted for tackling complex environmental problems. These may be multidisciplinary, in which different specialists combine to implement aspects of the solution according to their expertise, or interdisciplinary, where specialisms are merged to produce a more integrated solution. (The term 'concerted action' tends to be used in European contexts.) Although integrated management and planning has been the aim of environmental projects for many years, it is now often fashionable to refer to 'holistic' planning, in which problems are viewed in a seamless manner and team members are encouraged to leave their disciplinary perspectives behind. Characteristically, holistic approaches will also view projects on a life-cycle basis, so that solutions are worked out in terms of the total implications of the project, thereby (in theory at least) optimising its long-term sustainability.

In practice, natural resource planning has rarely if ever been simultaneously integrated at the policy, programme and practice levels

(Mowle, 1988). Potential solutions to this kind of disintegration are three-fold. First, it must be recognised that many conflicts between competing land-use objectives are intrinsically deep seated, and may sometimes be impossible to reconcile. However, usually there is some degree of common ground, or indeed positive mutual benefit, which can be achieved by the use of inclusive methods of consultation and design. Secondly, since sectoral administration hampers integrated practice, it must be assumed that some environmental conflicts are exacerbated by extrinsic factors associated with the bureaucratic infrastructure itself. The need to respond to issues on a problem-solving basis and by generic client group (of which the environment is but one) is gradually leading to more flexible modes of governance. Thirdly, improved understanding about the ways in which partnerships and networks operate can help deliver individual policies, programmes and projects on a horizontally integrated basis. Inevitably, scholars who subscribe to a 'political economy' analysis would view such solutions as deriving from a 'naive administrative' perspective; whilst they would be wrong to dismiss the importance of improved practical procedures, they would be right to recognise the deep-seated societal conditions which hamper the implementation of any environmental solutions, let alone integrated ones.

The plethora of departments and agencies with some environmental responsibility in Britain is further complicated by the fact that many national agencies tend also to have a regional structure for administrative convenience. This additional complication has been rationalised during the 1990s, first, by the creation of regional government offices (consolidating the regional offices of all government departments) and, secondly, by the creation of regional development agencies (RDAs) in England and a parliament and assembly for Scotland and Wales. Below the regional level, it is still often the case that administrative boundaries for different bodies rarely coincide, which can make partnership arrangements unduly cumbersome. Further, some topics, such as pollution control, are split between central state and local state.

One important approach to integration in recent years has been the reform of environmental agencies to take on board a wider range of responsibilities. However, some of the potential realignments were too difficult to contemplate, such as the inclusion of nature conservation responsibilities in the Environment Agency, which would have been allocated potentially irreconcilable duties for both pollution and countryside protection. There have been some concerns about the effectiveness of agency reform in these cases and whether or not genuine integration has occurred on an organisational basis, or whether the traditions and trainings of staff are too separate for joint policy and practice to emerge. It appears that, in the countryside organisations, there is continuing distance between recreation and conservation staff, whilst in the environment agencies, field staff (such as water bailiffs) do not always communicate well with each other or with higher-order staff. The extent to which organisational rationalisation genuinely assists effective integration is therefore debatable.

Complementary to the higher-order objectives of integrated environmental governance are the projects pursued at the local level. As an example of early moves towards 'integrated rural development' (IRD), various schemes were promulgated by the EC from the late 1970s, when projects on rural job combination and job flexibility ('pluriactivity') were initiated. A key conclusion was that, whilst land-use systems were themselves often socially and ecologically resilient, rural policies and measures were poorly integrated and legal instruments were too narrowly agricultural. Thus, it was argued that the prime initiative should lie with communities themselves rather than depend on the actions of distant bureaucracies. This concept was tested in the Peak District where a system of alternative agricultural grants was approved on a trial basis for two parishes. Here, farmers could obtain backing for maintaining wildflower meadows and other distinctive landscape features. Collaboration between private agencies also facilitated grant aid for a variety of social and economic projects, including village halls, playgrounds and tourist or light industrial facilities. The complexities normally associated with grant aid were greatly alleviated by providing 'one-door' access to funding and advice through locally based project officers. This experiment confirmed that most success was obtained when projects took account of the three 'I's' of IRD (Parker, 1985), namely:

- *individuality* of a particular locality as a source of economic strength, distinct social identity and environmental character, in contrast to the usual treatment of all rural areas in a blanket fashion by public authorities;
- *involvement* of local communities in thinking about their own future and in working out and putting into practice their own ideas for improving the future; and
- *interdependence* of rural areas as a whole, requiring a package of measures which eliminate harmful side-effects and encourage mutual benefit.

Despite the many encouraging features of this project, it is clearly unrealistic to assume that conflict resolution lies in the hands of local communities alone: the fundamental ground rules are clearly set by the EU, national and local government, public agencies and larger private landowners. Indeed, 'integrated' projects funded by the EU have come in for regular criticism on the grounds of still being too production orientated and of causing significant environmental damage, especially in the less developed member states of southern Europe.

Collaborative approaches to environmental management often stress the planner's role as a mediator and catalyst, a view which is clearly located in planning theories concerning 'brokerage' roles. Whilst these theories have their limitations, it is nevertheless true that planners often have access to relevant information, are experienced at consulting and liaising with community, policy and commercial interests, and have acquired many skills of negotiation and liaison. Whilst during the 1970s and 1980s planners tended to emphasise their experience in servicing partnerships and agreeing joint strategies, the 1990s witnessed a growing interest in the use of consensus-building techniques and mediatory methods to reconcile land-use conflicts. A particular expression of this has been the production of interdepartmental and multiagency strategies to help co-ordinate policy and budgetary expenditure across a range of environmental topics.

For example, it is now common to produce various types of rural strategy, including those for conservation, rural economy and community, landscape, nature conservation and recreation (CCW, 1995). Integrated strategies of this nature, which typically are prepared at the county level, can, according to their advocates:

- provide a forum for bringing together those with a responsibility and interest in the well-being of rural areas;
- increase knowledge of the existing state of rural areas, in particular their main assets and the threats to their well-being;
- develop understanding of shared objectives and integrated approaches;
- influence the drafting of statutory development plans covering rural areas;
- provide a vehicle for public involvement in establishing co-ordinated rural policies; and
- provide a framework for monitoring change, reviewing progress and ensuring adaptability to new circumstances and opportunities.

In most cases, for the purposes of decision-making, these are prepared by local 'countryside forums' which in turn tend to operate on the basis of 'working groups' or 'task groups', linking a wide range of organisations. Actions within these strategies tend to aim for: improvement of the information base; provision of advice; integration of objectives into statutory plans; networking; setting of targets; and implementation of specific projects. To date, however, is it probably fair to say that there is general disappointment at the slow rate of implementation of strategy objectives. One important constraint in practice appears to have been that of 'ownership' of strategies, so that contributing organisations may feel relatively unconcerned about their success or about phasing and targeting their budgets to assist collective objectives. It appears that, in situations where the local authority has been the prime author of the strategy with only nominal input from an advisory group, there has been little sharing of objectives and little desire on the part of others to take forward specific actions; conversely, there may be considerably greater success where task groups who jointly authored the strategy continue to meet on a regular basis to oversee the implementation and review new issues as they arise (Box 3.12).

More 'rationalistic' approaches to integrated planning have tended to be based on particular techniques and procedures. One method sometimes adopted is that of adaptive environmental assessment and management (AEAM), which involves having high-quality information available in a

Box 3.12 An example of a countryside strategy: the Rural Strategy for Gloucestershire

Purpose of strategy:

- Develop an agreed voluntary framework for co-ordinating action between statutory bodies, organisations and individuals;
- increase the awareness of the range of economic, social and environmental issues affecting the county;
- provide better understanding of the aims and objectives of the various statutory bodies and other organisations working in the rural areas;
- stimulate positive ideas and promote action to resolve the problems of the loss of rural services and environmental change, especially through innovation and multiagency projects; and
- provide a corporate approach to the integration of services and the resources of the various agencies working in rural areas.

Action plan topics:

- Environmental assessments to establish the effects on the environment of development proposals;
- enhancement and conservation of sites and their settings;
- support for voluntary and community groups in conservation and care for the environment;
- continuation of education and interpretation initiatives in the countryside;
- more careful use of water resources;
- careful working of mineral sites and waste disposal sites;
- encourage use of public transport, cycling and walking;
- assist businesses to find appropriate premises having due regard for the local environment;
- identify training and retraining needs of rural inhabitants;
- monitoring changes in the rural economy;
- appropriate opportunities for business, tourism expansion and processing facilities;
- surveying and responding to rural housing needs;
- provision of affordable housing;
- monitoring, integration and improvement of rural service provision;
- conducting 'village appraisals'; and
- multiple use of community facilities.

Steering group composition (for monitoring implementation):

- local authority representatives;
- government agency representatives;
- Rural Community Council and Association of Parish and Town Councils;
- Training and Enterprise Council; and
- voluntary sector representatives.

Source: Rural Strategy Advisory Group, 1992

responsive management information system such as a GIS, and having a range of experts on hand to adjust the design and implementation of the project as it progresses, rather than being bound by an original master plan. More generally, the term integrated environmental management (IEM) is used. Broadly speaking, the object of IEM is to integrate environmental considerations into all stages of a development process in order to achieve economic and social benefits at minimum environmental cost. Rather than seeking to stop development, IEM is intended to ensure that environmental factors are considered from the earliest stage of any proposal, and continue to be respected during its implementation and management. One prerequisite for this is that no expensive commitments are made in the early planning stages to a proposal which might have unacceptable external effects. IEM typically starts with a 'proposal generation stage', involving informal discussions between the developer and the regulatory agency. This enables exploration of environmentally acceptable ways of achieving the development objectives, and consideration of the planning and policy constraints before making major investments or commitments. Interested parties – the public, developer and government – then join in a search for viable alternatives, with the intention of identifying designs which are broadly acceptable. There is also an investigation of the possible effects of the proposal, at which point any potential drawbacks of alternative options can be identified. After this, formal environmental assessment may take place for those options which remain under consideration. The regulatory agency then reaches the 'decision stages', either approving the proposal (usually subject to conditions) or rejecting it (subject to appeal), and usually prescribing monitoring and auditing arrangements during operation.

Where 'rural' uses such as agriculture, forestry, conservation or recreation are the principal objectives, rather than built development, reconciliation of user conflicts is often related to the inherent capabilities and limits of the land under consideration. Thus, solutions are based on the potentials of the land to support one or more uses on a sustainable basis. A multiple or integrated-use solution may be achieved where conflicts between users either do not arise or can be reconciled through site management. Thus, the alternative outcomes are:

(1) exclusive use (often a socially and environmentally undesirable option) in which those with proprietorial interests reject the claims of other groups to influence their activity or share the resource;

(2) multiple use, in which secondary interests (e.g. water-based recreation) are tolerated and perhaps positively encouraged as a complement to the primary use (e.g. water impoundment);

(3) integrated use, in which equal land-use partners are managed in conjunction, generally to their mutual benefit (e.g. agroforestry); and

(4) compatible use, in which specified land-use activities are sanctioned provided they are not incompatible with acknowledged environmental constraints.

The shared use of resources can be accomplished by various options, or combinations of them. These include (Ridd, 1965) concurrent and continuous use of several resources obtainable on a land management unit; alternation and rotation of the uses of various resources or resource product combinations on a unit; or geographical separation of uses or use combinations so that multiple use is accomplished across a mosaic of units. An underlying principle, where 'compatible use' solutions are proposed, is that the management of a given ecosystem should not damage and should, wherever possible, yield beneficial side-effects to another ecosystem.

Conclusion

There are clearly many powers available for environmental management. These range from pollution and land-use control to nature and landscape protection. Some powers are exclusively within the statutory domain of town planners, some are principally the responsibility of other professions, whilst on occasion, planners can play an indirect role. Nevertheless, there is usually some basis in administrative practice for planners being involved – as controllers, consultees or enablers – in most areas of environmental concern.

An evident feature of environmental law and administration is its growing concern for integrated and holistic approaches. If environmental planning in the past has been characterised by interdepartmental rivalry and sectoral decision-making, then the current trend is unmistakably towards collaborative working for integrated and 'best overall' options. This inevitably greatly increases the complexity of decisions, and perhaps delays the time taken to arrive at strategic or local solutions, but may result in steadier progress towards crossing the sustainability transition.

A major effect on contemporary environmental management has been the growing influence of international bodies, impelling national and subnational administration towards standards of best current practice. Of particular importance have been the moral and legal pressures exerted by international conventions and protocols, and the binding requirements imposed by membership of the European Union. In Britain, the official attitude towards environmental issues has palpably changed from one of reluctant participation to championing good practice. Although land-use planning, being primarily local or subregional in character, experiences relatively limited international pressure for change, the wider issues which it seeks to address are starting strongly to align themselves with continental and global standards. This is likely to favour environmental planning solutions which are both administratively and conceptually integrated, and which resonate with universal concerns for sustainable development.

4 Planning and Managing the Natural Resource Base

The Changing Countryside

A major focus of environmental planning is the management of natural resources in the countryside. These resources can broadly be summarised as 'productive' and 'protective': productive uses are principally those of agriculture, forestry and minerals; protective uses include nature and landscape conservation. Despite its popular image of tranquillity and permanence, the countryside in most industrial and postindustrial countries is subject to continual flux. Characteristic landscape changes in Britain which have been revealed by repeated surveys comprise:

- an increase in cereals (especially wheat) at the expense of pastoral agriculture;
- removal and coniferisation of broadleaved woodland;
- considerable additional tree planting, of both conifers and broadleaves, but very different in quality and composition from old and ancient woodlands;
- removal and deterioration of hedgerows and stone walls;
- creation of extensive new farm boundaries, mainly with wire fencing;
- construction of new farm buildings, and increasing dilapidation and conversion of older ones; and
- a 'drying out' of much of the countryside, with loss of wetlands, lowering of levels in rivers and depletion of groundwater reserves.

The most recent Land Classification Survey in the UK (Walford, 1997) provides a range of statistics on current land-use composition and change (Box 4.1).

Key evidence on the countryside reveals that the rural land cover of Britain is made up of four main types; most important is grassland which covers 32%, arable comes second with 22%, heath and bog cover 14%, woodland 12%, whilst wetland and water make up nearly 2% with rocks and scree, and coastal habitats, 1% each. Thus, about 30% of Britain is semi-natural in character, whilst built-up land covers 8% of the rural landscape with recreational space at about 1%. It should be noted that the rarer land types are associated with relatively large standard errors, as this survey is based on a sampling frame rather than a comprehensive 'census' (Barr and Fuller, 1997). A variety of recent evidence on UK land-use

Box 4.1   Aspects of land use in the UK

- Over 10% of the land of the UK is built up;
- a third of England is covered by arable land, but the proportion of arable land in the remainder of the UK is less than one-tenth;
- land devoted to agricultural use (arable and grassland) makes up well over half (55%) of Britain (some of the remainder is moorland);
- almost half the land in built-up areas is residential, followed by transport uses;
- only 3.6% of land in the urban areas of Britain is now in industrial use; the largest regional proportion (8.6%) is in the West Midlands;
- the percentage of open space (9.6%) in urban areas is nearly three times that devoted to industry;
- the area of forest and woodland (12%) in Britain in 1996 is double the area recorded in the 1930s; and
- pylons or communications towers were identified as landscape features in 25% of the areas surveyed.

Source: Based on Walford, 1997.

change is summarised in Box 4.2 (Minor discrepancies between these figures can be accounted for by rounding and sampling errors). This chapter explores some of the causes and consequences of these changes.

Productive Uses of Rural Resources

Agriculture

Agriculture is no longer a leading industry economically in most western countries, yet it is still the dominant user of the countryside. Consequently, its activities are of major importance to the environment and it may maintain a strong political significance. Even though it lies outside planning control in most (though not all) legislatures, planners still inherit the consequences of its changes. Recent years have seen something of a retreat from agricultural intensification as the EU's Common Agricultural Policy (CAP) has been reformed, in response to chronic overproduction of many commodities, budgetary constraints, pressure from environmentalists and the need to comply with international trade agreements. However, the scale of this retreat is arguable, and farmers still find themselves having to achieve high yields within very constrained economic circumstances.

A particular interest of environmental planners is the type of *landscape* produced by farming. Very broadly, there are three main types of farm landscape in Britain: the enclosure landscape dating from the period of systematic land improvement of the eighteenth and nineteenth centuries; the extensive pastoral landscapes of the uplands; and the modern functional landscape related to widespread use of machinery and intensive

Box 4.2　Recent land-use change in the UK

Pre-1939 patterns of land-use change

One-third of farmland in the UK was covered by rough, semi-natural vegetation used largely for extensive grazing by sheep and cattle. In the 1930s, the only major transfers of use from grazing were to forestry and these were exceeded by gains due to the invasion of pastures by sedge and bracken, and to the direct reversion of arable to rough pasture. Within improved farmland, a depressed demand for agricultural products was encouraging the turnover of arable to permanent grass. Also, there was the rapid suburbanisation of farmland – from 1936 to 1939, each year an average of 25,000 ha of farmland in England and Wales was transferred to urban use.

1939–45 changes in land use

During the war years the most marked and rapid change in UK land use this century occurred – even more marked than during the 1980s. Ploughing grants and compulsory powers of the County War Agricultural Executive Committee led to a reversal in the decline of the arable area. Two enduring changes occurred:

- the transfer of land to the armed services; and
- extensive felling of timber.

Post-1945 land-use structure

Land use during this period has been the product of myriad management decisions. Ownership of land in the UK reflects 270,000 separate holdings in the rural sector; about half of these farmers are also landowners, the remainder being tenants whose management decisions are constrained by conditions of lease from the owner – so there are about 400,000 decision-makers involved in the use of rural land in the UK.

Agents of post-war changes in land use have been:

- growth of urban areas and their changing composition;
- changes in the agricultural sector (especially between rough land and improved land, tillage and pasture);
- extension of forest and woodland; and
- growing competition for rural land from quasi-urban uses (e.g. recreation and water-gathering).

The UK urban area has been advancing at the rate of about 10% per decade, though rates are less than two-thirds of the 1930s rate. This is affecting the better land, though the impact is being offset by gains in agricultural productivity.

Key rural land-use changes

Tillage and grassland have shown substantial increases, and major regional variations of change have served to enhance the contrast between east and west. The most notable changes have been an increase in the area devoted to barley, as a feedcrop for cattle, and a decrease in labour-demanding root crops and a decline in orchards.

Rough grazing – 6 million ha (a quarter of the farmland) contributes perhaps only 5% of the nation's agricultural product by value, but its aesthetic and ecological value is substantial. Between 1933 and 1980 about 1 million ha were transferred to improved farmland – in some areas reclamation was particularly extensive (e.g. in the national parks, especially the North York Moors, Dartmoor and the Brecon Beacons, where plateau areas were relatively easy to plough and where slopes are gentler.)

Woodland – over one-third of the land lost to farming has been converted to forestry. The UK area under woodland has almost doubled since the 1940s.

Changes to the rural landscape – the declines in broadleaved woodland, semi-natural vegetation and grasslands are largely accounted for by the expansion of cultivated land use by 25% during the period 1945–80. It is estimated that 22% of hedgerows in England and Wales were lost between 1947 and 1985, 50% of all ancient broadleaved woodlands, 50% of lowland fens and marshes, 60% of heathlands, 80% of chalk grasslands and perhaps over 95% of herb-rich grasslands.

Future climate change may increase the potential for growing a wider range of crops, and will have the general effect of lengthening the growing season in northern Europe but generally shorten it in southern Europe. Higher humidity may be as much a constraint as temperature, though if humidity were to decrease it could see crops such as sunflower thrive. Climate changes could shift the thermal limits of native species in the order of 300 km of latitude and 200 m of altitude per °C, and could radically shift plant distributions. Models, based on the Institute of Terrestrial Ecology (ITE) land classification, suggest a general simplification of land use.

Source: based on Parry *et al.*, 1992.

livestock husbandry. A great deal of sentiment has attached both to the infield and outfield systems of the first of these phases: the 'hand-built' and visually diverse appearance of the former contrasts effectively with the wilder, nutrient-poor, semi-natural condition of the latter. Also, by fortunate ecological coincidence, the wooded field boundaries of the enclosures have formed a rich network of corridors for wildlife, and these have been widely removed from functional landscapes. The future of the agricultural landscape is uncertain: plausible scenarios range from ranched, dismal vistas bearing no regional distinctiveness, to an economically cushioned countryside in which farmers promote the enhancement of plagioclimax vegetation and de-emphasise the role of conventional production.

The attrition of farm landscapes in the pursuit of technological efficiency has been well documented. Systematic field drainage has occurred (sometimes in association with arterial river management programmes), normally to convert pasture to arable land. Hedgerows have been neglected and grubbed up, both because of their maintenance implications and because small fields restrict the work output of modern agricultural machinery. Similarly, there has been a qualitative and quantitative loss of lowland farm woodlands. Semi-natural grasslands have been reclaimed and re-seeded with ecologically impoverished mixes of agriculturally nutritious grasses. On the base-rich soils of chalk and limestone areas, modern methods of cultivation have enabled grazing pastures to be converted to intensive leys or crops, causing a similar ecological loss. Of course, wildlife is adaptable and landscape tastes variable, so traditional sentimental attachments to past farm landscapes cannot be considered sacrosanct, especially as hedgerow and woodland removal has occasionally opened up fine vistas. However, there is no doubt that a serious net loss of wildlife habitat has occurred, and that former (labour-intensive) methods of field and boundary management, and traditional grazing practices, were ideally suited to maintaining the seral condition of plagioclimax communities. Dutch elm disease and severe storm damage have further compounded the attrition of tree cover caused by modern agriculture.

A further important consideration is the nature of the *farm business*. The major spatial trend in farm enterprise patterns has been towards greater regional specialisation. This is particularly manifest in the tendency towards arable farming in the drier eastern counties, dairy and stock rearing in the west and extensive production of prefattened livestock in the uplands. Specialisation has been accompanied by the shedding of labour for capital-intensive machinery and functional farm buildings. This process has been fuelled by a price–cost squeeze, in which inflation and a decline in the demand for food relative to income levels have caused the costs of production to rise at a greater rate than the price of foodstuffs.

In broad terms, it is also possible to distinguish between the prosperous agricultural lowlands, the marginal fringes (uplands) and the urban fringe. Whilst the uplands are economically marginal and vulnerable to farm abandonment, the urban fringe suffers from a combination of vandalism, trespass, neglect and development pressure, but also offers opportunities for selling conventional and unconventional farm products to large, accessible populations. In both fringes, farm fragmentation may be severe (Ilbery, 1991). Farm landscape change has thus tended to vary according to enterprise type, local considerations and farm business factors, with the most notable trend evidently being arablisation of the lowlands (Ward *et al.*, 1990).

Farm businesses are conventionally analysed in terms of their inputs, outputs and profit margins (Table 4.1). The total value of all outputs is known as the gross or total enterprise output. However, this merely reflects volume of turnover: if margins of profitability are required, two measures are commonly employed. First, the 'variable costs' which fluctuate with changes in the scale of the enterprise, such as fertilisers and sprays, must be subtracted in order to yield the *gross margin* (GM). From this in turn may be subtracted the overhead or fixed costs, such as buildings and machinery, which would not readily change with alterations to the area of the farm. These are commonly expressed in terms of depreciation on assets to yield an equivalent annual cost. Deduction of these would show the *net farm income* (NFI) (effectively, the annual salary of the farm family). A more sophisticated variant on these measures is 'whole farm budgeting', which identifies separately the different types of enterprise on the farm, and their associated input and output revenues.

A great deal of interest has focused on the processes of capital restructuring in agriculture, particularly those which tend to create a dualism between family farms on the one hand and large agribusinesses on the other. This in turn affects the direction and nature of capital penetration (subsumption) in the farm, producing various outcomes, which range from

Table 4.1   Diagrammatic structure of measures of farm profitability

	£	£
Total enterprise output (TEO)		_____
Minus: Variable costs (VCs) (e.g. seeds, sprays, veterinary expenses)	_____	
Gross margin (GM) (TEO – VCs)		==============
Minus: Fixed costs (FCs) (overheads, rent)	_____	
Net farm income (GM – FCs)		==============
Farmer's income + return on tenants' capital + profit		

the traditional farm family managing a single unit farm, to the corporately owned agribusiness handling several enterprise units and using only contract labour. The inevitable transformation of the family farm has created a number of responses, such as the search for diversified sources of income and capital, which in turn have contributed to a growth in part-time farming and a decrease in the economic centrality of the agricultural enterprise itself. Increasing subsumption of external capital in farm businesses has increased the influence of food processing and marketing companies on the management (especially mechanisation and use of biocides) of rural land.

Particular attention has turned recently to part-time farming, a phenomenon which was treated rather disparagingly in times when policies were directed to output maximisation, but which is now seen to have a significant role in rural diversification and environmental enhancement. Part-time farming is difficult to define, as it ranges from hobby farmers whose net income from farming, if it exists at all, may be unimportant to them, to those who maintain an extensive and highly profitable farming activity but who derive also a significant income from nonfarm sources. It thus comprises genuinely nonviable holdings, part-time working by farmers, dual job-holding by farmers and farm households with other gainful activities. It is also sometimes assumed that farmers who are not wholly dependent on farming for their living can better afford to protect the natural environment. Equally, those who are not economically or psychologically committed to maximum production may be more willing to provide 'staged' farming (contriving to produce a particular type of amenity scene) which society appears to demand. 'Hobby' farmers are especially significant around the edge of major cities and their objectives – probably not strictly economic – may be crucial factors in landscape planning, though it would be naive to suggest that subeconomic farming always produces positive environmental benefits.

Studies of agricultural change traditionally used to emphasise land and business factors, but there is now a greater recognition of the characteristics of the farming *community*. This is diverse, and comprises all members of the farm family (not just the farmer), full-time and part-time participants, and various approaches to investment and risk-taking. In most countries – not Britain, but many member states of the EU – the agricultural workforce is still large and thus it possesses a key socioeconomic role which ensures strong political support for farming and food policies. This partially explains the difficulties of reforming European agricultural policy. Considerations about the nature of farmers are crucial in understanding why some more than others alter the landscape and apparently respond to policy inducements; they are equally important when seeking to influence farmers to pursue new land management practices. Farmers' behaviour has been shown to vary according to the diffusion of information, age, education and personality; specific changes in landownership (i.e. when a farm changes hands or land is acquired or lost) are also critical triggers to investment and consequent land-use impacts. A number of broad generalisations may be made with respect to these factors. Notably, the more information available about progressive techniques, and the greater the possibility of seeing them demonstrated in practice, the greater the likelihood of adoption. Generally speaking, farmers tend to become increasingly conservative with age. Attendance at agricultural college increases the likelihood of 'modern' farming, and inservice training enables new ventures and innovations to be adopted. Finally, there is a spectrum from high-risk takers to risk-averse farmers, whilst wealthier farmers are typically more amenable to taking the risks associated with product specialisation. An example of the significance of personal factors is provided by the levels of participation in Environmentally Sensitive Area schemes. Key factors which led to farmers becoming involved, apart from the intrinsic extent of wildlife resources on their farms, were the payments offered by the scheme, information provided by government extension services, the 'success factor' and dynamics within the district; important secondary influences were age, education and length of residency (Wilson, 1997).

A seminal study of the behavioural factors affecting farm investment was conducted by Potter (1986), who found that pollution and various ecological effects were frequently associated with short-term (operational) management decisions, whereas actual habitat destruction and landscape change normally resulted from investment (strategic) decisions. Within this framework, it proved possible to identify three types of decision, namely:

- managerial decisions, i.e. those associated with land-use and husbandry practices on a daily, weekly and seasonal basis, whose adverse effects may, for instance, include increased fertiliser runoff and pollution from slurry and silage effluent;
- investment decisions, about the deployment of land and capital, including land improvement, and which may comprise hedgerow removal, pasture reseeding and field under-drainage; and

- strategic decisions, affecting the pattern of enterprises (specialisation) or scale of business operation, including conversions from grazing to arable systems.

In his survey of decision-taking and impacts on the unutilised agricultural area of a wide sample of farms he identified three broad categories of investment styles. The poles were represented by 'incrementalists' – who tended to produce landscape changes which were localised and intermittent, and probably less extensive – and 'programmers', who were systematic in terms of co-ordinated and carefully planned landscape change. Farmers adopting a blend of these approaches were termed 'mixers'.

Gilg (1996), reviewing behaviouralist perspectives on the farming community, noted that many studies have emphasised the idiosyncratic nature of farmers and the attachment to farming as a way of life, coupled with the fact that decision-making processes are based on 'satisficing' rather than 'profit maximisation'. Farmers are remarkably resistant to change and appear to espouse schemes which fortuitously fit in with what they were doing or planned to do anyway. Thus, it may be more important to relate the pattern of landscape change in a locality to the process of farm business growth and development rather than to specific farm or farmer characteristics. The process of countryside change has an inbuilt momentum and will continue to be driven forward by factors which are embedded within the present structure of the industry and the value systems of individual farmers, and these will be slower to change than policy-makers would often prefer. Evidently, farmers appear to want a very substantial inducement to switch land uses and are relatively resistant to policy-driven schemes, especially for environmental benefits. Those farmers who do behave in an 'environmentally friendly' fashion tend to be distinguished by behavioural or attitudinal traits rooted in their personal histories and circumstances.

Land *tenure* may be broadly divided into owner-occupation and tenancy (though various types of lands are held 'in common', including the Anglo-Welsh concept of commons and their associated commoners with 'common rights', and the Scottish crofters' common grazings and arable lands). A mix of tenure types may be found within individual enterprises, and the broad distinction between simple freehold and other tenure forms has become especially blurred and complex as a result of tax changes which have made it difficult to transfer substantial wealth from one generation to another. The relative stability and security of agricultural land as an investment have also made it a popular element of portfolios, especially as the investment organisations concerned do not die and are thus not affected by death duties. However, the scale of institutional farmland ownership has remained comparatively modest, and has never reached the scale which was being anticipated in the 1980s. Nowadays, owner-occupation is typically subdivided into owner occupation *per se*, joint ownership, family farming companies, institutional landownership and trusts. In areas where land speculation for urban or mineral development is endemic, notably the urban fringe, nonagricultural interests are further complicating the land-ownership arrangements. Here, property rights over land are often being divided into various forms of user and occupier rights, including option arrangements and conditional contracts, especially where construction companies seek to exercise a 'first option' on land which the planning system may at some future date make available for development. Similarly, mineral operators may seek to purchase the mineral rights associated with land in the first instance, rather than its freehold, and this may affect the achievement of environmental planning objectives.

Estates which have let tenancies may prefer to retain direct control once the tenancy is vacated, and thus farm the land 'in hand'. Tenancies may be leased from private landlords, public landlords (since many public organisations own farmland either for investment or intended operational purposes), companies, charities and institutions such as insurance and pension funds. Conservation organisations, for instance the National Trust and Royal Society for Protection of Birds, own large tracts of land and these are commonly leased back to farmers with restrictive covenants which ensure a balance between conservation and economic agriculture. Land may also be rented on short-term agreements, often on renewable licences of less than a year's duration. This may be useful in country parks or other amenity land, for instance where grazing is seen as a useful management tool but the landowner wishes to retain maximum flexibility, although farmers experience problems of insecurity and visitor pressure. Often, several farms may be managed as a unit, and policy on the out-farms will be dictated by the needs of the home farm. In north and west Scotland and some other marginal European areas crofting is practised, normally within a system of pluriactivity (multiple sources of income). Security of tenure and house-building right is enjoyed by the crofter, and both these factors, together with improved collective organisation, have helped it enjoy a modest resurgence in recent years. However, chronic problems of land subdivision and restrictive terms of tenure limit the rate of change.

Both private and public objectives influence the ways in which land is farmed: thus the internal constraints to maximise profit and minimise risks at the business level are strongly influenced by national (public) goals to secure an efficient agricultural sector, satisfy food self-sufficiency requirements and ensure that food is available in adequate quantities at affordable prices. Broadly speaking, the public goals of agricultural policy relate to the pursuit of 'equity' (fairness) and 'efficiency' (cost-effectiveness), which may be addressed by various instruments. The major mechanisms of policy in Britain have been capital grants to farmers for assistance in farm improvements, price supports (guaranteed markets), quota payments (permits to produce at stipulated levels, and tradable between farmers within a particular type of enterprise), direct subsidies on livestock (headage payments), advisory services, and research leading to technical improvements and more effective agrochemicals. More recently, direct payments have been introduced for farmers to provide environmental services and, significantly, these are not considered to distort the fair trading position of the farmer, and are thus inherently more acceptable to negotiators of international trade treaties. These various types of state subvention have provided farmers with relatively stable investment conditions and assisted them in optimising their investment and management. In industrialised economies accustomed to high wage levels, agriculture has tended to become inherently uncompetitive both in terms of rates of pay and capital accumulation. To counteract this, government policy has normally been traditionally involved in a high degree of protectionism aimed at inflating farm income levels artificially above world levels. As noted, various approaches to this have been adopted, but several are under threat as western agriculture once more falters towards deregulation and policy-makers acknowledge the need to dismantle protective tariffs.

Overall, the contemporary farmscape in Britain has been influenced by a strategic desire for self-sufficiency. The historical reasons why this was pursued in the 1940s have now passed. Self-sufficiency seemed necessary following wartime conditions of siege but not only does most of Europe seem to have passed into a genuinely postwar (rather than merely post-1945) era but, even if war were to recur, it is virtually inconceivable that it would last a quinquennium (indeed, it might be over in minutes). Moreover, the conditions of peace and economic exchange which now exist in Europe render considerations of national self-sufficiency increasingly almost irrelevant. The only arguments which would support national self-sufficiency in agriculture relate to bioregionalism (internalising the impacts of our ecological footprint), but these would point to an environmentally benign style of agriculture which minimised chemical inputs and maximised public confidence in food quality. Either way, the national policy which has driven maximum input–maximum output farming, based on agricultural units with a high economic centrality for traditional farm products, now seems rather anachronistic: this does not mean that industrialised farming will cease, nor that all farmers will seek environmental payments, but rather that national farm policy will diversify away from promoting maximum output.

The first major attempt to create an efficient, modern farming industry in recent British history was the Agriculture Act 1947, conceived in the wake of wartime blockades and near-starvation. This introduced three main mechanisms: an annual price review of farm products, supported by government subventions as necessary, to ensure fair returns for the farmers and 'cheap' food for the public; grant aid for capital investment; and subsidies for livestock in the hills and uplands in the form of headage payments. As the food supply situation eased in the 1950s, the emphasis started to shift from subsidies to investment and efficiency. Soil wetness is a persistent problem in much of Britain, and both arterial and field drainage were massively assisted by grant aid until the mid-1980s. High land values throughout the 1970s also enabled farmers to borrow heavily against the value of their farm in order to finance capital investments, although with static or falling land prices and enforced output reductions, many farms are now seriously overcapitalised.

Accession to the European Community in 1972 brought a fundamental change in the form, but not the essential purpose, of agricultural policy. The Treaty of Rome (1957) established five objectives for a Common Agricultural Policy (CAP), namely to

- increase productivity through technology and rational development;
- ensure a fair standard of living for the agricultural community;
- stabilise markets;
- assure availability of supplies; and
- ensure supplies reach consumers at reasonable prices.

Under this policy, subsidies and grant aid remained comparable in many respects, although higher rates were typically available under the EC-supported grants which required a comprehensive five-year farm development plan. Capital grants to increase the technical efficiency of farms have been paid under a succession of schemes, though these can no longer be justified and have been successively withdrawn. Even the Farm and Conservation Grant Scheme – which supported environmentally sympathetic measures such as hedge planting, effluent disposal facilities, shelter belts, woodland enclosure, repair of traditional buildings and energy-saving investment – has been phased out.

The major difference between the pre- and post-1972 arrangements lay in the price support system, which involved a change from normal price guarantees to intervention by centralised agencies, which purchased over-produced material ('intervention purchasing') and so stabilised prices at a relatively high level by preventing gluts. Excess would then be stored at the expense of the Community (or sometimes denatured and used as animal food). Complex arrangements had to be introduced to rectify farm price differentials between member states, in the form of the green pound and monetary compensation amounts (MCAs). This was understandable in the historical context of an inefficient agricultural industry and a need for self-sufficiency in temperate foodstuffs. However, the buoyant prices caused by government policy, coupled with various other modernising influences, have led to various less desirable side-effects.

Although the 'price' policies of CAP have tended to receive the greatest attention and fuel the principal controversies, there has also been an additional strand of policy concerned with farm 'structures'. Structural measures have sought to consolidate fragmented and uneconomic farm units, especially in the south of Europe, and to improve their efficiency and ability to add value to basic products. Many proposals for reform of European agricultural policy involve an elevation of the relative importance of structural measures, although these, too, have been associated with disruptive impacts on ecological and social systems.

The outcomes of the CAP have resulted in some rather predictable consequences, notably:

- an intensification of production;
- an increasing regional specialisation of production;
- an increasing concentration of production on the larger farms; and
- a range of environmental and socioeconomic disbenefits.

Consequently, the main thrust of the recent agricultural policy agenda has been the reform of CAP. This is not solely in the interests of environmental protection, but also of reducing the cost of agricultural support and of moving the position of European farmers into a fairer relationship to farmers elsewhere in the world. Some countries outside the EU, which have traditionally enjoyed high levels of farm support, have already had subsidies withdrawn and have increasingly had to compete at 'world prices' (though the concept of a world agricultural commodity price is itself something of a fiction). Moreover, with the prospect of an expanding EU, it will probably be impossible for the common budget to subsidise the huge additional numbers of farmers who will enter from eastern Europe.

Some researchers thus describe the change occurring in farm policy during the latter stages of the CAP as the 'postproductivist transition'. Ilbery and Bowler (1997) note that this stage has yet to be fully defined, but it displays a number of known characteristics, including:

- a reduction in food output;
- progressive withdrawal of state subsidies for agriculture;
- the production of food within an increasingly competitive international market;
- growing environmental regulation of agriculture; and
- the creation of a more sustainable agricultural system.

It also entails specific shifts from intensification to extensification, from specialisation to diversification and from concentration to dispersion. Thus the CAP was subjected to successive reforms during the late 1980s and 1990s, although these did not seek to dismantle its single market and joint financial responsibility, but focused mainly on measures for the control of production. The effect of these reform measures was differential, with larger farms better able to absorb the effects of a price–cost squeeze and smaller family farms experiencing a deepening recession, so that a difficult balancing act ensued between socioeconomic and food production objectives. The main elements of this comprised quotas in the milk sector (1984) and set-aside in the arable sector (1988). Subsequently, more decisive but politically contentious action was taken to reduce high intervention prices and, following the MacSharry proposals of 1991, the EU farm ministers agreed a range of measures, including:

1) a 29% cut in cereal support over three years;
2) compulsory set-aside (15% of arable area initially) if wishing to qualify for income aid (see below);

3) income aid to all farmers setting aside arable land (regionally based and payable on each hectare of land left in arable production, i.e. on 85% of original arable area);
4) compensation for rotational set-aside, i.e. on the 15% of arable land withdrawn from production;
5) the introduction of new nonrotational set-aside; and
6) additional aid for environmental protection, forestry and early retirement (the 'accompanying measures').

All these changes are controlled (in the UK) through an Integrated Administration and Control System (IACS), where farmers have to record details of all crops, livestock and set-aside on an annual basis. The main objective of the reforms was to break the link between farm incomes and the volume of food produced, by moving away from price support and towards direct income aid to farmers. The process of breaking the link between farm support and food production (a link which has been a cornerstone of agricultural policy since the mid-twentieth century) is known as 'decoupling', and the need to achieve a decoupled agricultural policy has been prioritised by countryside reformers. In 1992, set-aside became compulsory for all commercial farmers, who became eligible to claim 'income aid' through an Arable Areas Payments Scheme (AAPS) in return for setting aside arable land. Two main set aside options became available to farmers: rotational and nonrotational, the former requiring farmers to set aside up to 15% of their arable land on a rotational basis for each of six years, and the latter involving the same amount of arable land for five years. These changes achieved reasonable success in reducing surpluses and stabilising farm incomes.

Hard on the heels of the MacSharry reforms came the GATT Uruguay Round Agricultural Agreement (URAA) of 1993 bringing, for the first time, agriculture fully within the General Agreement on Tariffs and Trade. The essence of this was that subsidies on agricultural production in Europe ran counter to the international principles of free and undisturbed trading, promoting further pressure for decoupling. Direct payments to farmers, for instance those associated with conservation management, were more acceptable to the GATT framework and thus became more viable politically. From the environmental point of view, Potter and Lobley (1998) note that GATT negotiations signalled that environmental goods should be recoupled to agriculture (in exchange for the decoupling of food 'goods'), and so awarded environmental schemes 'green box' status: i.e. they recognised these as nontrade-distorting forms of support which could remain in place indefinitely. In 1996, the 'Cork Declaration' further promoted the decoupling of agricultural support and food production, and proposed that payments should be more generally directed at rural development. Moreover, the EU's *Agenda 2000* document (Commission of the European Communities, 1997), in its agricultural chapter, acknowledges the problems of the present basis of agricultural support and points towards a more balanced rural policy. Its ultimate effect on farming and the countryside remains to be seen: whilst it certainly indicates possibilities for promoting landscape variety and distinctiveness in the countryside, it also contains many features which suggest continued agricultural fundamentalism.

Farm businesses have responded in various ways to the withdrawal of support for mainstream commodity production. Some have sought simply to become more economically efficient, whilst others have pursued complementary enterprises to supplement the main farming activity. This latter is popularly referred to as 'diversification' and, whilst this strategy has not proved to be the panacea predicted during the 1980s, the take-up of diversification opportunities provides interesting insights into the nature of farmers' responses more generally (Box 4.3). For example, Marsden *et al.* (1986) identified four major strategies which farmers adopted towards diversification in London's Metropolitan Green Belt, where there is potentially a wide variety of income sources. First were those farm families who resisted the possibility of gaining alternative incomes, even though farm income levels were very low, either because there were few opportunities to diversify or because of strong personal commitment to farming as a full-time activity. Secondly were farm operations too small to serve as a fully economic unit, and where additional off-farm sources of income or farm-based activities (usually very closely related to the farming enterprise) were adopted in order to survive ('survivalists'). Thirdly were those businesses which consciously expanded in order to accumulate a well capitalised and diversified farm base ('accumulators'). Here, processes of merging and redirecting capital between agricultural and nonagricultural activities were well established. Finally, there were hobby farmers where most of the income came from off-farm sources. Often, indebtedness was the spur to diversification although the reduction in scale (or elimination) of many dairy enterprises by the introduction of milk quotas in 1984 was a specific additional imperative. Survivalists adopt a number of farm adjustment strategies to stay in business (Marsden *et al.*, 1989). These particularly tend to relate to increasing levels of owner-occupation and take-up of credit, reorganising farm businesses especially by establishing partnerships and substituting family for hired labour. Munton (1990) has also identified seven elements of farm adjustment (Box 4.4), four of which relate to the deployment of onfarm resources, whilst three describe potential sources of additional farm family income.

It is well known that, despite the commendable achievements of modern agriculture, the industry has been associated with a range of undesirable environmental side-effects. The most noted adverse changes have been:

- the human health effects of pesticide and fertiliser residues, heavy metal accumulation and other contaminants in soil, water bodies, food products and the food chain;
- the diminution and fragmentation of biotopes valued for nature conservation; and

Box 4.3   Farm 'diversification'

One opportunity which has been promulgated for farmers affected by declining income from the principal enterprise is that of farm 'diversification' (although this is a difficult and risky option). Some types of farming system make a consistently high demand on employees' time all year round. Here, the finite amount of time and range of skills which it is possible to master mean that alternative enterprises must be substituted for existing ones. On certain livestock farms, labour may be much more peaked, leaving substantial slack periods in the farming calendar when additional activities may be taken on board. Critical factors to consider in either event will be the land available (its location and quality), spare labour within the farm family and surplus buildings in usable conditions. Broadly speaking, four main groups of alternative enterprises may be proposed, the first three of which represent 'structural diversification' and the last, 'agricultural diversification' (Haines and Davies, 1988; Ilbery, 1991). First, there are visitor-orientated facilities catering for tourism or recreation (Slee, 1989). The former include bed and breakfast, cottages/chalets, caravans and camping or activity holidays; the latter typically comprise visitor centres, stabling and riding schools, golf courses, farmhouse catering, and shooting and fishing. These will draw heavily on new or converted buildings, spare land and managed habitats (woodland or wetland). They may also be very sensitive to farm location, so that sites which are distant from urban areas or tourist routes may be disadvantaged. Although farm tourism has been enthusiastically espoused by some, it is likely that participation rates by farmers are low, and it has been argued that, given the relatively static levels of domestic tourism in Britain, an expansion of farm accommodation supply would merely compete with existing stock (Hughes, 1989). Secondly, value may be added to conventional products, especially in the processing on site of animal produce (direct sales of meat and home-processed dairy foods, animal skins, etc.), and crops (milled cereals, pick-your-own and direct sales). Processing of produce will clearly often require significant investment in buildings and machinery, and may entail additional traffic generation and roadside marketing. Thirdly, greater use may be made of ancillary buildings for industrial purposes, either for conversion to light industrial premises if town planning considerations allow, or for craft work (Slee, 1989). Finally, from time to time farmers have been encouraged to consider producing unconventional livestock and crops. These include under-developed products which are still in shortage in the EC, and those which are organically produced and may therefore hope to attract a premium. Unfortunately, EC guarantee and structural funds do not extend to unconventional products, and so farmers may be wary of embarking on them, though there is a minor tendency to use set-aside land for 'newer' crops not intended for human consumption and thus not breaching set-aside regulations. 'Alternative' animal produce includes sheep milk, rare breeds, fish, deer and goats, some of which may be especially in demand amongst particular ethnic groups. However, low nutritional levels and adverse climatic conditions on the hills may preclude the husbandry of certain animals. Crop products include linseed, evening primrose, teasels and timber. It is worth noting the opportunity for producing high-quality timber on farms for craft or construction work, which can command a high premium, although of course the labour commitments and harvesting cycle may act as deterrents. Quick rotation, coppiced timber for firewood or charcoal production can also be attractive, but may be difficult to produce and sell. The success of many of these opportunities depends on the presence of effective local marketing mechanisms, as most farms would be unable to distribute and sell their products independently.

Pressures for farm diversification have raised particular issues for environmental planners, who are being encouraged to take more flexible attitudes to development in the open countryside, which traditionally has been subject to very conservative attitudes. Certain enterprises will require the availability of buildings, some of which will have to be purpose built, some of which may be adapted from existing ones. In general, the farmer finds the latter an easier option, as the only planning permission required will be for change of use, and consents for new buildings are often difficult to obtain. Tourist enterprises may also include caravan and camping sites, and if these are in use for more than 28 days in the year planning permission will again be necessary: it is likely that high standards of landscaping and facility provision will be required.

Box 4.4 Elements of farm adjustment

Deployment of onfarm resources leads to changes in:

labour – e.g. by substituting family for hired labour;
business structure – usually by changing from sole operator to a partnership to reduce tax;
tenure – e.g. buying formerly rented land or selling owner-occupied land for lease-back;
size – buying or selling land either to expand the farm business or to finance restructuring

Changing sources of family income leads to changes in:

farm enterprise – changing emphasis, e.g. expanding sheep while contracting the dairy enterprise;
economic centrality – increasing (or occasionally decreasing) the income from off-farm sources
diversification – e.g. bed and breakfast, farm shop.

Source: Munton, 1990.

- contamination of ground and surface waters and the eutrophication of surface waters by nitrates and phosphates.

However, an unsubsidised farmscape would also have had its drawbacks: overgrown hedges, clogged drains, dilapidated farm buildings, prairies and, in the uplands, extensive ranching systems and abandoned villages.

A balanced review of agrienvironmental impacts has been provided by Skinner *et al.*, 1997), who note that pesticide impacts have tended to be associated with the contamination of waters, impacts on fauna and flora, and human impacts. Influential factors are the type and rates of applications of chemicals, the accuracy of applications and the weather conditions. One area of concern is the extent to which pesticides reach rivers and lakes by leaching and runoff. Thus, pesticide levels are generally found to be much higher in surface waters than groundwaters, probably because sources include surface and drainflow runoff from farmland, as well as spray drift contamination, where soil contact and therefore degradation is reduced. Some (but not all) pesticides are strongly absorbed or degraded in soils, and degradation to less harmful compounds increases with rising soil temperatures and moisture contents. Pesticides may also affect humans, though, in Britain at least, there have been relatively few instances of harm to farm workers or to the general public by spray drift. Pesticide spray drift has also affected flora and fauna, and some studies have shown that pesticide flushes can occur at the headwaters of streams; presently, though, there is little evidence of long-term damage.

Nitrate concentrations in UK waters have increased over the past generation, and agricultural nitrate applications have led to increased leaching from the soils into both surface and ground waters (especially in areas such as eastern England where rainfall, and therefore dilution, is low). Ploughing up long established grassland, as occurred after 1945 in eastern England, also increases nitrate losses. In aquifers, high nitrate levels predominantly occur where there is no layer of clay between the soil and the water-bearing rock, though relationships between leached nitrate and applied nitrogen, atmospheric inputs, and the production of nitrate from organic matter in the soil under different land management systems are complex. Also, it should be noted that, whilst agricultural nitrates are often blamed as the villain of cultural eutrophication, phosphate rather than nitrate is often the limiting nutrient in freshwater systems, so that runoff rich in phosphates is a major cause of algal blooms. It has been estimated that agriculture only contributes about 20% of P inputs to surface waters, with sewage being the main source; thus sewage disposal, rather than agrochemicals, is generally the main culprit. In terms of biodiversity, the application of nitrogenous fertilisers significantly reduces species diversity in grassland as it stimulates the growth of competitive grasses at the expense of broadleaved herbs, and when a combination of agricultural inputs (fertilisers, pesticides, lime and manure) is used, the impacts on species diversity are particularly strong. Livestock wastes are similarly significant, and livestock farming accounts for about 80% of the UK's ammonia emissions, whilst odour associated with slurry storage and spreading may also be an issue, especially in populous areas. Finally, soil erosion in the UK is caused by the action of water and wind on arable lowlands, and also of frost and animals in the uplands. Soil may be lost entirely from a field or redeposited within it depending on topographic features. The coarse fraction of the soil is transported short distances whereas the finer silt, clay and organic matter can be moved well away from the site, causing the remaning soils to become progressively coarser grained and less fertile. Much of the phosphorus and pesticide losses from farm land to surface water are bound to eroded soil particles.

An important aspect of CAP reform has its implications for the environment, both those consequential effects of wider reform, and the effect of moving some of the mainstream farm budget to environmental provisions. Thus, various 'accompanying measures' to the main reforms have supported specific environmental benefits, whilst other European pressures (e.g. on pollution control) have also led to the introduction of farm environmental schemes (Parker, 1996). This complex of environmental land management schemes (ELMS) will be discussed subsequently, but they have introduced an important new principle of their own, namely, that of targeting support (Potter *et al.*, 1991). This has been an important departure in farm policy, where it had always been assumed that support measures had been widely available (and several were universal); indeed, targeting state payments to farmers in certain areas runs contrary to the original 'equity' principles of farm policy. In brief, ELMS cover the continuation of traditional farming (Environmentally Sensitive Areas), creation of new farm habitats (Countryside Stewardship), enhancement of wildlife interest on former set-aside land and SSSIs, and limitations on nitrate applications in sensitive groundwater zones (Nitrate Vulnerable and Nitrate Sensitive Zones). There is evidence of reasonable uptake of these schemes, despite some concern that they achieve limited 'additionality' (i.e. induce few additional changes beyond what fitted in with the farmer's existing plans). In addition to voluntary entry into ELMS, however, farmers face increasing mandatory controls on environmental impacts, especially through government guidelines on *Good Agricultural Practice*. Farmers who can be proved not to have adhered to these guidelines may face prosecution. For example, rules introduced under the EC Nitrates Directive impose stringent conditions on nitrifying land uses and compliance with these standards is mandatory and uncompensated.

Land has always been perceived as a scarce resource in a crowded country such as Britain, especially where it contains some special quality such as agricultural productivity, high load-bearing capacity or scenic attractiveness. Given the privileged position of agriculture since 1947, it is hardly surprising that good farmland has been considered almost sacrosanct, and has been treated as a major constraint on development. More recently, however, the case for an exclusively protectionist attitude to farmland has been undermined – the more so, since technological and managerial advances are leading to sustained increases in yield even on a declining land area. Moreover, whilst pressure for food production has receded, land of scenic or nature conservation importance has been perceived to increase in relative value. The official view is nevertheless that top-quality agricultural land should continue to be highly regarded, since it is highly flexible and thus most capable of supporting new crops in response to changing market needs. Dwellings on farms (for farmers and farm workers) are typically only approved subject to a condition requiring *agricultural occupancy*, although this requirement may on occasion be lifted at a subsequent date, especially if there is no longer agricultural or forestry need in the locality. Conversion of buildings to uses other than for agriculture normally requires permission, even if no structural alterations are intended, the critical considerations being acceptability of use for the particular site and the condition and nature of the buildings. Stabling for horses again requires consent, unless they are working horses for agriculture. Farm shops do not normally need permission if they are genuinely ancillary to the farm (though facilities for food processing and pick-your-own schemes would depend on the scale of operation proposed). Traffic generation, especially the ease of access and exit from the farm and the type and design of parking facilities, is usually an important factor, whilst even farm-side advertisements may need permission in certain cases. Planning policy guidance affirms the desirability of planners adopting a more favourable attitude to rural enterprise diversification, but this is often a fraught task in residentially desirable parts of the countryside where pressures from developers to exploit planning 'loopholes' are continually intensifying.

Forestry

Forests are climatic-climax ecosystems for much of the world and thus regenerate naturally *in situ*. Ecologists tend to distinguish between *primary* (continuously covered) and *secondary* (recolonised) sites, and conservationists attach an especially high value to the former: the terms 'primeval woodland' or 'wildwood' are commonly used. Peterken (1981) notes that primary woodland is a useful hypothetical concept rather than an identifiable state, and prefers to draw a pragmatic distinction between ancient and recent woodland, with a threshold date around AD 1600. This is typically the benchmark used to define *Ancient Semi-Natural Woodland* (ASNW).

Ecologically, there are major differences between self-sown and plantation woodland. In a woodland ecosystem, gaps left by the death of mature trees are colonised by replacement trees, often by a fairly complex sequence depending on the diversity of species available so that regeneration produces a mixture of tree species and age classes. Higher light penetration in self-sown stands also encourages an abundant ground flora, which is often characterised by slow methods of dispersal and may therefore take centuries to become established. By contrast, most plantations have a restricted range of highly productive but often non-native species and are typically even-aged; they also display unnaturally high densities, to encourage upright growth (as trees compete for sunlight) and produce suitably shaped specimens for machine processing. In countries which have an established woodland resource and a greater variety of native species which will thrive in local conditions, and even in some mixed woodland estates in Britain, timber is often extracted by methods other than clearfelling and replanting. Various systems of partial/selective clearance and self-seeding are often feasible, and this more closely mimics nature both in appearance and ecology. Contrasts between plantation and self-sown woodlands are summarised in Table 4.2.

Forestry, like agriculture, experienced an 'industrial' phase of growth in response to historical shortfalls, but subsequently entered a more mature 'multiuse' phase. In Britain, there are various characteristic types of

Table 4.2 Differences between plantation and semi-natural woodland

Plantation	Semi-natural
Trees have even ages and heights	Trees are uneven-aged and have irregular canopies
Single or few species	Mixed species
Planted, on open ground	Self-seeded, under the canopy
Harvested on large scale	Harvested selectively

woodland, which are associated with distinctive ecology and scenery. Original woodland cover was either mixed deciduous or conifer (predominantly Scots pine) forest, but very few fragments of this remain. However, quite considerable, though still patchy and vulnerable, expanses remain of ASNW. There was a modest renaissance of woodland planting during the eighteenth and nineteenth centuries as country estates invested in plantations for production and/or amenity. Whilst this did not contribute huge quantities of timber, it did introduce considerable knowledge about the performance of 'exotic' species (which had been introduced by scientist-explorers) in Britain. However, the general national trend was one of progressive deforestation since the Iron Age and this had, by the end of the nineteenth century, left the country largely denuded of woodland cover. The demise of traditional rural economies, in which woodlands provided important fuel and building materials, also accelerated the loss of management skills, and those woods which remained often became degenerate. Additionally, following the introduction of capital taxation measures in the early part of the twentieth century, the breakup of the great estates accelerated as freeholds were sold to tenants, and woodlands and species-rich turves were stripped.

It was wartime exigency, once more, which promoted the need for a policy response. During the 1914–19 war, Britain grew perilously short of timber, one particular problem being the scarcity of pit props to enable coal production. As we note below, forestry is scarcely economic in Britain and so, unlike agriculture, a key element of policy was to promote state entrepreneurship. Thus, the Forestry Act 1919 set up the Forestry Commission, comprising a Forest Enterprise (which was to acquire and plant land on a commercial basis) and a Forestry Authority, with the purpose of regulating the forestry industry, generating knowledge and advice regarding forestry and assisting the private sector. A priority task for the Commission (which is, in all but name, the government Department of Forestry) was to effect the creation of a new national forest on bare ground – a *strategic reserve* of 2 million ha – as quickly as possible. This role of a timber-producing agency is in distinct contrast to countries such as Australia and Canada, where substantial tracts of natural or semi-natural forest remain. In these cases, the role of the state agency is normally that of granting concessions to companies to extract timber, on payment of a royalty; felling activities are usually subject to re-planting conditions or else a requirement to extract only about a quarter of the standing timber so that natural regeneration will take place.

These priorities produced a familiar pattern of twentieth-century forestry in Britain. Given the tradition of safeguarding good agricultural land (the sentiments for which long pre-dated the planning legislation of the 1940s), planting by the state sector was until recently restricted to rough grazing. This land often produces good yields of timber, but only within a very restricted range of species able to withstand the poor soils, damp climate and high winds of upland Britain. Thus, plantations have tended to comprise just one or two species of exotic (non-native) conifers which display particularly good growth increments. The shape of plantations was typically determined by the legal boundary of the land which had been acquired by the Forest Enterprise or private owner, and the consequent straight lines often fitted uncomfortably into the natural sinuosity of the upland landscape as well as restricting recreational access on to the higher ground. The appearance and ecology of the early 'first rotation' forests were thus widely considered to be 'alien' to the landscape, but as early as the 1960s the Forestry Commission's landscape architects had developed ways of softening appearances, so that many of the newer and 'second rotation' forests were more sensitively designed. On the most marginal land, however, there is still limited scope for diversification.

Since the creation of the Forestry Commission there have been flourishing forest industries in both the state and private sectors. The balance between the two has varied, although during the 1980s the private interests, represented mainly by traditional landed estates and forestry companies, became dominant, especially in terms of planting bare ground as opposed to 'second rotation' forests. The Forest Enterprise itself has been subject to commercialisation and still remains a potential target for privatisation. Two main strands have existed towards the promulgation of a private sector: grant aid and tax relief. Financial grants are often heralded as the main method of influencing investment, but it was the hidden value of tax relief which did most to propel the ascendancy of the private industry. Prior to the Finance Act 1988, individuals could claim income tax exemptions on the sale of forestry land (the tax only being levied on increases in value of the underlying land), as well as decreased tax liabilities on final timber sales. These benefits led to widespread 'investment forestry', in which companies sought to attract clients from top income tax brackets, obtained land which they could plant on their clients' behalf and then contracted to undertake the planting work themselves. Once the initial, relatively high-risk, phase was over and the young plantation well established, these forests were often sold to quasi-public institutions (such as pension funds), who saw them as attractive investments in a situation where world timber prices were likely to continue to rise in real terms.

Although investment forestry, because of its sheer scale, made a major contribution to the attainment of annual planting targets and to supplying timber to forestry industries, it was a controversial activity. In particular, the massive plantations were often criticised for their alleged environmental damage and insensitivity, whilst centralised and purely commercial land management decisions were being imposed on remote areas in the interests of distant tax avoiders. Moreover, during the 1980s there was growing interest in farm forestry, as a strand of farm enterprise diversification, yet most farmers did not earn enough to take advantage of high-level income tax relief. Coupled with criticisms of the covert sanctioning of tax avoidance, and the uncapped cost to the Exchequer of forgoing taxes, it is not difficult to see why all but a few relatively minor benefits and transitional arrangements were removed by the Finance Act 1988. Estate planting has generally been more readily welcomed, and its wider acceptability hinges on a number of features which tend to distinguish 'traditional' schemes from those implemented by the new 'institutional' landowners on recently purchased sites. For instance, Mather (1989) noted that estate and farm planting tends to cause less displacement of livestock, and is more sensitive in scale, contains significantly larger proportions of broadleaves, and makes more use of permanent local labour than the mobile gangs favoured by the forestry companies. Presently around 15,000 ha of new planting is carried out annually by private landowners in the UK, more than half of it broadleaved. Second rotation planting accounts for about another 14,000 ha, though the Forest Enterprise contribution to this has fallen steadily since the 1980s. Statistics reveal a considerable interest in the planting of new pinewoods comprising native species.

In order to understand the reasons for 'industrial' style forestry planting, it is necessary to appreciate the economics of establishment and management. These make it clear that a rapid payback is necessary, which greatly restricts the manager's options. Given a situation (as has historically tended to occur in Britain) in which a forest has to be created afresh, so that there is no income accruing from previous felling to finance new planting, the investor will have to wait some 50 years before being able to enjoy the revenue associated with the final timber crop. It is thus necessary to estimate the terminal revenues from a forest stand and relate these to present-day values, so that its cost-benefit ratio can be compared more readily with land uses which yield an annual return on capital. Forests may yield some income before final felling – for instance immature timber taken out at the thinning stage, from recreation or even from Christmas trees – but this is a comparatively minor amount. Central to these calculations is the concept of discounted cash flows, and a brief explanation of this is given in Box 4.5. This account, though, reflects 'official wisdom' on forestry economics, and some advocates, especially of farm and community forestry, would argue that more favourable bases for evaluating economic returns should be applied.

Forestry policy agendas have thus changed considerably from the days of creating a strategic reserve, and now entail diversifying a still predominantly upland estate, creating more naturalistic forests in the second

Box 4.5 The effect of discounting cash flows

In general terms, if a company wishes to invest in a project whose full benefit may not be realised for many years, then this cost of waiting will entail risk, loss by inflation and opportunity costs of not being able to invest in other projects. Thus, anticipated returns must be deflated by an appropriate amount to reflect those factors. To assess the 'present value' of money in a year's time, we must find the amount that needs to be invested today to arrive at a given sum one year hence. Thus, if the project yields a cash flow of £110 in year one, then to arrive at this sum the company would alternatively have had to invest £100 today at 10%; the discount rate would express the reverse of this and deflate a return of £110 in a year's time to a present value of £100. In this way, the value of money is modified by time. Where r is a rate expressed as a fraction (e.g. 0.1 for 10%), then the gross present value (GPV) of a project yielding an annual return A over n years is:

$$ \text{GPV} = \frac{A_1}{1+r} + \frac{A_2}{(1+r)^2} + \ldots + \frac{A_n}{(1+r)^n} $$

Normally, investors are concerned with the net receipts, and thus subtract the cost of the initial investment from the gross present value to obtain the net present value (NPV). If the resultant value is positive, then a favourable cost:benefit ratio has been obtained and investment can be justified (the higher the ratio of benefits to costs, obviously the more favourable the investment becomes). Discounting is a technique which enables profits in n years time to be compared directly with profits today, and the rate r applied in the above example is termed the *discount rate*. It will be noted that the higher the discount rate, the lower will be the NPV, especially the proportion contributed by distant years. Thus, discounting imposes pretty brutal results on timber generally, and especially on hardwood species, where the main revenue will accrue in the distant future.

It is also important to recognise that in industrial or commercial situations, investors will be looking to achieve discount rates of around 10%, and even for socially desirable public investment an economic return of 7% is normally sought. By contrast, unsubsidised 'first rotation' forestry (i.e. where no previous forest crop exists to provide investment revenue) in Britain rarely achieves a rate of return of more than 1–3% and so cannot be considered economic by normal criteria. Originally, there was a need to create a strategic (i.e. potential wartime) reserve of 2 million ha, and this to some extent overrode economic considerations. However, once this figure was surpassed in the early 1980s, new planting had to be justified on other grounds. Thus, the extension of the national forest estate is now defended at least partly in terms of rural employment, landscape and wildlife benefits, recreational provision, alternative land use and contribution to the balance of payments.

rotation and encouraging multipurpose lowland woodlands (Box 4.6). Specific policy agendas have been introduced to create new community forests and a national forest, supplemented by urban and community woodlands. Hence, major principles of forestry policy are now those of 'sustainable' and 'multiple' use, and timber production is far from the sole consideration. In 1990, a House of Lords Select Committee called for a much more comprehensive forestry policy with clear and realistic planting targets, and a positive approach to the role of forestry on surplus agricultural land.

Nontimber benefits include pre-eminently recreation, for which an annual grant in aid is paid, as well as wildlife, landscape, employment, social cohesion, integrated land use, certain hydrological benefits and community involvement. Additional 'environmental services' appear to be CO_2 absorption, air filtration, control of microclimates, erosion control and absorption of agricultural nitrates (see Box 4.7). Environmental disbenefits of forestry, though, have included the loss of wildlife habitats, the scenic imposition of irregular plantation blocks or clearfells on the landscape, loss of archaeological features and settings, and effects on hydrology. It is also reasonable to observe that many critics (especially economists) argue that forest expansion is a waste of resources, and that state assistance could more productively be directed to other areas of investment.

Cultivation of a tall tree canopy is not always the most appropriate forestry strategy, and the main alternatives are coppice and woodpasture (Table 4.3). These entail taking advantage of the ability of many deciduous trees either to sucker or to regenerate multiple shoots from a lopped trunk. (This may have frustrating consequences for eradicating invasive species, such as sycamore, from conservation sites, where precautions have to be

Box 4.6   Key themes of Forestry Commission policy

Promoting multipurpose forestry
- planting of woodlands on set-aside
- Forest Enterprise Design Plans
- National Forest Tender Scheme

Protecting trees, woods and forests
- biodiversity
- deer initiative and squirrel forum
- felling permission
- plant health

Promoting woodland recreation and access
- Woodland Improvement Grant (WIG 1995–98 provided assistance to help encourage informal public recreation in existing woodlands)
- Walkers Welcome (encourages private owners to welcome people walking in their woods)
- Community Woodlands Scheme
- Community Forests (locational supplement)

Supporting the creation and management of native woodlands
Native pinewoods and birchwoods (e.g. Millennium Forest for Scotland Trust).

Enhancing the economic value of woodlands

Promoting health, safety and training

Promoting public understanding of, and participation in, the management of woods and forests
Forest Education Initiative.

Source: Forestry Commission annual reports and Forestry Commission, 1998a.

Box 4.7   Potential environmental services from large forest stands

Safeguard of water catchment – the cover of forest vegetation protects soils (often on steep, potentially unstable slopes) from erosion, reducing flash-floods and siltation of reservoirs, and maintaining water quality. (However, there may be some conflict with water supply objectives, due to the massive evapotranspiration losses from canopies, as well as exacerbation of soil and water acidification. Catchment studies are presently exploring the effects of various land-use mixes on water quantity and quality.)

In arid climates, forests may also reduce dry-land salting in areas of high salt accumulation.

Provision of plant and animal habitats, many of which depend upon specific forest environments for survival: the major interest here attaches to un-disturbed rainforests, which support an exceptional abundance of species and a great economic and scientific potential, although primary woodlands in temperate areas still contain rare genetic variants.

Possible benefits from integrated farm forestry schemes: shelter of fields and creation of shared tracks (albeit there is often antagonism between farming and forestry interests, especially when poorly planned, at the detailed level).

Leisure pursuits: tourism, recreation and environmental education.

Sequestration of CO_2 (though the role of this as a policy influence on new plantations is slight: Clayton (1994) notes that if the UK were to plant 1 million ha of three-year rotation poplars this would only account for 3% of the UK total CO_2 emissions and a mere 0.01% of global emissions).

Table 4.3   Coppice types and terminology

Type	Description
Simple coppice	Crop consists entirely of coppice, all of which is worked on the same cycle (even-aged)
Coppice with standards	Coppice (underwood) with scattering of trees (standards) being grown to timber size. Standards may be of seedling origin (maidens) or promoted from a stump shoot (stored coppice)
Short-rotation coppice	Coppice worked on a rotation of less than 10 years to produce stick-size material; also relevant to biomass production
Pollards	Trees cut off at 2–3 m above ground, to avoid grazing pressure. Formerly a component of woodpasture
Underwood	Coppice or scrub occurring under another tree crop

Source: Adapted from Evans, 1984.

taken to kill the bases of the tree.) Coppicing is a particularly efficient way of producing small roundwood for minor constructional purposes, or for firewood and charcoal. Suitable temperate zone trees include oak, hornbeam, alder and hazel. If the coppice is freshly planted then the first cut can be taken after 5–8 years. In former times, it was common to manage this as a productive understorey in a forest amidst high 'standard' trees, used for boat and house-building (hence the term 'coppice with stand-ards'). This system is also widely used in developing countries where wood-lands may be managed for fuelwood on a community basis to counteract the disastrous consequences of native woodland clearance.

In developed countries, outlets for coppice wood have now substantially disappeared, despite determined efforts by some to promote its use in energy cropping (especially willow) and charcoal production. Thus, much of the ancient coppice has been cleared, abandoned or become overgrown, and requires lengthy and skilled management if it is to be regenerated. Apart from the historical interest of coppice, it may have exceptional wild-life value (if not too clinically managed) associated with the relatively wide spacing of the trees (normally 3–5 m apart) and their low height, allowing high levels of light to penetrate to the shrub and herb layer. A management plan for a coppice should aim to produce a series of stands at different stages in the coppice cycle, enabling plants, birds and animals to migrate from one to another. Equally important for conservation purposes is to prevent invasion by non-native species and to exclude potentially damag-ing forms of recreational use (Rackham, 1976; Peterken, 1981). Coppicing is a labour-intensive activity and many attempts have failed through inade-quately sustained management programmes. Volunteer conservation la-bour is not always suitable because of the high skill levels required. It may sometimes be preferable therefore to reconvert to high forest by 'singling' out the strongest and straightest coppice stool, which then grows rapidly (having all the tree's resources at its disposal); in a commercial system, this may be an effective way of promoting hardwood production.

Similar to coppicing is the practice of pollarding, used in areas grazed by livestock, in which trunks are lopped above the level at which browsing cattle and deer would strip the fresh growing shoots. The regenerative properties of the pollard also provided an alternative to natural forest regeneration, as saplings would themselves quickly be grazed by livestock. This was typical of riverside willows and trees on commons, and is the central feature of 'woodpasture'; it thus forms a feature of exceptional historic landscape interest. Woodpasture has been defined as woodland permanently available as pasture (Peterken, 1981) and it developed out of the prehistoric practice of grazing cattle within natural woodlands, surviv-ing subsequently on commons and in forests, parks and chases.

Perhaps the closest present-day parallel to woodpasture is agroforestry, in which single trees are spaced at considerable intervals, although there the similarity ends, as they are fenced and allowed to grow to maturity. In tropical countries – with high solar radiation – it is common to grow crops amongst comparatively closely spaced woody perennials (trees, shrubs and bamboos). There may be beneficial economic and ecological interactions between the tree and nontree components of the system, as nutrient inputs are increased and soil erosion is reduced by the protective cover of foliage. In temperate latitudes, comparable practices are often based on conifer planting, with trees often widely spaced (perhaps no more than 100 stems/ hectare) in grazing pastures; they may be individually fenced and pruned until greater than *c.* 1.2 m high, but the good-quality timber and consequent premium may justify the effort. Some of the most promising temperate latitude experience has been gained in New Zealand, reflecting high timber yields and low ('world market') agricultural prices; consequently these conditions of relative profitability are not readily transferable to Britain, where there has been little interest. Nevertheless, Bullock *et al.* (1994) have suggested that more favourable grant aid arrangements could result in much wider uptake in Britain (woodland grant aid presently is linked to density of stems per hectare, for which agroforestry is inevitably very low). They argue that silvoarable and silvopastoral systems are potentially attractive, and could offer a wider diversity of landscapes and wildlife habitats. Silvopastoral systems based on sheep appear to offer the greatest potential.

Since the early days of policy support for private producers, various grant schemes have been available. These are now consolidated into the Woodland Grant Scheme (WGS), which offers a sliding scale of payments depending on the species and type of ground planted. A number of supplements can be added to the basic grant, for example, for lowland forestry, farm woodlands or native species. The Forestry Commission scrutinises schemes to ensure they are satisfactory in their design and implementation, and will normally pay grant on a phased basis as the plantation becomes established. Long-term woodland benefits are now aided through 'extended payments', whereby foresters receive continued subsidy for sound management during the lifetime of the trees. When tax relief was removed in 1988, grant aid was considerably increased by way of compensation and this, of course, benefited all potential investors rather than only those with high tax liabilities. Nevertheless, the scale of forestry expansion fell considerably, and it is widely felt that grant levels have been insufficient to maintain adequate planting rates.

Grant aid is only awarded after essential consultations, notably with the agriculture ministry, the nature conservation agencies (in SSSIs), the landscape conservation agencies (in designated areas) and local planning (including National Park) authorities (over sensitive locations or large schemes). Occasionally, other more specific interests may be involved. Prior to the abolition of tax relief (which was not subject to consultation or clearance), there was a 'gentleman's agreement' not to proceed where grant aid had been refused following objections by consultees. On a small number of occasions this was flouted, and was a contributory factor to the demise of tax relief. Clearance of agricultural land for planting has traditionally only been for rough grazing, but this was significantly relaxed in the 1980s to include a variety of medium-to-poor quality agricultural land classes, including some where pasture improvement schemes had previously been carried out. If objections from consultees cannot be resolved, the application is referred to a Regional Advisory Committee (RAC) and in the very rare instances where this is unable to decide, the relevant minister reaches a final adjudication.

Although forestry lies outside planning control, it is such a major user of the countryside, especially in the uplands, that it has been the subject of longstanding interest by planners. The most systematic means whereby planners seek to become involved is through the means of a 'woodland strategy'. Such strategies have been produced by many local authorities, often as a purely informal basis for conducting consultations, and providing guidance to investors on preferred locations and designs for future planting. The most official basis for this is the 'indicative forestry strategy' (IFS) which has been commended by government, and which identifies constraints on and opportunities for forestry activity within the local authority area. The principal intention of IFSs is broadly to distinguish between *preferred areas* for new planting (with land suitable for growing trees and where there are few competing interests), *potential areas* (containing perhaps one serious constraint which may be overcome by careful design) and *sensitive areas* (where the number, intensity or complexity of issues render large-scale afforestation generally undesirable). This analysis is based on a sieve mapping exercise of constraints imposed by suitability of land for tree growth, agricultural land quality, wildlife and landscape protected areas, mineral deposits, public recreation, archaeology and waste. The output is a *schematic strategy map* and a *supporting statement*. A number of IFSs have been produced, especially in Scotland, but their use has been relatively limited as they came onstream largely after the major phase of industrial planting had ended, following the Finance Act 1988 (SDD, 1990; DoE/WO, 1992b; Selman, 1997).

A major strategic dimension has been given to lowland planting by the former Countryside Commission's proposals for community and lowland forests. In mid-1989, the Countryside Commission and Forestry Commission announced a programme to create *community forests* on the outskirts of major towns and cities in England and Wales, with a comparable forest also being established in central Scotland. During the 1990s these have started to take shape, and are emerging as areas with high percentages of tree cover rather than continuous forest. In addition to conventional commercial operations, a substantial part of the planting is undertaken by farmers, and local authorities and other landowners. Community groups and amenity societies are also expected to play an active part in design, establishment and management, ideally using local people as wardens and rangers when the forests mature. Each forest has its own officer with support staff and budgets, facilitating the preparation of forest plans and business plans in consultation with local authorities, landowners, community groups, voluntary bodies and government agencies (Countryside Commission (with the Forestry Commission), 1989). Complementary to this is the creation of a major lowland forest in the Midlands (the 'National Forest'), again based on a mix of conventional plantations, farm woodlands and community-orientated public planting. Predictably, these schemes have multiple objectives, including those of alternative land use, recreation and tourism potential, landscape enhancement, employment provision and wildlife habitat creation. Although there is a strong reliance on voluntary participation by farmers in these projects, and compulsory purchase of land is not anticipated, the constraints imposed by complex urban fringe land-ownership, inflated land prices and current lack of interest in small-scale forestry are formidable. One important way forward will be to make positive use of the planning system, through development plan policies, conditional planning consents and planning 'gain' or 'obligation' (Bishop, 1991). Whilst most forests are broadly on target in terms of quantity of planting, this is not always of the type and in the locations most appropriate to the forests' multiple objectives.

In terms of the protection of existing tree cover, an important instrument of planning law has been the Tree Preservation Order (TPO). This may be applied to individual trees, groups of trees and, occasionally, woodlands, although not those under an approved Forestry Commission grant scheme. The sole principle involved in designating TPOs is that of *amenity,* and it is a moot point whether this can be interpreted to embrace ecological factors. Although TPOs have led to the retention of a very important amount of amenity trees, they have been criticised as being too negative (not promoting good management), too easily over-ridden by public bodies (such as highway authorities) and too laborious in their preparation. Revisions to planning legislation during the 1990s improved some of these aspects. In addition, controls are also exercised through the operation of felling licences under the Forestry Acts; whilst the permissible levels of felling are rather complicated, it is roughly the case that about one mature hardwood tree can be felled from land in each calendar quarter without the need for a licence. Licences, when granted, may have replanting conditions attached to them. An important supplement to legislation on woodland protection is the Hedgerow Regulations (1997), and it is now often necessary to obtain permission from a local authority to grub up (or otherwise damage) a hedgerow. Although the regulations are quite complex, the essential criteria for a protected hedge are summarised in Box 4.8. Whilst these provisions seem comprehensive, they are not exhaustive, and highly important hedges have been grubbed up with impunity because, for example, they cannot be demonstrated to form an integral part of a field system.

Conservation Resources

The need to protect key features of the countryside has been recognised for over a century. Some of this related to romantic complaints against urbanisation, but some was more specifically focused into purposive action to conserve places and species. The movement for national parks started in the USA in the mid-1800s, but in Britain this was restricted to voluntary

Box 4.8   Criteria for protecting a hedge

1. Marks a pre-1850 parish or township boundary;
2. incorporates an archaeological feature;
3. is part of, or associated with, an archaeological site;
4. marks the boundary of, or is associated with, a pre-1600 estate or manor;
5. forms an integral part of a pre-Parliamentary enclosure field system;
6. contains certain categories of birds, animals or plants listed by official conservation agencies;
7. includes:

   a. at least 7 woody species, on average, in a 30 metre length;
   b. at least 6 woody species, on average, in a 30 metre length and has at least 3 associated features;
   c. at least 6 woody species, on average, in a 30 metre length, including a black-poplar tree, or large-leaved lime, or small leaved lime, or wild service-tree; or
   d. at least 5 woody species, on average, in a 30 metre length and has at least 4 associated features.

   (There are minor variations in this according to part of the country, and lists exist of acceptable woody species).

8. runs alongside a bridleway, footpath, road used as a public path, or a byway open to all traffic and includes at least 4 woody species, on average, in a 30 metre length and has at least 2 of the following associated features:

   i. a bank or wall supporting the hedgerow;
   ii. less than 10% gaps;
   iii. on average, at least one tree per 50 metres;
   iv. at least 3 species from a list of 57 woodland plants;
   v. a ditch;
   vi. a number of connections with other hedgerows, ponds or woodland; and
   vii. a parallel hedge within 15 metres.

Criteria for selecting hedgerows for protection under the Hedgerow Regulations

Source: Wallace, P., 1998, Oxfordshire Branch of the Council for the Protection of Rural England (pers. comm.).

action by organisations such as the National Trust, Royal Society for the Protection of Birds, Council for the Protection of Rural England and (former) Society for the Promotion of Nature Reserves. The role of the voluntary sector (environmental nongovernmental organisations – ENGOs) has remained especially strong in British conservation and, partly through its wide membership base, continues to exert considerable practical and political influence.

As pressure grew to safeguard special areas and to provide greater access to the uplands, the need to create something akin to overseas national parks became irresistible. Nevertheless, it was unclear what form 'national' parks should take in a much-altered and generally privately owned countryside like

Britain, where wholesale state acquisition of 'primitive' areas was impractical. Official responses to conservation commenced with the Addison Committee, which argued the case for regional reserves serving general
countryside conservation purposes. Subsequent to this, the interests of landscape protection and scientific nature conservation began to diverge,
however, and this created a continuing tension in the purposes of countryside planning. Thus, the Dower and Hobhouse reports (Dower, 1945;
Hobhouse, 1947), which considered the purpose and administration of
national parks in England and Wales, were paralleled by the Huxley Committee on wildlife conservation which produced quite separate recommendations (Huxley, 1947). In the ensuing legislation – the National Parks and
Access to the Countryside Act 1949 – the scientists' aspirations for a specialist Nature Conservancy, acquiring and managing a national system of reserves, were largely fulfilled. However, the provisions for national parks and
broader recreational access resembled more of an administrative compromise. Whilst the Act included measures for the creation of a National Parks
Commission which, during the 1950s, oversaw the designation of ten national
parks in England and Wales, the diluted nature of the designation disap-
pointed supporters of the international concept. In Scotland, the Ramsay
Committee (1945, 1947) recommended the creation of parks much closer to
the internationally recognised definition, but this advice was ignored for a
variety of reasons. Direction Areas were defined in which development
applications would be notified to the Secretary of State for Scotland, but this
was a minor concession, and the arrangements were themselves terminated
in 1980 when a more general system of National Scenic Areas (NSAs) was
introduced (Selman, 1988). The NSA system has never worked terribly con-
vincingly, however, and national parks are likely to be an early outcome
from the Scottish Parliament (Box 4.9).

During the 1960s, nature conservation was increasingly seen to be
dependent on wider developmental and political forces, so that the Nature
Conservancy was increasingly required to provide policy and planning advice. This was seen to create tensions with its purely scientific role and so a
Nature Conservancy Council was formed in 1973, retaining estate management, advisory and policy functions, but with much of the fundamental
ecological research being contracted to the Institute of Terrestrial Ecology
(ITE). Similarly, following growing concern about the scale of countryside
recreation and tourism, associated with increased affluence and car ownership, the National Parks Commission was replaced by a Countryside Commission, with a wider remit, under the Countryside Act 1968, whilst a
Countryside Commission for Scotland was created under separate legislation the previous year. A complementary trend to this administrative diversification has been the development of ENGOs who between them play a
major role in site acquisition and management, training, research and environmental education (Box 4.10).

Thus, there has been a dual divide in the conservation organisations:
between the public and voluntary sectors and, in the public domain,

Box 4.9   National parks for Scotland

Consultation documents on the probable national park system for Scotland suggest that the country will not be looking to import a standard system from elsewhere but will develop one which is 'tailored to the Scottish situation and the special circumstances in each area'. The SNH has suggested that national parks in Scotland should:

- meet the highest international standards for the protection of outstanding areas of natural heritage;
- be places where people will have the opportunity to enjoy what the nation has pledged to protect on their behalf, while ensuring that the qualities which people come to enjoy are maintained;
- promote sustainable rural development, setting an example of how to integrate the rural economy with the proper protection of the natural heritage;
- address the different needs of different places around Scotland;
- address needs not met by existing policy approaches, especially where a more integrated approach to land management is required; and
- aim to increase local accountability while also meeting national expectations for the care of the best of Scotland's natural heritage.

Box 4.10   Some environmental nongovernmental organisations in Britain

- British Association of Nature Conservationists
- British Association for Shooting and Conservation
- British Trust for Ornithology
- Council for the Protection of Rural England/Council for the Protection of Rural Wales/ Association for the Protection of Rural Scotland
- County Naturalist/Wildlife Trusts
- Friends of the Earth
- Friends of the National Parks
- Learned societies (e.g. British Ecological Society, Royal Entomological Society)
- National Trust (and National Trust for Scotland)
- Ramblers' Association
- Royal Society for Nature Conservation (umbrella organisation for county naturalist trusts)
- Royal Society for the Protection of Birds
- Scottish Wildlife Trust
- Wildfowl and Wetlands Trust
- Wildlife Link/Scottish Wildlife and Countryside Link (co-ordinating groups for several conservation NGOs)
- Woodland Trust
- WorldWide Fund for Nature UK

between the nature and landscape conservation agencies. This latter distinction was abolished in Scotland and Wales in the early 1990s, when the NCC and Countryside Commissions there merged into Scottish Natural Heritage and the Countryside Council for Wales. In part, this reflected the growing acknowledgement by nature conservationists of the amenity and community significance of wildlife, and by landscape/recreation interests of the ecological processes underlying scenic quality and tolerance of visitor pressure. An early response of SNH was the designation of a small number of Natural Heritage Areas to reflect the combined concerns of the new agency. A possible merger in England was resisted, mainly on the grounds of the resulting scale of bureaucracy, but English Nature and the Countryside Commission were instructed to develop closer forms of collaboration. An indication of the priority tasks addressed by the Countryside Commission and English Nature is given in Box 4.11.

Box 4.11 Key programme areas identified in annual reports by English Nature and the (former) Countryside Commission

---

Nature conservation priorities:

- Implementing EN's contribution to the UK Biodiversity Action Plan;
- providing high-quality advice to government on statutory consultations and on the reform of the Common Agricultural and Fisheries Policies;
- securing the safeguard of wildlife sites important at the European level, especially the completion of SSSI notification and possible SAC rivers, and planning for the delivery of favourable conservation status on Sites of EC Importance;
- extending the positive management of SSSIs through the WES, etc.;
- continuing to review the SSSI series, seeking equilibrium by 2000;
- managing the NNR series and continuing to increase the numbers of NNRs managed by approved bodies;
- increasing further the Species Recovery Programme;
- continuing projects on habitat restoration, lowland heathland and veteran trees; and
- improving, integrating and disseminating data.

Countryside priorities:

- planning for sustainable development (e.g. transport, environmental capital, development);
- long-term benefits from farms and woodlands (e.g. influencing European agricultural policy, land management initiatives, analysing landscape change, promoting countryside products);
- promoting sustainable leisure activities (e.g. Millennium Greens, Parish Paths Partnership, national trails);
- encouraging local pride (e.g. village design, Rural Action);
- improving the countryside around towns (e.g. community forests, countryside management projects); and
- protecting and promoting areas of finest landscapes (e.g. review of AsONB, national parks, countryside management services, heritage landscapes).

The original proposals for national parks envisaged that each would have considerable autonomy within the local government system, and would be administered by an independent board with responsibility for planning functions. It was proposed that all the members of the boards would be appointed: two thirds by the constituent county councils and the remainder by the central government minister in charge, to ensure that the 'national' interest was properly represented. However, opposition from the county councils in whose area the new parks lay largely pre-empted this degree of freedom. Only the Peak District (the first to be designated) possessed a fully independent board, whilst the Lake District (the second) had a board with reasonable independence but reliant on county council administrative services. Elsewhere, boards were replaced totally by county council committees – the National Park Committees (NPCs). In advance of local government reorganisation in 1974, the Sandford Committee (1974) had investigated the case for reform of this inadequate system, and it brought about three important innovations: every park was required to appoint a National Park Officer (accompanied by an expert staff); each park had to prepare a National Park Plan setting out its management proposals and expenditure programme; and, to pay for these responsibilities and in recognition of their national importance, 75% of their budget was to be found from the national exchequer. Subsequent review by the Edwards Comitee (Countryside Commission, 1991) sought to affirm the over-riding importance of environmental conservation in national parks, with a subordinate objective to promote the quiet enjoyment and understanding of the areas. Their recommendation for a greater degree of autonomy of National Park Authorities from the local authorities, however, had to await the creation (in the mid-1990s) of 'Peak District'-style independent boards for all national parks.

The other key landscape designation in England and Wales is the Area of Outstanding Natural Beauty (Northern Ireland also has this designation, but its nature is somewhat different). In parallel with national parks, a large number of Areas of Outstanding Natural Beauty (AsONB) were established, with little original role beyond that of recognition of the value of scenery. However, latterly, greater emphasis has been placed on the management of AsONB, with an expectation that they should be more positively managed through the medium of a Joint Advisory Committee, AONB Officer and Management Plan (Box 4.12). Similar treatment has been given to the Heritage Coasts. A plethora of additional non-statutory designations (e.g. Areas of Great Landscape Value) is identified in structure and local plans and these are used to influence development control decisions and landscape management policies. Since the 1950s, the only additions to the national park family (basically similar in purpose but having rather different powers and functions) have been the Broads Authority and the New Forest. In Scotland, the Natural Heritage Areas have proved to be an interesting experiment in partnership approaches to the reconciliation of land-use conflicts and the realisation of sustainable development opportunities, though their heavy reliance on negotiation has probably been inadequate to safeguard the truly exceptional conservation resources within their boundaries.

In the national parks, National Park Plans (NPPs) have been produced, and these set out the park authority's framework for its expenditure

Box 4.12   Proposals for strengthening the management of national parks and AsONB

The Countryside Commission's 1998b publication, *Protecting our Finest Countryside*, focused on the management of National Parks and AsONB, and advanced the following proposals:

- Appropriate and effective management and funding arrangements are essential if protected countryside is to be secured for future generations;
- each area of protected countryside should have the management and funding arrangements appropriate to the varied needs of its locality;
- it should be recognised that the carrying out of statutory functions requires secure long-term support from public funds;
- the authorities responsible for protected countryside, in collaboration with each other, should continue to give high priority to securing funding from nongovernment sources. But nongovernment funds should be seen as supplementary to public funds and not as a substitute for them; and
- the policies and activities of all departments of central and local government should reinforce the statutory purposes of protected countryside.

These views are derived partially from a study of AsONB which sought to place them on a more equal footing with National Parks.

---

intentions and the ways in which it intends to influence land management. NPPs are prepared in addition to normal development plans, so that 'planning' refers to the implementation of policies and decisions affecting the environment, and 'management' comprises the organisation and provision of services and facilities, and the use and management of land and resources to serve national park purposes. In broad terms, an NPP can be considered to include four elements:

- delineation of zones with a strategic role, such as remote open country;
- transportation services, including recreational access;
- park-wide services, such as interpretation, rangers and visitor information; and
- strategies for dealing with seasonal pressures, e.g. peak demands for accommodation.

It has been noted that policies relate to all aspects of the landscape, from the major elements such as wild areas and forests, to the smallest detail of field boundaries and farm buildings. An important activity in national parks has been the mapping of designated 'moor and heath' which, in the past, has been important as a basis for commenting on agricultural improvement grants and is now gaining new significance in the definition of public access land.

Rights to permitted development may be removed by an Article IV Direction in particularly important areas, and certain blanket removals of permitted rights occur in National Parks (England and Wales) and National Scenic Areas (Scotland). Thus, in national parks, a Special Development Order requires applicants to give the planning authority specified information about a wide range of proposed developments in settlements and on agricultural land which would normally be permitted under the GDO. Rural planning controls may be assumed to be enforced with greatest rigour in areas covered by protective designations. Given the stricter regime, it would be anticipated that marked differences in development control decisions would arise inside and outside protected areas. One way of ascertaining the effect of planning designations would be to measure the ratio of planning applications which were approved to those which were refused: it is reasonable to assume that this ratio would be lower inside the designated area than outside. However, most research has confirmed that little quantitative difference can be discerned. This could possibly be explained by the fact that many applications outside protected areas were for larger proposals, such as volume housing estates, so that a simple comparison of numbers of applications and approvals was inappropriate. Even so, when development pressure is remeasured in terms of the area of applications, the results again indicate similar approval rates. However, statistical studies generally fail to consider the higher quality of applications submitted in protected areas, the number of speculative applications which may be deterred by perception of a more rigorous regime and the stringency of conditions which are attached to planning consents. There is, indeed, a more qualitative body of evidence to suggest that the quality of applications in protected areas has been consistently high and that great care has been taken over location and siting of new buildings.

Nature conservation sites, which also were introduced through the National Parks and Access to the Countryside Act, were more restricted than the large expanses designated for landscape protection purposes. The safeguard of wildlife assets has often been by acquisition of the freehold or leasehold of land rather than by designation alone. This is mainly because sites of nature conservation interest almost invariably require active management, especially in long-settled landscapes where abandonment of traditional farming may result in loss of the plagioclimax state. Moreover, whereas landscape is a culturally defined artifact and compromises over its future appearance may be considered socially acceptable, the scientific interest of sites requires a high degree of integrity (including fundamental properties such as drainage patterns) and cannot withstand manipulation in the same way as scenic quality; moreover, ecologically valuable sites may take centuries or millennia to acquire their critical attributes. Thus for various reasons, outright site safeguard has been a more prominent objective of nature conservation organisations. The principal feature of the post-1949 system has been the National Nature Reserve which, in essence, represents a progressively growing collection of biotopes throughout the country which are representative of the range of biological diversity and are refuges for the rarest species (some are for the safeguard of geological/ geomorphological features). Programmes of site acquisition are costly to sustain, however, and it is unlikely that much more than 1 or 2% of Britain's land surface will ever be covered by formal reserves. Other types of site were the Local Nature Reserve, intended to be an equally impressive collection brought about by the actions of local authorities, but which have only been a modest addition to conservation practice, and the Area of Scientific Interest, which was intended merely as a reflection of significant conservation sites outside the NNR network; subsequently, these were renamed Sites of Special Scientific Interest (SSSIs). By the late 1970s, widespread concern was being expressed about the attrition of wildlife conservation resources in the 'wider countryside', and the SSSI thus started to be seen as a principal instrument of environmental protection. The substantial number of private nature reserves – owned, leased, managed directly and influenced indirectly – of ENGOs should also be recognised.

As previously noted, the Wildlife and Countryside Act 1981 introduced a system of management agreements for SSSIs (in Part II), which could be accompanied by compensation payments. Controversial in the first instance, these payments have increasingly been used to provide management payments to landowners and tenants of SSSI land. In addition, legislative protection is extended to individual species, so that Part I of the Act seeks to prevent the intentional

- killing, injuring, taking or selling of specially protected wild animals;
- disturbing of protected animals in their places of shelter;
- collecting or harming of wild birds, their nests and eggs; and
- uprooting, picking or selling of specially protected plants, and uprooting of any wild plant.

A Site of Special Scientific Interest can be established on land which the competent authority (EN, CCW, SNH) are of the opinion proves of special interest by reason of its flora, fauna or geological or physiographic features. The SSSI is registrable as a local land charge so that new owners will be made aware of its existence. When an SSSI is designated, any affected occupier or owner of land has to be 'notified', i.e. receive a statement of designation, a large-scale map and a list of potentially damaging operations (PDOs). The appropriate local authority and Secretary of State also receive copies of the notification, and the relevant conservation authority must consider any representations or objections made by these parties during a specified period (normally three months). A farmer or forester subsequently intending to undertake one of the specified PDOs must give written notice to the competent authority who then consider its likely effect on the site's scientific interest. If permission is refused, the authority must offer the farmer a management agreement: this will have various 'heads of agreement' (summarised in Box 4.13). In the early stages of the Wildlife and Countryside Act some farmers abused the notification procedure, and destroyed the nature conservation interest of an SSSI before the three-month confirmation period had elapsed, but now it is possible to impose a Nature Conservation Order which gives interim protection. Various potentially damaging recreational activities could also be carried out for no more than 28 days in the calendar year on SSSIs, under the GDO, but these permitted development rights have been withdrawn in respect of SSSIs.

Box 4.13   Management agreements: possible terms

Length of agreement

Area (including large-scale plan)

Purpose of agreement (a brief summary, especially for legal and estate management staff)

Undertaking by the offeree:

- not to carry out (or to carry out only with permission) a range of specified PDOs, e.g. ploughing, drainage, application of lime/fertiliser/pesticides; and
- to use best endeavours to retain existing conservation features, to accept controls on shooting and allow access as appropriate to carry out positive management actions, for instance burning, fence maintenance, tree maintenance, grazing restrictions

Undertakings by the offerer:

- to pay compensation and reasonable legal and surveyor's fees; and
- to undertake agreed management and monitoring tasks.

The use of management agreements is of general interest, and they have an important role in securing the positive custodianship of valuable sites. Given that acquisition policies can only be a limited solution to heritage protection, and that control mechanisms can to some extent only be reactive measures against undesirable change, it is clearly necessary to have a more conducive instrument designed to encourage the positive management of private land according to the public purpose. Agreements between public bodies and private individuals, for the latter to manage the land in a particular fashion, have operated for many years. Provisions were included in the Town and Country Planning Act 1971, although their early use was on a fairly limited scale, with the recipient entitled to a small financial 'consideration' and perhaps assistance from countryside management staff, in return for specific land-use concessions (Feist, 1979). However, in 1981 the Wildlife and Countryside Act forced management agreements into the front line and introduced formal guidelines for compensation. The Act enables agreements to be struck by the competent authority including national park authorities and local authorities in England and Wales; the Countryside (Scotland) Act 1981 extended similar powers to Scottish local authorities. Comparatively little use has been made of this facility by local authorities due to its expense and cumbersome legal formula, though National Park Authorities (NPAs) used it fairly extensively, mainly to prevent agricultural 'improvements'. By far its most extensive, costly and controversial use, however, has been in defence of SSSIs, especially those threatened by agricultural intensification or afforestation.

If a management agreement is struck then compensation is usually payable for loss of rights to improve land and convert to a more profitable operation. This compensation takes the form of either a lump sum (usually covering a 20-year period) which is set at the difference between the restricted and unrestricted value of the land, or an annual payment reflecting the 'profits forgone' as a result of the agreement (Table 4.4). Tenant farmers are entitled to only the latter option. When applying for annual payments the farmer is required to complete a form specifying details of the cropping area, number of livestock and labour engaged on the farm, as well as the current operations and proposed improvements for the land in question. From this information, an estimate of the annual financial benefits of the proposed improvement can be drawn up, by subtracting the extra costs associated with improvement from its anticipated revenue. Current profits can be similarly estimated, and the difference obtained by subtracting the latter from the former. A significant problem with this lies in estimating future benefits from a hypothetical operation and various complaints have been made over the years that anticipated future benefit streams have been overestimated or have been inflated by hidden subsidies. However, with experience, English Nature's land agents have become increasingly shrewd bargainers and costs have been kept down. Moreover, the relative cost of 'profits forgone' payments has steadily fallen (from 81% of funding in 1992–93 to 57% of funding in 1996–97) as EU subsidies and UK grant aid have diminished, and as management agreements have been used to conserve SSSIs through partnerships between English Nature's 'local teams', owners and managers of land, and communities. Positive management is now progressively being based on Site Management Statements and assisted through the Wildlife Enhancement Scheme.

Table 4.4 Calculation of 'profits forgone' for a farmer denied the opportunity of converting from pasture to arable

	Store cattle	Wheat
Output	O_c	O_w
Variable costs	VC_c	VC_w
GM of output over variable costs	$GM = O_c - VC_c$	$GM = O_w - VC_w$
Incremental fixed costs	FC_c	FC_w
Extra capital charges (at x%)		
Roads		C_1
Field drainage		C_2
Buildings		C_3
Extra repairs on fixed assets		C_4
Extra machinery costs		
Fuel		C_5
Repairs		C_6
Depreciation		C_7

Thus, the net advantage for the arable system per hectare is $(GM_w + FC_w + C_{1-7}) - (GM_c + FC_c)$.

Although various additional costs are incurred in converting to wheat, this is generally considerably outweighed by the GM value.

The practice of nature conservation is thus a complex issue, and the corporate plan drawn up by the former NCC in the early 1980s recognised the need to combine resources with the voluntary sector, to address conservation issues in the wider countryside and, latterly, to take a more targeted approach to the recovery of dwindling species. Moreover, the importance of conditions in the 'wider countryside' to nature conservation means that local authority planners have a complementary role in protecting nature. Current policy emphasises planners' role in helping safeguard statutory sites, but also expects them to have regard to other types of area, such as ancient semi-natural woodlands, nonstatutory sites of nature conservation importance (perhaps identified by a local naturalists' trust or local authority) and linear 'wildlife corridors' within built-up areas, including watercourses and woodland belts. Nationally designated sites should automatically be assured a high degree of protection, and one of the most important roles of planners will therefore be to safeguard locally important areas which complement this national network (DoE, 1994b).

Internationally, there has been increased interest in achieving a co-ordinated approach to the recognition and protection of landscape and wildlife assets. Most notable is the IUCN's influential classification of categories of protection (IUCN, 1994), whilst, at a European level, the Pan-European Landscape Convention and Council of Europe's (1995) landscape strategy are significant. The last of these addresses all biological and landscape initiatives under a Europe-wide approach and promotes the integration of biological and diversity considerations into social and economic sectors. Its aims are thus:

- substantially to reduce and, if possible, eliminate current threats to Europe's biological and landscape diversity;
- to increase the resilience of Europe's biological and landscape diversity;
- to strengthen the ecological coherence of Europe as a whole; and
- to ensure full public involvement in the conservation of the various aspects of biological and landscape diversity.

The EU has also produced its own Biodiversity Action Plan (BAP), whose 'strategy themes' are summarised in Box 4.14.

Landscape quality does not, of course, end at the boundary of a national park or AONB, but is distributed across the wider countryside. Thus, it is important for planners to recognise the landscape values of 'ordinary' areas which matter to local people and contribute to a sense of local distinctiveness. A zone of especial policy importance has been the 'urban fringe'. In these areas, landscapes are typically highly fragmented and disturbed, comprising a range of landscape types which are aesthetically more or less desirable (Box 4.15). Responses to the uncertain future of pert-urban landscapes have been founded on a variety of innovative planning and management measures. These have sought to counter landscape neglect by the creative after-use of rehabilitated land, ecologically sensitive management

Box 4.14   Principal 'strategy themes' in the European Biodiversity Strategy

Theme 1: Conservation and sustainable use of biological diversity
- *in situ* conservation (networks of designated areas)
- *ex situ* conservation (captive breeding centres, zoos and botanical gardens)
- sustainable use of components of biodiversity (environmental assessment of strategies, plans, programmes, policies and projects)

Theme 2: Sharing of benefits arising out of the utilisation of genetic resources
- promotion of negotiative frameworks for genetic resources
- technology transfer of biodiversity conservation technology
- technical and scientific co-operation

Theme 3: Research, identification, monitoring and exchange of information
- role of community research framework programmes, information and observation networks, and annual work programmes.

Theme 4: Education, training and awareness
- programmes of public awareness and professional training
- monitoring, assessment and reporting capabilities.

Source: CEC, 1998.

Box 4.15   Types of urban fringe landscape

Urbanised landscapes
Include all residential, manufacturing and transport uses and public utilities; sometimes referred to as 'urbanic'.

Disturbed landscapes
In the process of change, such as active quarrying, rubbish dumps, bare ground, land being restored and construction land.

Neglected landscapes
Comprise low-intensity agriculture, e.g. horse pasture, scrub, weed-infested land, derelict sites, dying or dead woodland, derelict quarries.

Traditional agricultural landscapes
Incorporate the farming landscape as most people expect and hope it to be like: small enclosed fields, mixed farming, hedges, animals grazing outside and traditional regional building materials.

New agricultural landscapes
Arable prairies, modern factory-like buildings, battery houses, extensive views but with little visual, recreational or wildlife interest.

Amenity landscapes
Include woodland, water and playing fields. A distinction is made between publicly accessible areas and land/water over which private access rights exist.

Source: Blair, 1987.

of temporarily blighted land and cohesive landscape planning of major road corridors. In addition, beneficial effects can be wrested from land-use change, such as the creation of desirable after-uses on restored mineral land, and promotion of urban fringe woodlands and alternative farm enterprises. Within the urban fringe, recreation and access opportunities can be broadened by increasing the attractiveness of various facilities, and 'animating' the countryside by interpretation and other means of active demonstration.

Mineral Resources

Minerals are, from a planning perspective, perhaps the most important category of nonrenewable natural resources (Box 4.16). Their specialised treatment may seem rather curious initially, but development related to mineral working has distinctive characteristics which have led over the years to the introduction of a 'special regime' for mineral planning. These features comprise the fact that mineral working:

1) is a transitory activity;
2) simultaneously involves both destruction and development;
3) entails workings which are an end in themselves, rather than as a foundation for superstructures;
4) can only take place where a mineral is located, so that planners have little room for manoeuvre in influencing location; and

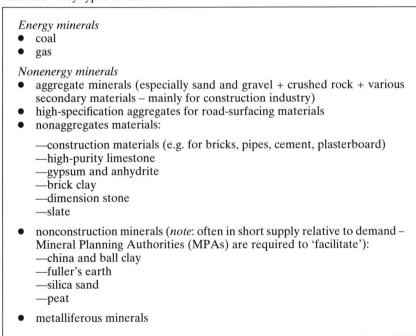

Box 4.16   Key types of minerals

Energy minerals
● coal
● gas

Nonenergy minerals
● aggregate minerals (especially sand and gravel + crushed rock + various secondary materials – mainly for construction industry)
● high-specification aggregates for road-surfacing materials
● nonaggregates materials:

—construction materials (e.g. for bricks, pipes, cement, plasterboard)
—high-purity limestone
—gypsum and anhydrite
—brick clay
—dimension stone
—slate

● nonconstruction minerals (*note*: often in short supply relative to demand – Mineral Planning Authorities (MPAs) are required to 'facilitate'):
—china and ball clay
—fuller's earth
—silica sand
—peat

● metalliferous minerals

5) is subject to changing economic conditions during the life of the quarry/ mine, perhaps making it subeconomic, whilst social expectations may vary over its lifetime, rendering it more or less acceptable to local residents and wider environmental interests (based on Stevens, 1975).

During the 1939–45 war, many mineral sites were hurriedly opened up to respond to the national emergency, but the controls over these were inadequate to meet the expectations of postwar society. Thus, immediately afterwards, the Waters Committee was set up (though it did not report until 1955) to consider the demand for aggregates (i.e. bulk minerals required for construction and road building). The first faltering introduction of effective controls occurred in 1951 when the Ministry of Town and Country Planning sought to

- establish satisfactory environmental conditions and relationships between mineral working and adjacent land uses;
- ensure that land was restored on completion of mineral working;
- ensure that mineral deposits were not unnecessarily sterilised by premature development; and
- ensure that necessary rights in suitable land were available to mineral operators to maintain production to meet national needs.

Nevertheless, planning controls over workings and the quality of site restoration remained shaky, and a concerted effort was made during the 1970s to respond to growing concern. The Verney Committee (1976) once more considered the vexed problems associated with 'aggregates' production, focusing primarily on the absence of reliable forecasts and lack of mechanisms to equilibrate supply and demand between different parts of the country. This resulted in the establishment of 'Regional Advisory Committees' comprising planners, central government representatives and members of the minerals industry, who subsequently provided forecasts which were interpreted by central government to try to achieve some sort of consistency in preparing mineral plans. About the same time, the Stevens Committee (1975) considered more generally the issue of planning control over mineral working, and this led to the Town and Country Planning (Minerals) Act 1981, which introduced some significant additional controls: these are now consolidated into the main 1990 Act. Official central government guidance on mineral planning was supplied in the so-called 'Green Book', but this single pamphlet has now given way to a more comprehensive series of Mineral Planning Guidance Notes (MPGs), and to a single National Planning Policy Guidance Note on minerals in Scotland.

One of the most difficult environmental considerations with respect to mineral planning is that of interpreting 'sustainable development', as normally the term relates to resources which are innately renewable. However, the content of sustainability must be reinterpreted for nonrenewable (or only partially recyclable) resources, and the official definition of 'sustainable' minerals development reads:

... making the best and most efficient use of all available resources, so that minerals extraction, processing and consumption are limited to what is necessary to meet the needs of the current generation; and that the overall quality of the environment affected by the mineral extraction should be conserved or improved over time, so that future generations are not disadvantaged by the activities of the present one.

(DoE, 1996d, p. 4)

This translates into specific objectives, namely: 'MPAs will wish to satisfy themselves that a balance is struck between the aim of ensuring a supply of minerals and the capacity of the environment to absorb the impact of extraction (*ibid.*)' The objectives for sustainable development for mineral planning are to

- conserve minerals as far as possible, whilst ensuring an adequate supply to meet the needs of society;
- minimise production of wastes and to encourage efficient use of materials, including appropriate use of high-quality materials, and recycling of wastes;
- encourage sensitive working practices during minerals extraction, and to preserve or enhance the overall quality of the environment once extraction has ceased; and
- protect designated areas of critical landscape or nature quality from development, other than in exceptional circumstances.

Increasingly, there is pressure from environmentalists to introduce taxes to encourage mineral users to make greater use of recycled minerals for a wider range of purposes. Many people would also draw attention to the economic benefits of mineral working which, for countries reliant on commodity production, are very considerable. Even in more diversified economies, mineral working can yield 'multiplier' effects – i.e. demand for additional goods and services related to mineral working, processing and transport, and supply of housing and services to employees and their families – though especially in rural areas and for smaller-scale developments, most of these benefits leak outside the local economy.

Many issues associated with mineral extraction impinge on the regional scale, and this is especially true in the case of aggregates, as certain regions with high development pressures are 'demand hungry' whilst others with lesser pressures but particular types of geological deposits are 'supply rich'. Because of the geographical imbalances between supply and demand, an important feature since the early 1970s has been the Regional Aggregates Working Parties (RAWPs), whose results have latterly been published in *Mineral Planning Guidance* (MPG) 6. The process is controversial, because environmentalists feel the RAWP forecasts to be too high, the minerals industry considers them to be too restrictive and the forecasts on which they are based are unreliable due to the volatile fortunes of the development sector. The government recognises that MPG6 is based on the 'best information currently available', but that this is subject to changes in demand, technology and environmental expectations. Despite investing in sophisticated economic reviews and an appraisal of the RAWP process, it is unlikely that really significant improvements to the current system will be found unless substantial policy and financial support is given to the use of secondary aggregates.

The regional model is also capable of making allowances for supplies from other sources, notably coastal superquarries and marine-dredged aggregates. The former was for a time seen as a major contributor to regional demand. The concept rests on massive quarries (with an output of at least five million tonnes a year), located on the coast, for which (marine) transport costs would be very low (Black and Conway, 1996). Typically, they would exploit hard rock resources for crushing and distribution to distant markets. Despite the early enthusiasm, only one has been developed in Britain (Morvern, Argyll) whilst a second in the Outer Hebrides was abandoned due to mounting public opposition. Marine aggregates are often of high quality and may be peculiarly appropriate to particular engineering structures, such as soft coastal defences. However, their removal can be very disruptive to pipelines and cables, as well as to natural assets and processes such as coastal sediment transport and inshore fisheries.

Given the specialised nature (both procedural and technical) of mineral planning, it has been treated as a 'county matter', with county councils having responsibility for both production of the relevant local plans and the execution of development control. Given the reorganisation of local government, and the creation of many new unitary authorities, this situation has been considerably modified. The strategic framwork is, as usual, contained in the structure plan, which sets out the broad policy framework for minerals within the national and regional context and provides for the co-ordination of mineral working with other elements of strategic planning. The operational framework is established in specific *mineral local plans,* which give detailed expression to structure plan policies and proposals and relate them to identifiable areas of land. Mineral policies in development plans have a number of functions which are summarised in Box 4.17.

This approach sounds logical enough, but it is very difficult to adhere to in practice because the minerals industry is very closely related to the development sector, which is itself a sensitive barometer of national economic performance. Thus, gauging the 'reasonable' extent of permissions or rates of extraction is an intractable task, which invariably upsets either the green lobby or the industry.

Complementary to spatial and strategic planning for minerals is the exercise of development control. Policies in Mineral Local Plans (MLPs) set out the criteria against which individual applications are judged and conditions are framed, and these criteria typically include:

- the nature of working programmes (e.g. the phasing of site extraction and the logistical possibilities for progressive restoration in the wake of recently completed extractive plans);

Box 4.17 Key concerns of mineral local plans

Ensuring supply:

- make adequate provision for supply;
- provide an effective framework within which the industry can make applications; and
- recognise that the county must make a realistic (perhaps more than its fair share) contribution to regional and national demand.

Safeguarding potential future mineral extraction areas:
- define *mineral consultation areas (MCAs)* in local plans, to facilitate liaison between counties and districts over 1) protection against development; and 2) extraction prior to other forms of development taking place.

Identification of landbanks in order to sustain continuity of production:
Demands are cyclical, and commissioning of extraction sites can take a number of years; thus, operators generally seek to confirm 'banked' areas of future working to respond rapidly to surges in demand.

Identification of areas for future working:

- specific sites
- preferred areas
- areas of search.

(This requires knowledge of mineral availability, the landownership situation and consideration of the planning/environmental factors.)

- the nature and extent of environmental impacts;
- transport considerations, and the capacity of country roads to accommodate site traffic;
- the potential for, and nature of, restoration, after-care and after-use;
- any ancillary development which will be necessary on site; and
- the need for widespread consultation (as mineral planning can affect water supplies, land drainage, farmland, nature/landscape conservation, archaeology, listed buildings and linear services such as railways and power lines).

In general, conditions imposed on planning consents need to meet a standard set of 'tests of acceptability' (they must be necessary, relevant to planning and to the development in question, enforceable, precise and generally 'reasonable'). These apply to the determination of any planning applications, but specifically with regard to mineral workings, planners are expected to consider:

- the supply and demand of the industry (i.e. the volatile nature of mineral economics);
- the topography and geological structure of the site (which may make certain patterns of working infeasible);
- the method of excavation; and
- the buildings and equipment necessary to make the site viable.

Thus, the conditions imposed on a mineral planning consent should reflect a programme of working designed to accommodate the operator's needs while at the same time paying due regard to minimising the effect on the environment both during and at the end of the mining operation. Typical aspects covered by planning conditions are shown in Box 4.18. Mineral applications are often associated with substantial types of environmental impact, and may require the submission of an Environmental Statement.

Box 4.18 Typical topics covered by planning conditions on mineral planning consents

- time limits;
- access, road safety, lorry routes
- working programme (hours, direction, depth, limitation on production, topsoil and subsoil preservation);
- environmental protection (dust, fumes, noise, waste disposal, blasting);
- surface water, drainage, pollution control;
- landscaping (e.g. TPOs, screening);
- boundaries and site security;
- restoration and after-care; and
- subsidence and support.

A controversial issue related to the winning of minerals has been the consents given to certain sites as a result of wartime exigencies, but which subsequently remained legal. These 'Interim Development Orders' (IDOs) have been the subject of widespread debate, but section 22 and Schedule 2 of the Planning and Compensation Act 1991 gave new procedures for dealing with permissions for the winning and working of mineral waste originally granted under IDOs (between 21 July 1943 and July 1948). These require various actions to be taken if permissions are to continue to have effect, namely that

- the operator must reapply for planning consent if the working is still active;
- if the site is dormant, a new approval scheme must be obtained;
- if the site has not been worked since 1979, the planning consent is deemed to have lapsed; and
- the operator is entitled to limited compensation if the MPA requires modifications to existing extraction arrangements.

These remedies, though, trade heavily on 'goodwill', and the guidance indicates 'realistic' types of extra conditions which may be imposed. Generally, the best prospects for bringing IDO sites in line with current expectations exist where a new extension is sought by the operator, possibly creating an opportunity to negotiate in the context of the whole site.

A defining characteristic of mineral workings is their longevity, so that public attitudes towards the acceptability of a scheme may change. Consequently, a permission (with its particular set of conditions) which seemed acceptable at the time of granting consent, may not meet amenity expectations 20 years later. Thus, mineral planning legislation provides for a periodic *review* of mineral workings, enabling the MPA to modify existing consent conditions. According to this the industry must accept 'reasonable additional costs' arising from the modernisation of old mineral planning permissions, whilst the MPA must undertake periodic reviews of sites, with the aim of ensuring that conditions are consistent with current minerals planning practice. In effect, conducting this review is one of the most time-consuming tasks facing MPAs, as it involves ascertaining the status, condition and rates of production of all mineral workings and giving detailed consideration to the most appropriate revisions to their control framework. Moreover, as the industry is only required to absorb 'reasonable additional costs' and the MPA is unable to pay significant levels of compensation, the degree of any modifications to existing planning consents must be very astutely judged.

Conclusion

The ways in which we use natural resources clearly creates major impacts on the environment. Although agriculture still dominates the land surface, major changes in land cover associated with other land uses have occurred during the twentieth century, whilst agriculture has itself been transformed almost beyond recognition over the same period. Tree cover has been significantly regenerated, even if it is of very different quality and composition from more ancient woodlands.

Yet the land-use planning system can exert relatively little control over these major land-use changes. Indeed, some key transformations in the use and management of rural land have effectively fallen outside any form of statutory control. But between them, planning, agriculture, forestry and conservation agencies can exercise a range of direct controls and indirect influence over many environmental changes. Planners are widely concerned with these, even if only because they inherit many of the consequences of change.

One area of natural resource management in which planners are centrally involved is in the development of minerals. Significantly, this is a principal category of nonrenewable resources and thus raises distinctive and intractable issues of sustainability. Indeed, the tenets of sustainable development are forcing us to reappraise more generally our use of natural resources.

Our understanding of what is optimal in terms of resource use is being transformed from one of maximum sustainable yield and supply management, to one of multiobjective planning within a policy context of demand management. This transition is as yet nowhere near complete, and policy-makers have taken only fairly faltering first steps. However, it is likely to be a dominant policy theme of the early part of the twenty-first century.

5   Landscape Ecological Planning

Principles of Landscape Ecology

We have noted that historical approaches to managing and planning the landscape have been based on site safeguard and designation of wider areas. This, for all its limitations, has been hugely beneficial, and despite the environmental impacts that have affected areas of 'critical natural capital', far worse would have happened had key sites and areas not been covered by protective designations. Consequently, key nature conservation sites have been protected from destruction and they have yielded much scientific evidence, whilst scenic areas have maintained much of their cultural interest and have generally been safeguarded against the worst excesses of recreation and development.

However, approaches based purely on site-based safeguard and passive protection are now seen to be necessary but not sufficient. Despite the application of scientific expertise and careful planning, there has been continued attrition of biodiversity and landscape quality. Species continue to be threatened by extinction, whilst everyday and even rare landscapes are becoming progressively punctuated by intrusive change to the point of decomposition. Both in theoretical and policy terms, the separation between ecological and scenic value is itself becoming increasingly sterile as ecologists recognise the cultural systems underlying nature conservation value and planners appreciate the bio-physicochemical dynamics of amenity landscapes.

The focus in recent years has thus shifted away from safeguarding individual sites according to their significance in terms of specific wildlife or scenic values, and towards the management of the 'wider countryside', acknowledging that the 'rest' deserve attention in addition to the 'best'. Thus, it is recognised that species loss may be heavily dependent on what happens outside nature reserves, and that scenic quality may be contingent on the fate of 'ordinary' landscapes in which people live and work. In this respect, all land is deemed to have distinctive 'character', and the process of 'characterising' all areas is seen as a way of protecting and enhancing those attributes which make them distinctive. An important, emerging basis for understanding the nature and dynamics of the wider countryside is now provided by *landscape ecology*. This appears to be able to help us explain, predict and plan change in the wider countryside, focused as it is on the patterns and processes within entire landscapes, rather than just in protected enclaves.

Theoretical research from the 1960s began to relate species viability to habitat size, and this has been influential in guiding the acquisition and design of networks of nature reserves. The concept of island biogeography (MacArthur and Wilson, 1967) was especially important. This relates the number of species and success of their populations to the size of islands and their isolation from neighbouring land masses (The flora and fauna of islands are typically impoverished because of their isolation, but may also develop unique species, because of the genetic separation of populations of initially identical species.) MacArthur and Wilson's theory enabled ecologists to relate island size to the range and viability of indigenous species through the production of 'species–area curves', which indicated that larger habitats (islands) would be likely to sustain a larger number of species. It was also highly influential on nature conservation policy, where it led scientists to debate the respective merits of protecting several small sites as opposed to a single large one within a particular area (the SLOSS concept – 'single large or several small').

The principles of island biogeography could be extended to terrestrial situations, in which isolated ecologically rich habitats were separated from each other by inhospitable zones (e.g. urban, intensive arable, monocultural forestry). This is an appealingly simple idea, but in reality the relationships between the population dynamics of species, and the qualities of core and intervening habitats, are far more complex. Consequently, the study of island biogeography has been elaborated into sophisticated studies of gene flows, migration patterns and life-cycle characteristics of interdependent species between ecotopes of differing characteristics and sizes. As each ecotope can be represented as a 'patch' of vegetation likely to favour particular types of species, island biogeography was reinterpreted in terms of theories of 'patch dynamics' (e.g. Wu and Levin, 1994).

For many ecologists, this theory is sufficient. Others, however, have favoured its further extension to 'landscape ecology' (Forman and Godron, 1986; Farina, 1998). This latter subject is controversial, especially in Britain, although it has proved very influential internationally. It is presented as a basic model here because its vocabulary helps to connect planning requirements with scientific principles, especially in relation to the development of a 'wider countryside' (including urban greenspace) framework for the conservation of 'critical' natural capital and creation of compensatory 'constant' natural capital. Planners should be aware, though, that whilst landscape ecology is strongly supported by some scientists, it is treated sceptically by others.

Briefly, the science of landscape ecology can be seen to revolve around a number of key themes. First, it is assumed that there is an *ecological infrastructure* of ecotopes conducive to different levels of species diversity. In particular, certain terrestrial and aquatic habitats (ecotopes/ biotopes) are relatively undisturbed and likely to support 'interesting' or rare species. These are typically semi-natural remnants ('islands') likely to be protected by designations. Between them are tracts of more intensively altered land (the 'oceans' of island biogeography) which are relatively impassable to many species because they are inimical to their survival or do not meet their life-cycle requirements. These tracts may, however, contain isolates of semi-natural habitat – such as field corners, hedgerows, ponds or urban wilderness – which may connect semi-natural ecotopes or provide refuges for rare species. Collectively, these features form an infrastructural network which underpins various biological processes. The two main considerations relating to this network are the *pattern* of biotopes and adjacent areas; and the *process* of species movement and associated gene flows (and other environmental dynamics, such as soil development and water movement) across the wider countryside.

A principal interest of landscape ecology is the population dynamics of key species. In particular, it departs from traditional ecology, which focuses on the life-cycle (birth–immigration–death–extinction) processes of individual populations, by stressing the interactions between individual populations across the wider landscape. Thus, individual populations often interact, increasing survival prospects and enhancing gene flow; apparently separate groups may thus interbreed and share other survival strategies as part of a *metapopulation* (i.e. population of populations). The *genetic viability* of apparently isolated and vulnerable populations may in turn be sustained if they are able to interconnect with other members of their species across a relatively hospitable countryside. Where populations become too isolated, 'genetic drift' may occur, in which the genetic diversity within a population starts to decline, so that the species locally becomes less resilient and adaptable to environmental change, thereby accelerating the likelihood of local extinction. The likelihood of population survival is affected by a number of *patch characteristics*. Thus the size and shape of an ecotope will influence species' prospects, with larger and rounder patches reducing the likelihood of species becoming vulnerable to 'edge effects' (e.g. predation from hostile species or drift of agrochemicals). Excessively elongated habitats may disfavour species by creating a disproportionate length of 'edge', whereas relatively circular ecotopes will create 'interior' conditions which maximise the preferences of species characteristic of the habitat and minimise their likelihood of disturbance.

Between the patches lies the *matrix* of interconnecting town and country, often profoundly altered by human intervention and even hostile to all but a comparatively few resilient species which may thrive in simplified environments. The pattern of ecologically valuable patches lying within this wider matrix is often termed the *mosaic*. One way for species to transfer between patches across the matrix is via *corridors*, such as hedgerows, verges and drainage ditches. It should be noted, though, that hardly any species depend on linear habitats such as these for fundamental life-cycle processes; moreover, movement along corridors may be hazardous, as it may expose species to the attention of predators which may stalk such passageways. However, species movement can be observed along corridors, whilst linear features are often valuable for their 'patch' characteristics, as they may represent the only remaining relics of semi-natural habitat in an area and thus provide important wildlife refuges and nuclei to which new habitat may be adhered. Key corridor characteristics are their length, unbroken continuity, width, age and diversity. The existence of corridors promotes 'connectedness' (where physical links exist between linear and patch features) and 'connectivity' where movement along (i.e. functional use of) corridors can be demonstrated. Where there are no connecting corridors (or even where there are), species must attempt to negotiate passage across intervening hostile territory in order to reach another conducive patch. This may be relatively successful, depending on conditions of *permeability* (where the matrix is not wholly hostile) and *porosity* (where, despite inimical matrix conditions, there are 'stepping stones' of favourable habitat, enabling 'percolation' of individuals).

Other key considerations in landscape ecology include:

- the possibility that connected landscapes may help species respond to *climate change*. Thus, individual species survive within a given 'range' of climatic and physicochemical conditions and, if these change and species are unable to migrate, extinction may occur. If, for example, landscapes display high levels of connectivity, it may be easier for species to respond to climate change by migrating northward or southward, or uphill or downhill, and thus modify their range;
- the prospect for creating new patches within the ecological mosaic. Where insufficient 'natural capital' exists, *compensation* measures may be used, that is, new development may be able to create or transfer habitats to complement landscape structure, frequently to offset the losses associated with urban development or agricultural intensification. Careful design and location of newly created features may be extremely valuable in supplementing existing landscape mosaics;
- the opportunities for analysing landscapes at a macroscale. Here, it is often possible to identify gaps in the current framework of protected areas, where desirable mosaic or corridor features are either absent or inadequately safeguarded. This principle underlies the method of *gap analysis*, which has been used to supplement protected landscape networks in some countries;
- the scope for maintaining semi-natural habitats, which are the result of centuries of low-intensity human use. Landscape ecological management thus rests upon the retention of *cultural systems* of land use and implies that means of support must be given to certain traditional systems of farming, woodland management and fishing. Similarly, as many plagioclimax ecological communities, which have arisen from an intimate relationship between people and land, seem to be intrinsically appealing – perhaps by virtue of their detail, complexity and human scale – there is generally a strong association between scenic beauty and ecological value. (This is not always true, as nature is generally 'untidy' and perhaps threatening, although many principles of scale, surprise,

seasonal variation, etc., connect human experience and conservation value in a positive way.) Thus, landscape ecological planning can generally integrate considerations of *aesthetics* with those of a more scientific nature; and

- that much of the science of landscape ecology has been made possible by the emergence of techniques enabling us to *perceive and analyse large-scale landscape patterns*, such as remote sensing and geographic information systems. These methods help us to assimilate copious spatial data quite rapidly and are now transforming the ways in which we understand environmental change in the wider countryside.

Whilst some of the suppositions underlying landscape ecology are thus quite controversial, the subject does contain many useful environmental concepts, and provides a powerful vocabulary for 'joined-up thinking' in environmental planning.

Environmental Planning and the 'Wider Countryside'

The British countryside is dominated by productive land use, and the area which is given over principally to amenity management is very small. In most instances, therefore, the main aim is to extract concessions from commercial enterprises towards multiple use, especially to nature and landscape conservation. Whilst this reflects the current state of play, it is pertinent to note that many environmentalists still see this as an unacceptably human-centred philosophy and look towards systems with a greater equality of economic and ecological objectives. Sustainability, they argue, cannot be achieved by cosmetic treatments of the countryside, only by fundamental shifts away from productive systems based on agrochemicals and monocultures. This account considers the reconciliation of productive and protective land uses in the wider countryside, rather than their treatment as essentially separate activities.

Farming in the Wider Countryside

We have already noted how the nature of modern farming has transformed the ecology and scenery of the wider countryside. This has arisen from a technological and managerial revolution, based largely on the intensive use of agrochemicals, machinery and, increasingly, genetically modified organisms (which may facilitate some reduction in the previous two). Farm policy is gradually favouring more environmentally friendly modes of production, but this is, as yet, having little effect on the general pattern of the industry.

In the longer term, it is conceivable that farming will become a radically different, low-intensity land use. This is unlikely for the foreseeable future, however, and so the principal ecological interest of farmland will be likely to be sustained in unfarmed patches or low-intensity pastoral use. Thus, the principal habitats of potential interest on farmland comprise woodlands, hedges, heaths, semi-natural grasslands and wetlands (see Table 5.1). Arable land is rarely of interest, although a restricted ecology has established itself and sympathetically managed field margins can be very important. Certain linear features are highly valued, especially hedgerows and stream edges, but also including conservation headlands, wildlife fallow margins and grass margins. Headlands occur where the expanded field margin is also part of the cereal crop and can continue to be harvested, albeit at a reduced yield. Certain 'weeds', especially broadleaved ones, are tolerated to encourage insects as a food source for partridge and pheasant chicks; they may also contain rare flowers and butterflies. (Some use of selective

Table 5.1 Values of existing habitat: a guide to assessing priorities for wildlife on the farm

Outstanding value	Great value	Lower value
Woodland Old (shown in first OS map) deciduous woodland of more than 1 ha especially native broadleaf trees and those with trees of different age classes	Smaller (less than 1 ha) old woodlands, secondary woodland and plantations of more than 1 ha	Plantations of conifers and small plantations of broadleaf trees
Hedges Thick old hedges (i.e. those with four or more species per 10 m)	Poorly managed old hedges. Thick younger hedges (less than four woody species)	Poorly managed young hedges
Heath and moorland Large (more than 8 ha) with both dry and wet areas, not intensively grazed	Smaller heaths and moors. Large heaths very heavily grazed	
Grasslands Unimproved pastures and meadows with numerous plant species. Flood meadows	Unimproved pastures and meadows with few plant species	Grasslands improved by fertilisers. Older grass leys
Wetlands Large, well established ponds and lakes with at least one edge free of trees. Unpolluted rivers, streams and ditches containing permanent water	Smaller ponds and totally shaded larger ponds. Permanent rivers, etc., showing signs of pollution or which are treated chemically	Small polluted shaded ponds. Ponds and ditches which are dry for some of the year

Source: FWAG, 1982; Tait *et al*., 1988.

herbicide sprays may be permitted.) Fallow creates ideal conditions for wild flowers which thrive on disturbed ground, many of which have become rare although large numbers of their seeds remain dormant in the soil. If a weed barrier is required between the margin and the crop, a sterile strip or grass strip may be used. In general, positive conservation action on farms relies on three complementary strategies: identification and protection of pockets of semi-natural habitat; positive utilisation of agriculturally unproductive areas for landscape enhancement; and modification of management regimes to retain traditional methods wherever appropriate.

There is now a good deal of evidence of growing environmental sympathy amongst farmers, as reflected, for example, in membership of a conservation group, planting of broadleaved trees, and creation and protection of wildlife habitats. This is clearly not a universal view, though, and many farmers are temperamentally disinclined or economically unable to make concessions to wider countryside needs; indeed, many would see this as 'farming badly'. Generally speaking, traditional estates and family farms, together with smaller-scale (including 'hobby') farms, are often comparatively favourable to conservation, whilst heavily capitalised farm businesses, sometimes owned by large corporations and often contracted to ('horizontally integrated' with) a major food retailer may have little sympathy for wider countryside matters. Also, many farmers who do foster habitat resources are those with little economic interest in agriculture and may only be farming on a small scale (wildlife, of course, will not be concerned about the economic status of the provider of the habitat, but will merely take advantage of ecotopes however they arise).

One approach to enhancing the landscape ecology of farms is to encourage farmers to take a more active interest in conservation and to make concessions to wildlife in the use of their land. A principal influence has been the provision of conservation advice through the Farming and Wildlife Advisory Group (FWAG) (Cox *et al.*, 1990) and the Farming and Rural Conservation Agency (the government's agricultural extension service, whose free conservation advice to farmers still continues despite the commercialisation of most of its activities). The Countryside Commission also aided our ability to communicate environmentally sensitive practices to farmers through an experimental Demonstration Farms scheme which provided practical examples of conservation measures integrated into the agricultural business; its lessons were disseminated more widely through an associated network of Link Farms.

Although the original national FWAG was established in 1969, it had a rather obscure existence until the early 1980s when MAFF started to see its potential for advancing their new statutory responsibilities towards the wider countryside, and when the Farming and Wildlife Trust was established. Most counties/regions now have their own FWAG, with a committee and full-time adviser. Its aims are to:

- provide practical advice and encouragement to farmers and landowners who wish to undertake conservation measures;
- provide for contact and discussion between farmers and conservationists;
- develop understanding between farming and conservation interests through practical demonstrations, publicity, talks and conferences; and
- identify opportunities, conflicts or developments which require study and research.

Although FWAGs receive grant aid from various public and voluntary services, they are also dependent on private subscription from farmers and, more recently, on charging for their advice. Those taking out membership are then entitled to request a visit and report on the conservation potential of their farm, receive FWAG publications and information, and be informed of functions and events organised by the local FWAG. Despite a somewhat precarious financial position and periodic problems of organisation and viability in the less prosperous farming areas, FWAG does continue to have an important catalytic role.

As the pressures for chemical and mechanical intensification lessen, farming may be moving slowly into a postproductivist era. Certainly, recent agricultural policy initiatives have emphasised the scope for farmers to engage in tasks aimed at producing 'conservation' outputs rather than agricultural outputs. In this sense, the countryside has become increasingly 'commoditised' and it has become respectable for farmers to be seen as producers of amenity goods, which society appears to demand in addition to its food supply. (There remains a possibility that exposure to world farm economics may yet produce a homogenised and substantially depopulated landscape, over a relatively short period, which is given over predominantly to commercial cereal and livestock production.)

Consequently, since 1985, various environmental land management schemes (ELMS) have been introduced, and these have complemented pre-existing initiatives to integrate farming and wildlife. International challenges to the EU's Common Agricultural Policy have focused on the production subsidies which are used to underwrite farm outputs, but do not threaten direct payments to farmers as these do not immediately affect the prices of products on world markets. Increasingly, therefore, direct payments have been used to reward farmers for the production of non-agricultural goods, including those of an environmental nature. A major aspect of these grants is that, despite having been piloted by conservation bodies, several of them now have been transferred to the much larger, mainstream agricultural budget. Their purposes are essentially two-fold: to introduce or retain landscape and wildlife features associated with traditional methods of farming; and to mitigate the pollution impacts associated with the application of agrochemicals, especially inorganic fertilisers. Research into agrienvironment schemes has shown how difficult it is to analyse their effect, because of the delayed response of landscapes and habitats to the various packages of financial support. Researchers therefore tend to measure responses by farmers such as quantities of fertiliser applied, but there is little direct evidence on habitats and landscapes themselves (Whitby, 1994). Equally, it is difficult to apply 'policy on–policy off' methods as the control situation is difficult to define: moreover, there appears to be a 'halo effect' where farmers who are being paid to abstain from particular actions may divert their payments into funding other unwanted activities, though there is little firm evidence of this. The principal ELMS are summarised in Box 5.1.

In addition to the schemes with explicit environmental components, it is also important to note the set-aside scheme, which was introduced in August 1988, initially on a voluntary basis whereby farmers received payments for setting aside at least 20% of their land currently in use for various surplus crops. Subsequent regulations enforced a proportion of set-aside for farmers participating in the Arable Area Payments Scheme, but in the late 1990s participation once more became voluntary, though with the prospect of reintroduction dependent on the recurrence of significant overproduction. Any form of agricultural production is prohibited on set-aside land, which means that farmers may not graze farm livestock or produce fodder crops either for sale or for use on the farm (other than for horses) though the land must be kept in 'good heart'. However, land and buildings may be used for a variety of nonagricultural activities such as tourist facilities, caravan and camping sites, car parks, game and nature reserves. Some crops, such as linseed, which are produced for industrial use rather than human or animal consumption, may also be grown, as may timber.

It has been suggested that for the less scenic, 'factory floor', agricultural lowlands, this may be the only realistic form of diversification. Set-aside has a number of disadvantages, including that of paying farmers for doing nothing, but it is favoured by some as a comparatively effective way of reducing surpluses whilst ensuring that land is available to be rapidly returned to productive use if necessary. The levels of compensation paid to farmers clearly have to be competitive with the profit margin attributable to the crops they would have grown, although a slightly lower figure would be acceptable as the farmer is adopting a no-risk and minimum-effort strategy. The supporters of set-aside point to its potential to achieve food and environmental objectives simultaneously, given that it could reverse the intensification responsible for much of the habitat destruction associated with modern agriculture. However, this is a rather simplistic view. For instance, food objectives are most likely to be met if land supporting a particular crop is idled, whereas conservationists would prefer the policy to be directed at land with a specific location, climate or ecosystem. ELMS such as the *Habitat Scheme* have helped to ensure that at least some long-term ecological benefit is secured from set-aside.

Farm forestry has a potentially significant role in deriving alternative uses for farmland but, if it is to succeed, requires grant aid, advice and co-operative organisation. Different forms of grant aid are available to plant trees on farmland, but the role of advice and support may be equally

Box 5.1 The principal ELMS

Countryside Stewardship

The *Countryside Stewardship* (CS) scheme was piloted by the Countryside Commission between 1991 and 1996 and transferred to MAFF thereafter. It provides a means of paying farmers and other land managers to enhance and conserve specified landscape types and, in contrast to most ELMS, is not confined to designated areas. The scheme's objectives are to

- sustain landscape beauty and diversity;
- protect and extend wildlife habitats;
- conserve archaeological and historic features;
- restore neglected land or features;
- create new habitats and landscapes; and
- improve opportunities for people to enjoy the countryside through the provision of new or improved access.

Whilst CS is not confined to designated areas, it is restricted to the creation or restoration of certain landscape types, namely, chalk and limestone grassland, waterside land, lowland heath, the coast, the uplands, old meadows and pasture, historic landscapes, old orchards, field boundaries, field margins on arable land and countryside around towns. Detailed objectives for each particular area are agreed through a process of consultation and targeting at county and regional level, and agreements (which usually run for 10 years) with farmers are drawn up individually to address particular management objectives and local circumstances. Public access provision is a common requirement of CS schemes, and priority tends to be given to sites with existing access or which are close to centres of population. In 1998, a further scheme (Arable Stewardship) was piloted within CS to assist arable farmers who wished to manage their land in ways more sympathetic to wildlife.

Environmentally Sensitive Areas

Given the need to practise conservation extensively on farmland in heritage landscapes, and the impossibility of paying farmers full economic compensation to do so, the alternative approach has arisen of making payments to manage land in accordance with generally desirable conservation prescriptions. This concept was included in EC Structure Regulation 85/797 and led to the designation of a number of Environmentally Sensitive Areas (ESAs). In the UK, these have been selected in accordance with four criteria. That

- they should be of national significance;
- conservation must depend on the adoption, maintenance or extension of a particular form of farming practice;
- traditional farming practices would help to prevent damage to the environment; and
- the areas should comprise discrete and coherent units of environmental interest.

The designation of ESAs resulted in a sharp policy departure from payment of universal benefits to selective geographical targeting on areas where they were likely to have greatest effect, and this targeting was based on quite sophisticated analysis.

As a result of these measures, it is now possible for farmers within designated ESAs to enter into schemes tailored to the environmental attributes of

(continued over)

the area within which their land lies. Detailed environmental objectives are set for a five-year period in each ESA and are agreed through a process of consultation with farming and conservation bodies. Individual agreements with farmers run for ten years, and annual payments are made on each hectare of land entered into the scheme. The schemes are entirely voluntary and all owner-occupiers, tenant farmers, trusts, partnerships and, in certain circumstances, landlords are eligible to join such schemes. Prescriptions for ESAs vary individually reflecting different habitats and environmental conditions. Farmers in ESAs receive their grant as a flat-rate payment for compliance with a fairly basic set of requirements, but those agreeing to more onerous restrictions or converting land to conservation uses can usually qualify for a higher rate of payment. Payments are based on measures enshrined within a 'whole-farm plan' to minimise the risk of an adverse 'halo effect'.

Tir Gofal

An imaginative approach to agri-environmental integration is represented by Tir Gofal in Wales. Covering the whole of Wales, this scheme is aimed at encouraging farmers to maintain and enhance the agricultural landscape and wildlife, and improve public access. It replaces existing agri-environment schemes and, being a 'whole farm' scheme, ensures that conservation on one part of the farm is not offset by intensification on other parts. The four elements in Tir Gofal comprise:

- mandatory compliance with whole farm principles, management of key habitats, and optional restoration or creation of certain habitats or features;
- voluntary options for creating new access routes or areas;
- payments for additional works to protect and manage habitats and features and to support new access provision; and
- training for farmers, including courses on managing specific habitats, such as wetlands and woodlands and practical skills, such as drystone walling and hedge laying.

Farm Woodland Premium Scheme (FWPS)

This scheme (which is part of the programme under EC Regulation 2080/92 on forestry measures in agriculture) replaced the pilot Farm Woodland Scheme in 1992, and aims to encourage farmers to replace productive agricultural land to woodlands by providing annual payments for 10 years (for predominantly conifer woodland) or 15 years (for predominantly broadleaved woodland) to help offset the agricultural income forgone. The scheme's purpose is to enhance the environment through the planting of farm woodlands (to improve the landscape, provide new habitats and increase biodiversity) and, at the same time, encourage land managers to realise the productive potential of woodland as a sustainable land use. FWPS is linked closely to the Woodland Grant Scheme in order to ensure high silvicultural standards, and the land must not be returned to agriculture within 20 years (for mainly conifer woodland) or 30 years (for mainly broadleaved woodland).

Habitat Scheme

At the time of writing, the Habitat Scheme was in its pilot stage. It is aimed at creating or enhancing certain valuable habitats by taking land out of agricultural production, or introducing extensive grazing, and managing it for the benefit of wildlife: it targets selected 'water fringe' sites, farmland previously in the Five Year Set-Aside Scheme and coastal saltmarsh. The general purpose of the scheme is to continue to safeguard and enhance the wildlife potential of sites of long-standing (or which, through being set-aside, have acquired) conservation value. Farmers enter into long-term management agreements to pursue prescribed agricultural practices.

Moorland Scheme
Introduced in 1995, this scheme aims to protect and improve the upland moorland environment, by encouraging farmers to undertake a range of positive measures designed to conserve and enhance the rural environment. In particular, it enables farmers to improve the environmental value of their moorland by grazing fewer sheep and by improving its management, principally by setting maximum winter and summer stocking density limits for heather moorland. Farmers are compensated for this 'extensification' by receiving an annual payment for the number of ewes removed from the flock in order to meet the lower densities.

Nitrate Sensitive Areas (NSAs)
The NSA scheme enables farmers in 32 selected areas of England, covering 35,000 ha of eligible agricultural land, to receive payments in return for voluntarily changing their farming practices to help protect valuable supplies of drinking water. All the NSAs fall within the areas designated as Nitrate Vulnerable Zones (NVZs) under the EC Nitrate Directive (91/676/EEC). The aim of the scheme is to help reduce or stabilise rising nitrate levels in key sources of public water supplies, through voluntary changes in farming practices, which should help ensure that groundwater abstracted for drinking water supply meets the 50 mg^{-1} limit for nitrate laid down in the EC Drinking Water Directive (80/778/EEC). The scheme operates through three different types of voluntary measures, namely:

- the Premium Arable Scheme, requiring conversion of arable land to extensive grass;
- the Premium Grass Scheme, involving extensifying intensively managed grass; and
- the Basic Scheme, allowing low-nitrogen arable cropping.

Organic Aid Scheme
This scheme, which commenced in 1994, is voluntary and offers financial help to those conventional farmers and growers who wish to convert to organic farming methods. Existing organic farmers may also benefit when converting new land. Various rules apply to participating farmers to ensure that food meets genuine standards of 'organic production'.

Source: Information mainly obtained from MAFF's website: //www.maff.gov.uk/
As these schemes change regularly, up-to-date information should be sought by visiting this website.

important. Isolated sellers are in a weak position, and are normally dependent on the specialist technical and marketing skills of timber merchants, who can be highly selective about purchasing only those trees which they consider to be of suitable age and condition. Consequently, interest has grown in collective organisation amongst farm woodland owners. One of the most widely acclaimed woodland regeneration schemes is Coed Cymru (Welsh Wood) which has particular significance for farms on which timber, if exploited commercially at all, would probably be felled for low-grade uses and not replanted. The operation is based on a combination of appropriate technology (e.g. diesel-powered mobile sawmills), free advice, production of management plans with farmers, and balanced conservation and development objectives.

Forestry in the Wider Countryside

The design factors influencing the visual and ecological qualities of plantation forestry may be considered in relation to the typical sequence of management operations. Normally, these commence with initial ground preparations (e.g. fencing, liming, drainage) and the planting of saplings (deciduous) or transplants (conifers). In the early years careful tending, especially against competition from weeds, is necessary as neglect can lead to high failure rates. Where grant aid is available, a proportion will normally be withheld until this stage is satisfactorily completed, whilst 'extended' payments are offered to ensure that good management is continued. The young thicket will eventually require pruning and/or thinning if an acceptable final crop of trees is to mature. Finally, felling of the forest stand will be followed by replanting to a similar or revised design in the 'second rotation'. If large areas of forest comprising native species are available, traditional systems of limited extraction may take place: timber is either taken as single trees chosen over a wide area, or by taking all trees from a more restricted compartment or *coupe*. Alternatively, a 'selection' system may be used, removing the older trees: whilst this approximates most closely to the dynamics of a natural forest, it is technically difficult to operate. A 'shelterbelt' system is more practicable, and comprises a number of variants, but in essence groups of poorer trees are felled and surrounding trees are allowed to seed in their gaps; the sheltering trees are then felled when the new crop is well established. The easiest system to operate is 'clearfelling', and this has dominated commercial practice in Britain; however, natural regeneration is restricted by exposure and high weed growth and replanting will be necessary. In a clearfelling system, the main concession to amenity is the size and design of felling coupes. If crops of mature trees are to be taken from amenity woodland, a selection felling approach is usually preferred.

The multiple use value of commercial forestry may be enhanced by the adoption of appropriate design practices. Such principles have been progressively developed since the 1950s when serious concern first started to be raised about the unsympathetic appearance of plantations. In general, good practice now involves relating forest shapes to landform, reflecting the scale of the landscape in the forest and creating as much diversity as possible. Shape is considered to be the most powerful and evocative factor in landscape perception. Whereas natural shapes tend to be irregular, curved, asymmetric and diffuse, artificial ones are typically geometric, straight edged, symmetrical and well defined. It is up to the landscape designer to achieve a forest form which is sensitively massed, although this may be difficult where the ownership boundary is straight edged. In attempting to relate forests to landform, the main principle observed is the tendency for the eye to be drawn downwards on spurs and convex slopes and upwards in gullies, hollows and concavities. Thus, if forest boundaries rise in hollows and fall on convexities, they respond visually to the shape of the ground. This also mimics nature, where trees survive in fertile, moist, sheltered hollows at a higher level than on dry, exposed ridges. The 'scale' of the landscape can be related to the expanse which is visible, with scale increasing in broad, open moorland vistas and where there are large differences in elevation. As the scale decreases, so smaller plantations and felling coupes become necessary. Diversity thus depends both on careful design of major features and varied species and age classes, and on the retention of naturally occurring elements. These latter include water, views (especially focal views up valleys), vegetation, walking routes, wildlife habitats, archaeological sites and open areas (Figure 5.1).

With respect to nature conservation, it is clear that plantation forests can become mature and diverse habitats in their own right, subject to careful design and leaving significant areas unplanted. Many schemes have rightly been criticised for their ecological monotony, and current good practice involves a range of planting and management modifications to produce a satisfactory result. Sympathetic designs include the positive use of ancillary habitats, such as fire ponds, fire breaks and roadways, and the wider spacing and earlier thinning of stands, to allow greater light penetration and establishment of ground flora. However, it is salutary to bear in mind that the qualities of ancient woodland can rarely if ever be totally regained, and its retention is therefore paramount. Even where fragments of ancient woodland are enclosed within commercial units it should be possible to exclude ploughing, drainage, fertilising and unsympathetic planting. This practice is assisted by the Forestry Commission's network of Forest Nature Reserves (FNRs). Grant aid is available through the Woodland Improvement Grant, not only for the regeneration of woodlands, but also for the specific promotion of biodiversity (with additional 'Challenge' funds for target areas).

Peterken *et al.* (1992) have also suggested that, in addition to conventional strategies of retaining and creating habitats on unplanted land, benefit could be obtained from assigning 15–25% of the plantations to long rotations containing small permanently uncut cores, while shortening the rotations of the plantations not assigned to long rotations. This should allow significant 'old-growth' habitats to be created, and increase the extent of temporary open space, apparently without a disproportionate sacrifice of wood production. Variations in the detail of tree management are recommended according to location and species type, but the general practice would be for long rotation stands to be located on the lower slopes of valleys, because:

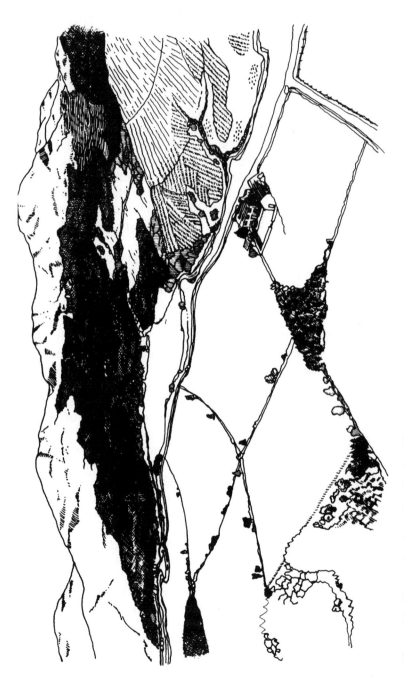

Figure 5.1 Forest planting related to land form
Source: Forestry Commission, undated

- these offer the best chance of protection from wind, which would enable trees to stand for 100 years or more;
- soils are mostly mineral, deep and fertile by forestry standards, thus offering firmer rooting and better stand growth, and providing better prospects for developing a broadleaved admixture;
- the locations are typically more humid, which will enhance the woodland interior microclimate; and
- there may be nature conservation benefits in having short rotation stands at the higher elevations, where the clearfelled compartments can be linked to unplanted moorland within the forest, and to moorland above the forest.

Both visual and nature conservation considerations point to a broad pattern in which the low ground is devoted to long rotation plantations, with a substantial broadleaf component, and managed by a small-scale patchwork of felling and regeneration associated with a small-scale matrix of complementary habitats. The high ground would then be devoted to short rotation plantations, with broadleaves mainly associated with the crags and moorland fragments; there would be a large-scale patchwork of felling and regeneration, and a felling sequence designed to maintain open space linked with moorland. However, general design principles may have to be modified locally to take account of additional factors, such as the close proximity of farms which would be a source of disturbances to the forest. The key fauna of old growth often needs remoteness, which implies that eventually old-growth retentions should be distant from recreational centres, contain few or no trails, rides or roads, and utilize physical barriers such as cliffs, rivers and wetlands as boundaries.

A good deal of controversy has surrounded the gains and losses to bird populations caused by the planting of bare ground. In general, there is considerable evidence of higher density of bird pairs compared to pre-afforestation, but these are mainly the commoner species, although both the crossbill and siskin (conifer-seed eaters) have greatly extended their range as a result of softwood cultivation. However, the losses of a smaller range of rarer moorland species might outweigh these benefits. On the initial *enclosure of land* the removal of grazing pressure, combined with draining and short-term nutrient replenishment, causes a flush in plant growth, mostly associated with grasses and heather sustaining large insect populations and increased numbers of small rodents. Thus, high densities of small insectivorous birds (e.g. meadow pipit) thrive, as do predatory birds (e.g. short-eared owls and hen harriers), whereas wading birds are adversely affected by the depth of vegetation. Of these latter, only curlew appear to persist, and serious losses may occur of greenshank, golden plover and dunlin. Later, as the *trees grow*, predatory birds fail to find food or nesting shelter and leave. The development of scrubby conditions favours willow warbler, reed bunting and redpoll, but the open-ground species and those which used the emerging trees as songposts (e.g. stonechat, whinchat) depart. On the *development of a dense thicket*, the structure of the bird community is drastically altered, favouring canopy feeders such as goldcrest, chaffinch and coal tit. When *brashing* of the lower dead branches occurs this may, if they are not removed, provide valuable foraging surfaces and nesting cover for robin, wren and dunnock. By the end of the polestage and *before closure*, the forest would be expected to hold a number of the species already mentioned, together with treecreeper, woodcock, siskin and sparrowhawk. Forest structure will affect species densities and numbers: especially in deciduous forests, with a multilayered structure, canopy, middle-layer and ground-nesting birds can be represented. Hole-nesting and shrub-nesting birds may be limited by the availability of nest sites, especially in managed forests where dead timber and scrub are often removed (Bibby, 1987).

One of the principal 'wider countryside' benefits of forests is that of recreation. Visitors on foot are normally permitted access to all the Forestry Commision's forests provided this does not conflict with management and protection of the forest and provided there are no legal agreements which would be infringed by unrestricted public access. For instance, forest paths may be closed if logging operations are underway which pose a threat to public safety. Access is, however, more generally restricted in many private forests as owners fear their liability for injury to members of the public, although the private sector claims increasingly to honour a wide range of amenity and recreation obligations. Support for private foresters is available through the Woodland Improvement Grant to facilitate public access. In Forestry Commission forests the main emphasis is on the provision of facilities for day visitors, notably car parks, picnic places, viewpoints, forest walks and visitor centres. These are usually free, although some charge may be made for car parking and publications, whilst a few of the Commission's vehicular tracks may be used additionally as 'forest drives', and a toll is generally charged. Holiday chalets, camping and caravan sites, may also be provided on a profit-making basis. Some sporting activities are considered to be compatible with the forest environment and these include fishing, shooting, horse-riding and orienteering. The value of recreational benefits from forestry has been estimated by economists as a 'consumer surplus', and this is very substantial (e.g. Benson and Willis, 1992). As a reflection of this, the Forestry Commission receives annually, as part of its grant-in-aid, a recreation and amenity subsidy.

Since forests are amongst the main regulators in catchment systems, it is to be anticipated that they would have major hydrological effects, both on water quantity and quality. Water supply authorities, in particular, have noted that in humid windy climates a tall tree canopy cover will increase evaporation, and thus lead to a reduction in available water. Afforestation practices typically required in wet, cool and windy (i.e. British upland) conditions have yielded increased runoff, discoloration, a tendency to acidification and, over longer timescales, increased sediment supply to channels. Impact on sediment load, and thus on fish spawning and nursery areas, may be severe: whilst much of this may be restricted to the early stages of ploughing, insertion of drains and road construction, etc., other impacts are long term. In order to mitigate hydrological impacts, the forest industry tends to employ either 'accommodation' or 'allocation' measures, the former modifying design and the latter creating no-go areas where planting would not be allowed (Newson, 1991). Not surprisingly, the Forestry Commission prefers to rely on accommodating hydrological considerations within existing plantations and a great deal of faith has been pinned on its *riparian policy*, which emphasises the use of buffer zones alongside watercourses. It is argued that serious soil losses should not occur if cut-off drains are sufficiently frequent and aligned so as to prevent erosion. All drains are then designed to stop short of watercourses so that the water they discharge is filtered on its passage across the undrained area – a practice which also obviates the need for drainage machinery to come right up to stream banks and possibly cause their collapse. Planting should not extend right up to stream edges for a variety of reasons, but of particular importance is the potential effect of excessive shading by conifers, largely eliminating ground cover and leading to accelerated bank erosion. Some planting of smaller deciduous trees is, however, desirable to provide nutrient input and dapple shade. Valley topography, soil and aesthetic factors help define the area and shape of the buffer strip, but there is some agreement that it should be no less than 5 m wide and preferably 2–3 times as wide as the stream bed. The policy of varying ploughing contours for aesthetic reasons, or interrupting them on fragile areas, also diminishes risk from erosion.

Landscape and Nature

We have previously noted a variety of mechanisms for protecting wildlife and scenic resources, some of which were very site specific and some of which cover large areas (Box 5.2). In densely settled countries, the task of conservation is largely that of protecting *cultural* landscapes; these, rather than being pristine wildernesses, contain a long history of human usage conducted at an intensity sympathetic to the maintenance of interesting features. Various general approaches have been suggested for the management of cultural landscapes, typically involving the enrolment of the indigenous population.

Conservation practice has developed incrementally in Europe over the past 50 years or so, and has resulted in a wide variety of protected area systems. However, these have not achieved all their policy aims, and there has been an undoubted loss of biodiversity and landscape quality. In relation to the UK, the principal limitations of protected area approaches have been identified as:

Box 5.2 Mechanisms for protecting wildlife and scenic resources

- World Heritage Sites (global)
- Ramsar Sites (global)
- Biosphere Reserves (global)
- Biogenetic Reserves (Council of Europe)

Protected areas created under EU legislation:
- Special Protection Areas
- Special Areas of Conservation
- Environmentally Sensitive Areas
- Nitrate Sensitive Areas

Protected areas found throughout the UK
- Forest Park
- Forest Nature Reserve
- National Nature Reserve
- Marine Nature Reserve
- Local Nature Reserve

Protected areas found at the country level:
- Regional Park (S)
- Natural Heritage Area (S)
- National Scenic Area (S)
- Marine Conservation Area (S)
- Area of Special Scientific Interest (NI)
- Site of Special Scientific Interest (E, S, W)
- area of special protection (E, S, W)
- National Park (E, W)
- Heritage Coast (E, W)
- Area of Outstanding Natural Beauty (NI)
- AONB (E, W)

S = Scotland; NI = Northern Ireland; E = England; W = Wales.

Source: Bishop *et al.*, 1995

- the separation of nature conservation from landscape protection and enjoyment;
- the evolution of partially different administrative and legal arrangements in each part of the UK;
- the separation of nature conservation at sea from that on land;
- the development of certain types of protected area types within the system of town and country planning, and development of a different set of protected areas by conservation agencies;
- the creation of protected areas to encourage environmentally sensitive management of farmland; and
- the increasing number of protected area types established under European and international agreements (Bishop *et al.*, 1995; 1997).

Not surprisingly, this situation displays a range of strengths and weaknesses (Box 5.3), so that protected areas may be viewed as a necessary, but not a sufficient, aspect of conservation policy. For example, the practice of nature conservation has been evolving away from the protection of species towards the protection of their habitats and, indeed, away from the protection of species and habitats towards conservation strategies which are based on the protection of the natural processes upon which they depend. Moreover, there is a recognition of the increasing need to replace a 'self-contained' policy approach to nature conservation with an integrated approach, so that it forms part of the planning and management of the terrestrial and marine environment as a whole, and is placed into each economic sector. The narrow, reductionist principles on which scientific arguments for conservation have been based are also starting to be supplanted by a recognition of the importance of biodiversity (i.e. ecosystems, species and the variety within species) as a component of sustainable development. There is also less emphasis on 'protection' and more on creative conservation, both to restore lost features and to create new ones. Similar trends in thinking can be detected in the area of landscape conservation, where there is a retreat from an almost exclusive concern with the protection of the 'best' towards an interest in 1) the diversity of the entire landscape, and 2) local distinctiveness. Reflecting the incrementally blurred distinction between science and aesthetics, there is a drift away from an essentially visual approach towards a deeper appreciation of the ecological, historical and cultural values of landscape and the way in which these are interwoven.

Box 5.3   Strengths and weaknesses of the UK system of protected areas

---

Strengths:

- safeguard places which are outstanding in terms of natural wealth, natural beauty and cultural significance;
- maintain the life-supporting diversity of ecosystems, species, genetic variation and ecological processes;
- protect species and the genetic variation that humans need, especially for food and medicine;
- provide homes for human communities with traditional cultures and knowledge of nature;
- protect landscapes reflecting a history of human interaction with the environment;
- provide for the scientific, educational, recreational and spiritual needs of societies; and
- provide benefits to local and national economies and are models of sustainable development to be applied elsewhere.

Weaknesses:

- the tendency to treat protected areas as 'islands' set apart from the areas around;
- the tendency to see protected areas as an alternative to, rather than one element within, a national strategy for conservation;
- the failure to integrate protected areas requirements into policies for the sectors (e.g. agriculture, tourism, transport) which affect them;
- the inadequate recognition of the needs and interests of local people upon whose support the long-term survival of protected areas will depend; and
- limited public and institutional support for protected areas.

Source: Bishop *et al.*, 1995.

Protected areas, whilst still the mainstay of conservation practice, are clearly coming under increasing scrutiny. It is not yet clear whether superior alternative systems exist, but it is apparent that protected areas alone, especially those which make strong distinctions between nature and landscape conservation purposes, are insufficient to address the landscape ecological issues affecting the wider countryside. For example, Bishop *et al.* (1997) argue that protected areas negate a holistic approach to action: whereas the environment is all-embracing, the practical effect of over-reliance on site protection results in a disaggregated policy response. The 'designation' approach encourages the view that conservation is a sector or a land use, rather than a fundamental principle underlying all sustainable development. Protected area boundaries are also defined by relatively arbitrary lines on maps, which is probably essential for firm planning decisions, but a misrepresentation of the subtle gradations which exist within scenery or ecosystems. Similarly, lines on the ground create a 'boundary effect', leading to artificially separated management regimes, yet in reality the integrity of a protected area cannot be divorced from what goes on in its surroundings. Neither environmental problems – such as, for example, long-distance transport of air pollutants or polluted rivers – nor biological phenomena (for example, animal migrations, and animal and plant eruptions and dispersals) – recognise protected area boundaries. Moreover, the pattern of 'designations' has become excessively complex, having been developed to protect differing environmental resources at different scales, through various agencies, using different mechanisms and employing different names. More speculatively, physical changes such as global warming could make protected areas obsolete, and render it impossible to protect biodiversity within areas, as species struggle to establish new geographical ranges. Nevertheless, whilst conservationists would welcome more flexible and dynamic systems of ecologically based land management, it is difficult to conceive of an administratively workable alternative to site designation which would also ensure the practical defence of critical natural capital against strong development pressures.

An important complement to protected area systems is the current interest in treating these as 'core areas'. In practice, therefore, protected areas will continue well into the future, but it is likely that they will be embraced within spatial strategies which reflect modern theories of landscape ecology. Thus, protected areas can be viewed as 'core patches', to be linked up through ecological networks. A good example of this is the EECONET (European Ecological Network) of The Netherlands and Flanders, which seeks to reverse the fragmentation of habitats into small, isolated islands by establishing and developing a coherent European network of habitats, based on four kinds of action:

- better protection of core areas;
- the development of support zones around these;
- the creation of corridors between these; and
- the restoration of damaged habitats.

In order to effect EECONET it is envisaged that the protection of existing parks and reserves in Europe will be strengthened as the cores of 'nodes' of the network, and that CAP reforms will be used to strengthen the network.

More generally within Europe, the European Community Biodiversity Strategy (CEC, 1998) provides a framework for developing policies and instruments to comply with the Convention on Biological Diversity, in particular the proposed production of action plans. Four key themes pervade this strategy, namely:

1) conservation and sustainable use of biological diversity – *in situ* and *ex situ* conservation (e.g. the Natura 2000 network and gene banks, respectively), and the sustainable use of components of biodiversity (e.g. environmental assessment and ecolabelling);
2) sharing of benefits arising out of the utilisation of genetic resources – multilateral frameworks, bilateral co-operation and technology transfer;
3) research, identification, monitoring and exchange of information – e.g. through the Multi-Annual Work Programme of the European Environmental Agency; and
4) education, training and awareness.

These themes are in turn related to key policy areas, comprising the conservation of natural resources, agriculture, fisheries, regional policies and spatial planning, forests, energy and transport, and development and economic co-operation.

A further reflection of the ways in which nature and landscape conservation are both pursuing increasingly mutual objectives and addressing 'wider countryside' issues is through the process of *characterisation*, involving the production of novel maps of land resources by English Nature and the Countryside Commission. Thus, English Nature has defined a system of *Natural Areas*, to ensure that nature conservation policies now affect the whole countryside (EN, 1998). This is not only a basis for more co-ordinated local action and administration, but provides a framework for ensuring that landscape ecological objectives cover all areas. Natural Areas are zones defined not by administrative/governmental boundaries, but by their wildlife and natural features, and often landscape homogeneity and land use and human history. Each Natural Area is supported by a description, or profile, of its ecological character, and a set of long-term visionary objectives which have resulted from discussions between English Nature staff and local nature conservation bodies. They provide a framework to link local and national priorities, at a scale which helps local decision-makers to understand the wildlife resources in their area, relative to their place within the country as a whole. In many parts of England, Natural Areas have been subdivided in the hope that this will enable people to gain a 'sense of place' within the larger Natural Area (Box 5.4). The Countryside Commission's *Countryside Character* approach is based on a similarly comprehensive analysis and understanding of the character of the

English landscape, and reflects the current view that landscape character exists everywhere in equal measure, even if for certain purposes we value some characteristic landscapes more than others. Countryside Character descriptions are comparable to Natural Area profiles, and provide summary information, review past and recent changes to the landscape, and identify the main forces for such change. For example, the Countryside Character approach is intended to furnish a backdrop against which development planning and development control can be exercised (Box 5.5). Thus, the method should help planners devise how best to enhance and respect local distinctiveness and, in this regard, performs a comparable function to Natural Areas (CC and EN, 1996).

An integrated map of these two approaches – the 'Character of England: landscape, wildlife and natural features' – has been assembled as a means of defining natural heritage in a single framework. The framework is intended to assist local authorities and area officers by forming a bridge

Box 5.4 The North Northumberland Coastal Plain (an example of a Natural Area description and its objectives)

The North Northumberland Coastal Plan Natural Area comprises the belt of low-lying land running northwards from the Coquet Valley to the River Tweed, and westwards to the Fell Sandstone moorland edge. The area is overlain by glacial till and, more locally, blown sand or peaty deposits. The Natural Area is characterised by a diverse coastline emcompassing the high sandstone cliffs north of Berwick, the low-lying whinstone and limestone cliffs south of Bamburgh, and the extensive dune complexes found from Lindisfarne to Amble. Inland the area is characterised by an open agricultural landscape with Whin Sill outcrops and the river valleys of the Coquet, Aln and Tweed. Broadleaved woodlands are now largely confined to these river valleys.

Nature conservation objectives are thus to:

1) maintain all semi-natural habitats in the Natural Area and to enhance the most important and characteristic types such as dunes, intertidal flats, rocky coasts, estuaries, fens, Whin grassland/heath, rivers and woodlands;

2) maintain and expand the populations of internationally and nationally important species which are characteristic of the Natural Area;

3) enhance the wildlife potential of the wider countryside throughout the Natural Area and to replace habitats lost, degraded or reduced and achieve Local Biodiversity Action Plan targets;

4) enhance populations of nationally declining species characteristic of lowland farmland habitats within the Natural Area, particular emphasis to be given to those areas which still maintain relatively good populations of such species;

5) ensure that the geological and geomorphological features of the Natural Area are maintained for future research and education;

6) establish a framework of adequate legislation, liaison, influence, information and understanding through which objectives 1–5 can be achieved.

Source: CC/EN, undated.

Box 5.5 The North Northumberland Coastal Plain (an example of Countryside Character)

The North Northumberland Coast Plain comprises a long linear plain which fringes the North Sea. The low-lying area rises inland to the Cheviot Fringe, Northumberland Sandstone Hills and Mid-Northumberland. The southern boundary is formed by the South East Northumberland Coastal Plain.

The complex underlying geology of volcanic Whin Sill outcrops and sedimentary rocks including shales, sandstones, coal seams and limestone measures gives rise to an exceptionally varied coastline. Inland a thick fertile blanket of boulder clay has resulted in an open agricultural landscape. The Middle Coal Measures found in the southern part of the plain were previously important for their rich mineral wealth and were exploited as part of the Northumberland coalfield.

The dramatic and varied coastal landscape of cliffs, long empty golden beaches, rocky shores, mudflats and saltmarsh, is backed by open windswept farmland and intimate river valleys. Agricultural use of the coastal plain is generally mixed arable on heavy loam soils, in combination with beef cattle and sheep grazing. Fields are generally large, bounded by low-cut and often fragmented hedges, and fences. Hedgerow trees are scarce and woodland cover is low, except for mixed coniferous shelterbelts adjacent to farmsteads and remnant broadleaved woodland in river valleys. The River Tweed acts as a natural border between England and Scotland, and has a long history of medieval crossborder warfare. The prominent line of strategic coastal defences, medieval castles, fortifications, religious buildings and the town of Berwick upon Tweed contributes to this rich heritage. Fishing villages and coastal towns with well established caravan and chalet sites provide traditional seaside attractions for tourists. Elsewhere, the dispersed settlement tends to consist of small villages and isolated farm hamlets. The A1 road and the east coast main line railway, which run roughly parallel to the coast, are prominent in the landscape.

The dramatic and diverse coastal scenery is a prominent feature of the area. To the north of Berwick the coastline is characterised by the spectacular high sandstone cliffs, while whinstone and limestone cliffs can be found south of Bamburgh. Offshore islands, rocky headlands, raised beaches and wide sandy bays backed by sand dunes, intertidal saltmarsh and mudflats also occur. The coastline provides an important habitat for a large number of sea birds including eider. The cliffs with patches of heather, sea thrift, hairy thyme and spring squill attract puffins, kittiwake, guillemot and several species of tern. The sand dune complexes with lowland heath, scrub, grassland and mires harbour a number of rare species such as the narrow-lipped helleborine, variegated horsetail and ringed plover.

The Holy Island of Lindisfarne, linked by a tidal causeway to the mainland, is a prominent feature that retains a remote and spiritual quality. This island, together with the Farne Islands, contains outstanding historical features and provides important habitats for nesting seabirds and wintering waterfowl.

Mining and quarrying of rich mineral deposits have greatly affected both the past and present landscapes and semi-natural habitats. In the past coal mining provided a valuable source of revenue. This has now been superseded by the quarrying of limestone and Whin Sill.

Numerous rivers and watercourses meander across the coastal plain to the North Sea. Small remnants of broadleaved ash, oak and alder woodland, often with pied flycatchers, occur along the Aln and Coquet River Valleys, while the other water courses are generally open in character. The rivers are

especially significant for their water crowfoot beds, otters, salmon, freshwater
crayfish, mayflies, stonefly and caddis fly populations.

Whin Sill outcrops and glacial landforms and deposits provide dramatic
landscape features that are also of considerable geological interest. The outcrops are particularly important for dry heath and grassland communities that
support species such as maiden pink, hairy stonecrop and seaside centaury.

Small areas of base-rich fern characterised by bulrushes and grass of Par-
nassus provide an important habitat for invertebrates.

Source: CC/EN, undated.

between national policies and programmes and local interpretations of
needs and demands. In this way the map will help deliver national consistency by serving as a starting point for developing more informed policies
to shape the landscape, wildlife and natural features of the future. Examples of how this joint approach will be used include: identifying where
and how agrienvironmental schemes might best be extended and tailored
to specific parts of the country; and helping in decisions about the siting
and composition of major new woodland in ways which respect local
character.

A further approach to countryside planning, influenced by landscape
ecological principles, is that of gap analysis (Scott and Csuti, 1997). This
uses remote sensing and GIS to assist in the assessment of the status and
distribution of several elements of biodiversity. Although not a substitute
for traditional biological surveys, it can provide a preliminary, landscape-
scale assessment of the distribution of both species and ecosystem diversity
that can be used to guide future field research and to provide a spatial
framework for a preliminary national biodiversity conservation strategy.
Burley (1988) identified four steps in gap analysis:

1) identify and classify biodiversity;
2) locate areas managed primarily for biodiversity;
3) identify biodiversity that is un- or under-represented in those managed
   areas; and
4) set priorities for action.

Gap analysis requires three primary GIS data layers: vegetation cover,
species range maps, and the location of land managed primarily for native
species and natural ecosystem processes. Where available, information on
other environmental factors – such as elevation, slope, aspect, soils, aquatic
features and climate – can be used to improve the accuracy of maps of
vegetation and species distributions. Additional information on socioeconomic attributes of landscapes can be examined to refine approaches
to land-use planning. Gap analysis assumes that plant communities serve as
integrators of many physical factors that interact at a site and that floristic
composition provides a common denominator for the description of plant
communities. Although there are many small-scale factors which the mac-
roscopic approach of gap analysis fails adequately to consider, it is assumed

that species maps based on wildlife–habitat relationships can best be used to predict the occurrence of species across general vegetation types and environmental conditions (Morrison *et al.*, 1992). Maps of species distribution generated for gap analysis are intended to be used and validated at 'landscape scales'. Gap analysis may thus be thought of as a coarse filter approach to conservation strategy – it identifies areas in which selected elements of biodiversity are represented; once those areas are identified, other principles of conservation biology, such as population viability analysis, ecosystem patch dynamics and habitat quality can subsequently be used to select specific sites and determine appropriate management area boundaries.

The increasing emphasis on biodiversity has latterly been a major influence on conservation planning. In particular, 'action plans' have been prepared to arrest the loss of biological diversity from the countryside as a whole. In some ways, this appears to revert to a reductionist, habitat and species-specific approach; in other ways, though, it affirms the importance of all biological diversity wherever it may be found. Within the UK, for each key habitat (listed in 'Biodiversity', UK Steering Group, 1995, Volume 2), a steering group has given strategic focus to implementation; groups were formed from relevant sectors, ranging from government departments to land managers, and undertook national evaluations of the resource, including areas under greatest pressure and opportunities for conservation action. Overall, a *three-step process* has been established for biodiversity planning:

1) review of species and habitats – species audit, habitat audit, potential for reintroduction or re-creation, additional data (require data/expert opinion on extent, quality and national significance);
2) evaluate and prioritise (UK priority species, significance of local resource in national context, opportunities to enhance the local resource, local decline rates/rarity/threats/fragmentation/importance for key species/distinctiveness); and
3) set local targets (realistic but ambitious, same measurable parameters as in the national targets, set against clear timescales with milestones).

These national plans are being cascaded into Local BAPs, organised on a partnership basis at county level, so that national plans can be interpreted by local professionals and volunteers to implement action on the ground.

Planning Catchments and Rivers

The quantity, quality and distribution of water resources have become a major concern in recent years. As a generalisation, it could be argued that we are moving away from an engineering solution to water use, in which water was impounded and supplied on demand in ever-increasing quantities to end-users ('predict and provide'), to one in which water is treated as a valuable, multipurpose resource requiring comprehensive planning and the management of demand. More generally, in addition to specific approaches to water management, the focus on catchment planning opens up further possibilities for integrated countryside action.

A starting point for understanding water habitat planning is to consider the major types of freshwater habitat found in Britain, namely:

- the fast-flowing upper sections of rivers, often associated with areas of high precipitation; especially in the north and west of Britain, peat may form above the water table leading to raised and blanket peat bogs;
- lowland areas with slower-flowing rivers often fed by streams draining soft sedimentary rocks, generally leading to nutrient-rich conditions (sometimes including fen and swamp); and
- sites of industrial or agricultural origin, such as canals, flooded mineral workings, subsidence flashes and farm ponds.

Conditions in the first of these categories may require little active management, and the main conservation objective is often to safeguard extensive areas (large enough to retain overall hydrological systems intact) from forestry, drainage and commercial peat cutting. In the lowlands conservation objectives tend to be more complex. Often, these habitats display a spatial zonation of different plant communities, comprising free-floating plants in the open water, floating plants in the shallows nearer the bank and emergent aquatics such as reed mace on the bank. Traditional management of wetlands and poorly drained meadows often involved the maintenance of a complex and delicately balanced system of dykes to control winter and summer water levels and these have often been in conflict with the needs of modern agriculture or have fallen into disrepair (RSPB, 1994; Lewis and Williams, 1984; Environment Agency, undated).

More extensive losses to conservation have arisen through the straightening and clearing of rivers and streams by drainage engineers and farmers, who have often assumed that watercourses must be reduced virtually to culverts to provide efficient flow conditions. Fortunately, it is now acknowledged that careful design frequently enables the retention of many of a river's wildlife and landscape features without seriously adversely affecting its drainage functions. Two factors influence the desired hydrological properties of rivers – bankside vegetation and channel form. Whilst the tendency has been to make both as uniform as possible, this is highly undesirable from an amenity standpoint and is also increasingly incompatible with integrated land drainage objectives. It is recognised that bank vegetation may be helpful in preventing scour and protecting bed and banks, but may also increase flood risks, reduce channel capacity and contribute to channel obstructions through jams or fallen trees. Traditional methods of bank management used vegetation positively, to stabilise banks and deflect flows, but regular manual maintenance was necessary and the skills have largely been lost. Nevertheless, traditional methods can be adapted to current requirements, and a diversity of materials and techniques can be used for bankside protection including faggots, hurdles and the planting of alder, willow or reeds. These are not all universally appropriate but may be selected for particular circumstances and may be combined with modern materials such as gabions, revetments and geotextiles.

The starting point for bankside management is to undertake a survey of physical and vegetative features present in each reach, plants which may disappear as a result of river works and those plants which may be suitable for replanting during restoration. Although some suppression of excess bankside vegetation may be feasible by shade control (i.e. planting trees to restrict sunlight penetration), removal of vegetation and realignment of banks must often be undertaken mechanically. Here, the key conservation objective must be to select stretches for clearance on a three to four-year rotational basis rather than clearing the whole channel at once; riverbanks can then recover naturally after disturbance, having a reservoir of plants and animals available for recolonisation. Where engineering works affect the river bed, channel design should aim to retain variation in depth, substrate and velocity, as smooth, uniform bed profiles tend to be associated with low species diversity. If a channel is to be regraded, special features such as steep banks, pools, riffles and islands – as well as the hydraulic mechanisms which led to their original creation – should be retained wherever possible, and various channel design options are available which combine flood control capabilities with conservation requirements.

Large lateral variations in habitats are found in natural river systems, ranging from threadlike, braided gravel-bed highland channels to broad periodically inundated river floodplains with extensive backwater systems. These each support characteristic species, and even provide special conditions for important life-cycle functions; the continuum of habitat types is summarised in Box 5.6. Channel form is influenced not only by management and design factors, but also by land-use planning, and so it is now normal for water and planning authorities to liaise over matters such as:

- urban expansion and drainage;
- urban expansion on the floodplain;
- waste disposal;
- use of natural flood areas;
- use of retention ponds for runoff; and
- encouragement of groundwater recharge by land-use controls.

Rivers are dynamic systems, continually modifying their form and creating new habitats. However, their ability to rejuvenate themselves and make these responses has been hampered by channelisation (e.g. resectioning and the construction of embankments); and such activities have resulted in changes in the frequency and magnitude of flooding (altering seasonal patterns of flows and hydrograph form) as well as natural patterns of sediment transport and nutrient exchange. Reduction in the link between the river and its alluvial zone also affects the relationship between surface

Box 5.6 Channel control options combining flood control with nature conservation and recreational opportunities

Floodbanks These are raised embankments designed to retain floods. Their height depends on the distance they are set back from the riverbank. In general, least damage to wildlife occurs if they are set back from the river channel (although more agricultural land is sacrificed) and if constructed from spoil imported on to the site rather than material dredged from the river.

Flood storage areas This increasingly popular technique creates a washland area; in the most ambitious schemes this has incorporated multiple use (e.g. a country park, nature reserve). However, it does alter conditions downstream, so that riverside wetlands are flooded less often.

Flood by-pass channels Here the original channel is retained untouched, and continues to take normal flows and small flood discharges. This is good for conservation and game fishing as it enables habitats to remain undisturbed for long periods. A new channel is created, but it must be maintained carefully so that colonisation and silt deposition do not occur.

Multi-stage channels These involve cutting berms at progressive heights to accommodate high flows; however, it leads to a loss of bankside vegetation and so should preferably apply only to one bank.

Partial dredging If dredging does prove necessary, it should preferably only take place over two-thirds or less of the width; this will retain the variety of conditions (i.e. a part of both edge and middle communities remains undisturbed).

Source: Derived from information in RSPB, 1994; Lewis and Williams, 1984; Environment Agency, undated.

water and groundwater in the aquifer underlying the floodplain. Despite the historic impacts on river channels, there is increasing recognition of the role of fluvial processes in creating sustainable river habitats, so that river managers now acknowledge two principal approaches to stream rehabilitation. The first is passive, in which disturbance is reduced and the stream is left to readjust naturally; in some cases, the morphology of some artificially widened or deepened channels has been observed to adjust without intervention. The second involves specific rehabilitation measures, dependent on the intrinsic 'energy' of the river, so that the lower the stream power, the more necessary will be active physical rehabilitation of morphological features.

The water's edge is an inherently attractive recreational venue, and serious visitor pressures may impact upon the lakeside or riverbank, where visitors are drawn by the venue's inherent attractiveness but where substrates are often poorly drained and easily eroded. Access to lake shores is often impeded by woodland or shrubs, so that recreation tends to be heavily concentrated on sites which have car parks, picnic sites or other forms of public access. The recreational attractiveness of a lakeside site (Table 5.2) is influenced particularly by the type of shore, presence or absence of a beach and the nature of shore vegetation. Pressures tend to be concentrated in a fairly narrow zone comprising the shore and adjacent water, and conflict of use can arise from different patterns of movement, namely, those on to the beach, from the beach into the water and linear use of the beach. This creates mutual hazards between users and presents an intense but localised impact on the site itself (Countryside Commission for Scotland, 1986). Liddle (1996) has confirmed that the main effects of recreation at the water's edge are turbulence and turbidity, propellor action (cutting), the direct contact of boats and the eutrophication of waters from sewage release.

Table 5.2 Factors affecting attractiveness of lakeside venues for recreation

User	Main requirement	Foreshore: best type	Backshore: best type
Sitting, picnicking, walking, beach play	open aspect; beach and water should be visible	open shingle beach or low rock outcrops; no vegetation	open grass/ woodland, permitting view of lake
Water-based sporting pursuits	easy access to water; correct offshore topography for specific sport	open shingle beach; no woody vegetation	low backshore; easy access
Wildfowling and fishing	diverse habitat	herbaceous/woody vegetation	variable

Source: Adapted from Countryside Commission for Scotland, 1986.

Six major types of recreational impact occur in littoral zones. Trampling can lead to elimination of ground-cover vegetation and destabilisation of soil, especially on the vulnerable backshore edge, which may collapse and expose tree and shrub roots. Rolling by vehicle wheels may similarly compact soils and reduce the vigour of vegetation. Burning can lead to direct damage, for instance by the use of tree trunks as fireplaces. Digging by children or for fishing bait may lead to undercutting and collapse of the cliff (similar effects arise from animals burrowing). Abrasion caused by the launching of boats can damage emergent vegetation; whilst litter can contribute to unsightliness or pose more serious danger (e.g. broken glass and nylon fishing line). Nonrecreational impacts may also be important, such as wave action, puddling of ground by cattle, and sand and gravel working (sometimes in contravention of planning consents). Wave action is especially damaging where backshore vegetation is absent or retreating. Various types of physical management may be practised (Countryside Commission for Scotland, 1986). Where erosion of the backshore edge is taking place, engineered bank protection may be necessary, making use of various materials: geotextiles, rip-rap (random-laid stone on a graded slope), gabions, walls and revetments are appropriate to different settings and differing degrees of intensity of wave attack. In some of these, tree cuttings may be inserted to soften their appearance, provided they are trimmed to restrict pressure from roots. Protection through revegetation of the backshore (e.g. by alders, reeds) may help speed up recovery of the damaged area, but this is difficult to achieve if root exposure is already occurring.

At the seashore, considerable legal protection against development is available in Britain, where the foreshore is owned mainly by the Crown. Only the beach above the mean high-water mark (in England and Wales) or high-water spring tide level (in Scotland) may be owned and developed subject to planning conditions. The overall situation is complex, though, as both the legislation and the government departments and quangos have varied coverage. The offshore zone is technically the responsibility of the Crown Estates Commissioners and the marine and fishing interests, headed up by the MAFF for fishing and pollution control and the DTI for offshore transportation generally. There are no formal planning controls over activities in the offshore zone, though planning devices such as environmental impact assessment apply for activities covering dredging and oil/gas installations.

Traditional approaches to coastal protection have, not surprisingly, relied on acquisition and designation of linear areas. Thus, much coastline in England and Wales has been protected by the National Trust's *Enterprise Neptune* programme of acquisitions, whilst in 'links' areas some inadvertent protection is afforded by golf courses, which originated on the close-grazed turf of dune systems. Slow progress continues to be made on marine nature reserves (e.g. around Lundy Island) and marine parks (e.g. Isles of Scilly). The most extensive form of coastal designation is the Heritage Coast, covering about a third of the shoreline of England and Wales. Principal planning objectives are to conserve the quality of the coastal scenery, to provide careful management and enhancement where applicable and to enhance and facilitate recreation and enjoyment of the coastline. As well as pursuing rigorous development control, local authorities are encouraged to prepare Heritage Coast Management Plans as a mechanism for guiding countryside management projects, and promotion of compatible recreation and tourism projects. A Heritage Coast Forum promotes these objectives, and acts as a focus and liaison point for organisations.

In terms of environmental systems, the coastal zone represents an area of dynamic transition where land and sea interact and which includes both landward margins and inshore waters. Although it has no precise definition, it includes:

- the adjacent land, including both developed and undeveloped areas;
- estuaries, tidal inlets and the intertidal zone; and
- inshore marine zones.

Given the complexity of this environment, land-use planning alone will be insufficient, and so must be complemented by a process of coastal zone management (CZM) which brings together all those involved in the development, management and use of the coast within a framework which facilitates the integration of their interests and responsibilities to achieve common objectives. It has not been deemed practicable to propose a single framework for CZM, as the environments and administrative situations are so varied, but various 'good practice guidelines' have been commended (Box 5.7).

Given the dynamics of the shoreline, coastal management is increasingly influenced by process models, which reflect sediment transport and changing sea levels. Coastal geomorphologists have long recognised that coarse sediment movement is largely restricted to specific areas, and only on extreme conditions does it move outside these areas (French, 1997). Work by the Hydraulics Research Institute has demonstrated how the coastline can be split into a number of units termed *coastal cells*, defined principally by coarse sediment movement, and these can be used as a basis for coastal management. A cell boundary will occur where longshore transport is zero, and this often occurs at headlands or across estuaries, though their delineation is relatively fluid and thus maps of their distribution cannot be treated as definitive: under extreme wave conditions coarse sediment can cross these, and the pattern of light sediment transport (silts and clays) is even more diffuse. Whilst these do not cause major problems for CZM, managers must recognise that their plan boundaries are permeable.

According to the 1995 report of the Intergovernmental Panel on Climate Change (IPCC, 1996), sea levels around the world have risen by between 10 and 25 cm over the past 100 years, and it is plausible that 2–7 cm of this is due to thermal expansion of the oceans linked to increased global temperatures (O'Riordan and Ward, 1997). Some 2–5 cm of this may be caused by ice melt, though the precise role of the major ice sheets in Greenland and the Antarctic remains a mystery. The IPCC also report circumstantial evidence of more extreme weather patterns, including an increased likelihood of coastal storm surge events. Coming on top of modest but measurable sea-level rises, associated with already established climate change, this combination suggests that shorelines generally will be subject to greater threat of flooding, erosion and accretion of sediment in the future. The Climate Change Impacts Review Group (1996) estimated that, for the UK, sea levels may rise by 19 cm over the average for 1961–90 by the 2020s, by 37 cm by the 2050s, and by 42–63 cm by the 2090s. The IPCC (1992) recommended three complementary approaches to shoreline policy:

1) *managed retreat* by progressively abandoning land and protective structures, and by creating or restoring appropriate alternative habitats or landscapes to replace especially prized areas that will be lost as a consequence;
2) *accommodate* by continuing occupancy and by applying a selection of adaptive measures including adjusting to periodic flooding by raising the level of buildings and access routes, or accepting temporary inundation; and
3. *protect, or hold the line*, by providing robust and reliable defences.

Box 5.7   Good practice guidelines for CZM and a framework for the integration of management plans

The 'best practice' manual produced by Nicholas Pearson Associates (1996) for the former DoE proposes the following key points:

- seek sustainable use and development of the coastal zone;
- work with natural processes, aiming to conserve biodiversity;
- take a long-term, strategic approach to guide activities at the local level;
- encompass both land and sea;
- integrate plans, strategies and activities, whilst respecting individual responsibilities;
- establish a clear vision which identifies genuine issues and real needs for action;
- maximise the benefits of the work, experience and connections of existing initiatives;
- promote consensus through early and comprehensive participation and partnership;
- be democratic and open in decision-making;
- choose simple working arrangements and communicate clearly throughout;
- provide reliable and necessary information, available to all partners;
- achieve action through an attainable timetable supported by clear responsibilities and adequate capital and revenue funding; and
- set up monitoring of agreed parameters to measure success from the start.

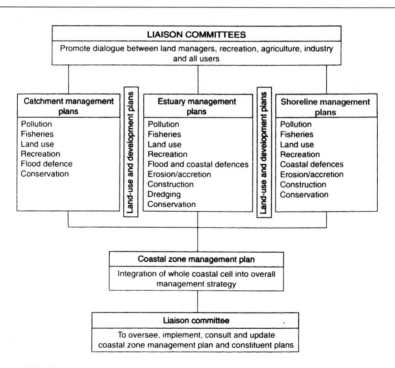

Figure 5.2   Framework for the integration of various management plans into an overall coastal strategy
Source: French, 1997, p. 208

Of these, the concept of 'managed retreat' is perhaps the most pressing in terms of planning priorities. Its appeal to conservationists lies in the fact that, without such a strategy, crucial habitats will be lost in a 'coastal squeeze', as encroaching tidal limits press ever more closely against landward defences. However, farmers in particular are concerned that the implications of breaching sea walls have been given insufficient consideration.

Recreation Ecology

A major concern of recreational managers has been to moderate the effect of visitors on sensitive rural sites. Whilst scenically attractive and ecologically valuable sites tend to attract visitors, many of these venues are fragile and are vulnerable to damage by recreational use. Countryside managers have, therefore, sought mechanisms through which potential conflicts can be anticipated and reduced. Thus, a popular concept since the 1960s has been that of recreational 'carrying capacity': this is loosely comparable to the notion of ecological carrying capacity (which defines the maximum population of a species supportable within a given habitat), as it reflects the maximum levels of use sustainable within a particular location before intrinsic site qualities start to deteriorate. Recreational carrying capacity has, though, proved a very difficult principle to implement in practice, as it is much less easy to define than ecological limits (though these are far from straightforward). Typically, it has been customary to refer to four types of site capacity (e.g. Goldsmith, 1983), namely:

- ecological capacity, i.e. the level at which unacceptable change starts to occur in floristic composition, soil structure and bird and animal populations;
- physical capacity, i.e. the point at which site facilities (such as car parks, visitor centres) or access routes become congested;
- perceptual or social capacity, i.e. the point at which the recreational experience starts to deteriorate; this varies according to the ability of different types of environment to absorb visitors (e.g. open heathland is low whilst forest is high), and the psychology of the visitor (e.g. solitary or gregarious); and
- economic capacity, i.e. the threshold beyond which the economic returns of the enterprises are diminished (e.g. by ramblers crossing a farm).

Such are the difficulties of defining 'carrying capacity', however, that it is more often treated as a useful conceptual framework than a practical management tool.

Many studies in recreational ecology have assumed a very experimental stance. For example, Liddle (1996) has shown how recreational activities have both a horizontal and vertical component of energy in terms of their ecological impacts. Mechanical forces are created by various recreational activities, such as trampling, horses, trail-bikes, four-wheeled vehicles and other machines in which people may be transported over a rough and sometimes fragile terrain. Also, boats (propelled by various means) cause impacts especially in lakes and rivers where the natural waves are normally quite small. The most widely studied impact is that of trampling, whose principal effects are found to be:

- soil compaction;
- bruising of vegetation, leading to a reduction in height and flowering frequency;
- reduction in species diversity (plants unable to adapt to the changed conditions will disappear, and be replaced by resistant species); and
- possible creation of bare ground with subsequent soil erosion.

Characteristically, trampling results in a negative curvilinear relationship with regard to ground cover and biomass. This response is probably because there is usually an initial sharp decline in cover as the more valuable plants are eliminated by trampling, and then a slower attrition of those resistant individuals that are left. There appear to be two genetic strategies for plant survival in trampled situations and these both allow the production of small, often prostrate individuals. One is to have small prostrate forms which are genetically fixed and the other is to have sufficient phenotypic plasticity so that the adult plant can exist and flower as a small reduced individual.

Most interest in site capacity has focused on ecological impacts, though the pursuit of an elusive concept of ecological capacity has been displaced by an acceptance of more general principles of recoverability and vulnerability. Vulnerable sites are generally those with fragile vegetation, thin soils or waterlogged soils. Whilst many sites appear to suffer significant damage from visitor impact, they often recover and thus display resilience. Most past research has been on higher plants, but an increasing amount seems to be on primitive plants (e.g. mosses, lichens) perhaps reflecting the growth of 'adventure tourism' in fragile/cold/montane terrains. Stocking (1994) has expanded on the concepts of resilience and sensitivity, noting that the former reflects the capability of a land system to absorb and utilise change (i.e. resistance to a shock), whilst the latter reveals the degree to which a land system undergoes change due to natural forces, following human interference (Table 5.3).

A starting point from which to consider impacts on recreational ecology would be to imagine a hypothetical untrampled, lightly grazed point. This would tend to be dominated by tall grasses but, with the commencement of light trampling, the tall grass stems and leaves would be bent and broken and the vegetation would become more open. The lower-growing broadleaved species then start to receive more light so that invader species could also join the community. At this stage dominance would switch to the broadleaved species of low-growth forms (e.g. *Plantago*, *Bellis*,

Table 5.3 Resilience and sensitivity (after Stocking, 1994)

		Sensitivity	
		High	Low
Resilience	High	Easy to degrade, but responds well to land management that restores capability	Only suffers degradation under very poor management and persistent mismanagement
	Low	Easy to degrade, unresponsive to land management and should be kept in as natural a condition as possible	Initially resistant to degradation, but after severe misuse land management has great difficulty in restoring capability

Taraxacum), and this dominance would only be lost when the trampling intensity increased and narrow-leaved species were the only survivors. Soil organisms show a general decrease in numbers of individuals and species, reflecting changes in habitat conditions, as well as direct injury and death. Removal of surface layers of litter and soil clearly destroy habitat, and even compaction of litter by light trampling can have a major effect on animal numbers. Compaction, and the consequent hardness of the deeper layers of soil, is also associated with a decline in the numbers of individual species of the deeper-living organisms, probably due to the physical difficulty of burrowing, but also possibly associated with changes in soil moisture. Birds show different sensitivities (some are very easily disturbed) so results are species specific. The most vulnerable sites (which also often display poor recoverability) are those which are at an early stage in the successional sequence, composed of delicate vegetation types and/or are subject to limiting site factors such as highly erodible soils. These generalisations, however, must always be balanced with local considerations of visitor activities and weather conditions. These various terms are included under the general term of 'trampling' which, despite sounding somewhat nontechnical, is a major obsession of researchers concerned with countryside recreational impact (for a wide-ranging review of these issues, see Liddle, 1996).

Recreational managers may not be too concerned about changes in sward composition over part of the area, especially if there are no important nature conservation objectives for that part of the site. It is difficult to ascertain when a critical level of visitor impact has been reached because of variations in the stage at which advance warning 'indicator species' appear. Restoration of sites usually involves fencing, and thus a polite and informative explanation will need to be given to visitors of the purpose for their access restrictions. Improvement of footpath surfaces (e.g. duckboards, geotextiles) may be attempted, but care must be taken to ensure that these do not simply funnel movements to other sensitive areas. Since visitor movements are strongly related to access points from car parks, relocation of these can sometimes provide a solution. Site treatment with topsoil, fertilisers, turves, etc., may also be appropriate as long as management objectives do not seek to maintain existing ecological and edaphic conditions (Figure 5.3). For all such operations, it is essential to have a sympathetic contractor with small and light machinery, and this is not easily guaranteed.

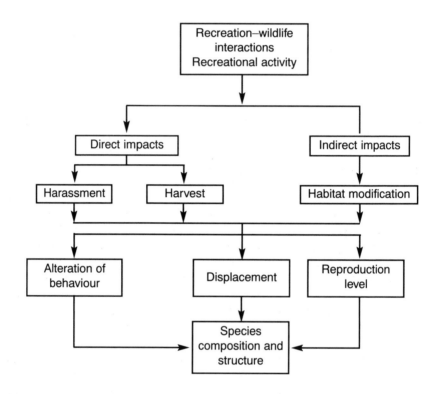

Figure 5.3   Summary of the main effects of recreation–wildlife interactions
Source: Hammitt and Cole, 1987

A key feature of recreation ecology in upland areas is the variability and sensitivity of the environment as influenced by slope, rock and soil type, vegetation, drainage, climate and local topography. Diverse topographic forms are thus often present, and may be overlain by different soils and sediments; soils tend to be thin and infertile, whilst vegetation cover is typically low and sparse. There is in addition an important distinction between the process of trampling by visitors (abrasion) and the subsequent natural processes of erosion, which may be initiated or exacerbated by the effects of abrasion. The severity of physical impacts to vegetation and soil is related to both people and the environment. The former determine the frequency and intensity of visitor use, and the shear stress imparted by various types of footwear, whilst the latter is conditioned by the vegetation community and soil properties (Aitken, 1986). Given the diversity and fragility of the natural environment, and the linked roles of abrasion and erosion, it follows that footpath formation is highly variable and unpredictable except within very broad limits.

Where serious erosion of upland vegetation has occurred, remedial treatment may be attempted and increasing success is now being obtained in the reinstatement of vulnerable habitats. Attempts to restore eroded heather moorland have involved study of experimental plots, in which different regimes of fencing, grazing, seeding and fertilising have been tested. Typical results (cf. Tallis and Yalden, 1983) indicate that:

- if moorland vegetation (especially on peat) is being disturbed through construction works the top 5–10 cm layer should be carefully stripped off, stored and subsequently respread;
- commercial seed sources will usually be acceptable, but regular lime and fertiliser will normally be necessary;
- on a very mobile substrate a nurse crop (e.g. *Agrostis tenuis*) plus lime and fertiliser will normally be necessary to secure initial plant cover;
- it is usually beneficial to exclude grazing at the establishment phase;
- blanket treatment is not sufficient in the context of large areas of heterogeneous moorland, and due account must be taken of critical topography and differing seasonal growth patterns of plants; and
- any treatments can be ruined by chance erratic events, so managers should not be deterred by failure on one occasion.

However, re-establishment of moorland vegetation is generally difficult and a key recommendation would be to avoid damage in the first place if at all possible. A major instance of intensive upland recreational use in Britain, which serves well to illustrate many points of ecological interest, has arisen in connection with downhill skiing in the Scottish Highlands, particularly the Cairngorm massif (Box 5.8). Although it has considerable regional economic importance, skiing has been associated with serious environmental impact. Part of the problem (or opportunity, from the economic point of view) is that skiing comes as part of a package, including tourism promotion, hotels and new roads. It has been argued that the Scottish climate, which is more akin to arctic than alpine conditions, maximises the likelihood of damage to vegetation and minimises the potential for repair or prevention. Wood (1987) notes that similar problems are not encountered in alpine ski developments where ski runs can be bulldozed, drained, graded, sown and hydroseeded to restore a stable vegetation cover suited to snow retention and skiing.

Another environment which is especially vulnerable to recreational impacts is the coast, especially the 'soft' (i.e. readily erodible) coastlines of sand and mudflats, saltmarshes and dunes. Sands and mudflats teem with worms, snails and molluscs and thus may attract huge populations of

Box 5.8   Seven main spheres of skiing impact in upland areas

Construction works, such as excavation of foundations for chairlift pylons, may damage vegetation and soil, as can the use of inappropriate techniques such as bulldozing to create pistes. Above about 900 m, accelerated erosion of bare soil is increasingly likely, and its deposition downhill as a layer of sediment over vegetation can rapidly enlarge the degraded area. Most mountain plants are also unable to grow through more than about 6 cm of sediment. This effect is particularly noticeable along the sides of constructed vehicle tracks, potentially leading to gullying and erosion, whilst infiltration is reduced over the compacted surface of the track.

Vehicles may cause persistent damage to vegetation, and bulldozers have been known to create bare tracks by a single passage. Even piste machines, which have a relatively low ground pressure, can cut up the vegetation and loosen the surface soil, and operators typically agree not to use them when there is under 10 cm of snow.

Skiers themselves may trample and churn vegetation, in addition to the slicing effect of their skis. Damage is most likely to occur following a thaw at the end of the season when snow cover is often incomplete. Vegetation may recover naturally, or be helped by reseeding, and impact may be minimised by directing skiers away from areas of patchy snow cover. Use of ski tows is undesirable when snow cover is thin, and operators sometimes provide artificial ski matting.

Walkers, either participating in skiing or taking advantage of the rapid access to high ground provided by chairlifts, have caused widespread impact. Footpaths act as foci for erosion, especially where taking a direct uphill route. If a path deteriorates badly walkers may keep to the area alongside it, thus spreading the erosion further: detectable, though slight, damage to lichens has been recorded 50 m from heavily used paths.

Alteration of site hydrology may occur through the construction of car parks and other facilities, and the cutting of drainage channels. This leads to faster runoff, although it is unclear whether this is of particular significance. Water quality is of greater concern, being affected by effluent from septic tanks and nutrient-rich leachates from fertilised reseeded areas.

Damage and disturbance to wildlife may occur. Until the late 1970s there was no evidence of disturbance to birds and, indeed, grouse and ptarmigan were reported to have become remarkably tame. Subsequently, however, ptarmigan became locally extinct, and this has been attributed to their colliding with overhead wires along the ski tows and chairlifts. Walkers, and especially their dogs, may disturb birds during the nesting season and dogs may kill both adult birds and chicks. Litter and waste food attract scavenging birds, some of which are also inveterate nest robbers, and this has reduced the breeding success of ptarmigan and dotterel.

Reduction of scenic/landscape quality is caused by ski centre paraphernalia, but careful design and site tidiness can help mitigate these effects.

Source: Highlands and Islands Development Board, 1987.

waders and wildfowl, whilst saltmarshes are important grazings for geese and duck. They are susceptible to disturbance, particularly from bait diggers and to a lesser extent from wildfowlers and water-sports participants. Soft sandy coasts are generally most vulnerable to off-road vehicles, trampling, fires, and camping and caravanning. Adverse impacts on cliffs are most often caused by climbers. The greatest concern has tended to focus on dune systems and their erosion (e.g. Boorman, 1987). Coastal dunes commonly comprise a sequence commencing with the strandline and small foredunes, through white dunes bound by stabilising vegetation (such as lyme and marram grasses), fixed grey dunes (taking their name from their lichen mats), to low-lying dune slacks where the freshwater table may reach the surface, forming open-water and marsh communities. Landward of this is often a pine plantation to stabilise the sandy soils, perhaps dating from the turn of the century. The system is inherently unstable, being a seral stage on primitive soils, and is thus vulnerable to human disturbance. It is also especially sensitive to erosion at particular points in the sequence and at particular times of year (Ranwell and Boar, 1986) (Figure 5.4).

Much effort has gone into describing and quantifying the recreational impact on dunes although it is still difficult to generalise from the various highly empirical individual studies. Damage may arise from vehicles, typically parking and caravanning at the landward edge or the use of motorcycles and four-wheel drive vehicles to cross the dunes. This reduces the height and density of vegetation, decreasing the surface roughness and allowing winds of higher velocity nearer to ground, as well as opening up the area to increased trampling and grazing. Loss of vegetative cover precipitates disruption of the thin humus layer, exposing the sand and promoting wind erosion. This both widens the damaged area and mobilises the sand, smothering vegetation elsewhere and reducing its stabilising effect. Experience at Aberffraw (Anglesey) showed that passage of 200 cars in summer led to reduction in dune grassland of over 50%, with even worse effects in winter; other studies have demonstrated the disproportionately serious impact of the first few vehicle passes. Quinn (1970) found the distribution of visitors at Brittas Bay (Co. Wicklow) to be 380 persons/ha on the beach, 392 on the yellow dunes, 82 on the middle dunes and 32 on the back dunes: whilst absolute numbers will vary greatly between areas, this does confirm the relative popularity of the sandiest and highest points, and the pressure that is likely to be concentrated on the access points to these. Several empirical studies have demonstrated reductions in dune turf height or complete erosion of vegetation with varying levels of human trampling, and the preferred solution is often a surfaced (e.g. duck-boarded) path on heavily used routes.

In some cases, dune recovery may be constrained by geomorphic factors, such as exhaustion of sand supply, but where this is not the case restoration may be achieved by physical and/or vegetative aids. All these make intensive use of labour and materials, and it may be impossible to achieve adequate levels of action on large dune systems. Most emphasis has been on dune planting, especially by marram and sand couch, whose long narrow leaves are resistant to abrasion and moisture loss, and which display unlimited horizontal rhizome growth. Ranwell and Boar (1986) have advocated a variety of planting designs depending on local conditions.

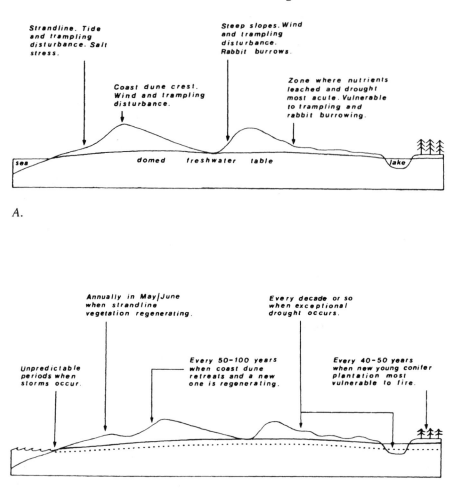

Figure 5.4 (a) Locations in space; (b) phases in time when dunes are especially vulnerable to erosion
Source: Ranwell and Boar, 1986

In general terms, the assessment of recreation sites to sustain tourist or recreation usage requires the collection of pertinent information on the biophysical and socio-economic environments. The former will be governed by the type of site being developed, and the latter includes the current levels of development, infrastructure, visitor facilities and users of the area. Secondly, an assessment should be made of the major projected recreational facilities, so as to match site characteristics to visitor demand. Following this, the current recreational pressure needs to be evaluated in terms of the seasonal population influx and participation rates for recreational activity. Resource demands must then be projected on the basis of estimates of additional visitor numbers and their participation rates in relevant activities. These then have to be converted into associated space requirements, such as parking areas, access roads and specific user facilities. The ecological carrying capacity and acceptable levels of crowding should then be considered. Both are difficult to establish, but ecological capacity may be inferred from the expert opinion of experienced ecologists about the resilience of different systems; acceptability of crowding can be determined from questionnaire surveys or structured interviews. This information can then be used to derive the optimum level and types of recreational use. In practice, it is suggested that the appropriate level will be defined by those constraints which cannot be overcome even through planning and management strategies and which thus impose the most severe limitations for further recreational development (Sowman, 1987).

Although recreation ecologists will generally tend to seek solutions through methods of site management, it must be recognised that much of the basis of success will rest on the subtle control of visitors. Visitor management techniques are often classified into direct and indirect approaches. The former seek to modify visitor use and behaviour by limiting freedom of choice. Management techniques are generally explicit and obvious, but may intrude into the recreational experience and are thus not always appropriate. Indirect approaches, conversely, seek to modify decision factors but leave the final decision to visitors, and are thus much more subtle in their nature. Direct management involves regulation (such as bye-laws) or explicit requests, possibly accompanied by physical barriers to prevent access to sensitive sites or areas undergoing recovery. Indirect management includes the provision of information about the area's recreation opportunities, its uses, its problems and the effects of certain visitor actions. It entails careful design of site installations and access routes (paths, surfaces, gates, location of car parks, etc.). Lucas (1985) further draws attention to the possibilities of 'use limitation' employed in American wilderness areas. For instance, in some areas, camper (i.e. overnight) use is restricted by issuing permits. Redistribution of use depends partly on indirect management, usually supplying information to people in an effort to influence their location choices, although direct management (rationing of permits) may also be included.

An integral agent in site and visitor management is the ranger. Strictly speaking, local authority rangers can only operate on land over which bye-laws can be made, i.e. country parks, certain common land and land subject to access agreements. In practice, however, their sphere of operation is much wider. A ranger's functions can broadly be grouped as:

- enhancement of the visitor's enjoyment of the site by information and interpretative duties;
- protection of sites through their conservation or better management;
- protection of the visitor from known hazards and possibly from the activities of others, both by good management and the provision of information; and

- surveillance, i.e. ensuring that visitors behave in a way sympathetic to the site and to other people's enjoyment of it, and seeing that by-laws are observed.

Also, rangers should maintain good relations with the local community and public agencies, and liaise closely with the police and emergency services. Like project officers, they may also have to execute small-scale landscape improvements, footpath schemes and recreation facilities. Increasingly, both rangers and project officers must become experts in attracting financial backing – typically 'core' funding from public agencies and 'project' funding from commercial sponsors – and their success is commonly measured in terms of the gearing/leverage ratio (i.e. the proportion of commercial to public funds) which they achieve. Many agencies make use of part-time and volunteer rangers, particularly to cope with seasonal and weekend patterns of workload as well as to involve the community (e.g. Bromley, 1990).

Another essential ingredient to visitor management is interpretation (Figure 5.5). This is the art of explaining the character of places to the visiting public so that they become aware of the significance of sites. Interpretation combines a type of informal education with a conservation 'message', aimed at emphasising the inter-relationships within the environment. Through telling a story, interpretation seeks to create an acceptance of the need for conservation generally and, in relation to particular sites, can raise awareness of the need to behave responsibly and reduce the likelihood of unintentional damage. It provides an enjoyable experience in itself but may also inform visitors about other opportunities in the area. Interpretative facilities can increase the number of tourists and perhaps extend their length of stay, and may also encourage visitors into areas not traditionally benefiting from tourist expenditure (Countryside Commission, 1979; Bromley, 1994). Various options are available, including visitor centres and museums (incorporating exhibitions, audio-visual programmes and publications), interpretation boards, leaflets and ranger services (providing talks and guided walks). Selection of a suitable approach depends upon knowledge of the likely visitors, the depth and level of detail appropriate to their needs, and the purposes, design and optimum location of interpretative facilities.

Natural Resource Management Plans

Given the lack of planning control in many 'wider countryside' situations, extensive use has been made of a variety of nonstatutory and semi-statutory land-use plans. Some of these are directed at sites, some at areas such as the urban fringe and uplands, whilst others deal thematically with topics such as biodiversity and forestry. The larger-scale, thematic plans are often prepared on an 'indicative' basis, that is, they indicate planners' views on the appropriateness of certain types of land use, even though there is

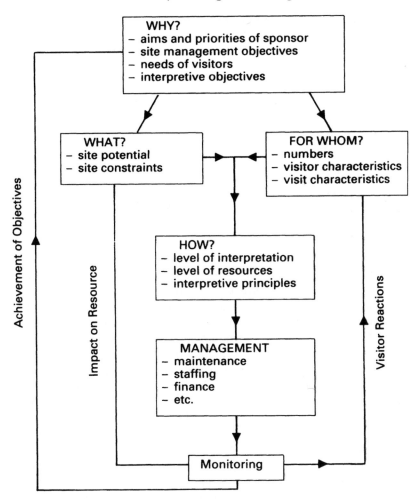

Figure 5.5   The interpretative planning process
Source: Countryside Commission, 1979

little or no statutory power to intervene. This, of course, contrasts mark-edly with development plans which are statutory formalisations of land-use zonations.

At the site level, detailed management plans are often prepared. These are typically used for facilities such as country parks and nature reserves, though the concept originated in commercial forestry. Where amenity is the primary land use of a site it will almost always be the case that management inputs are essential to the retention of those qualities on which site interest depends (e.g. Green, 1996). Ideally, a management plan should be prepared with the intention of assisting various functions, namely:

- helping in decisions about complex problems by clarifying objectives;
- serving as a checklist of actions;
- satisfying public accountability (i.e. demonstrating the ways in which funds are being spent);
- acting as a guide for new staff;
- providing an historical record (of site surveys, past management and experience); and
- serving to link local practice with national policy for reserve acquisition and management.

The typical horizon for such plans is five years. English Nature, for example, proposes a standard format for adoption by conservation agencies, including methods for site and species description, evaluation of nature conservation values, clarification of management objectives and the proper planning and execution of necessary works. Procedures for monitoring are also provided – especially important for nature conservation, where habitats may need to be maintained, created or enhanced, or breeding successes of species assured (Nature Conservancy Council, 1988).

The major division of a site management plan is into the 'descriptive' (survey) and 'prescriptive' (plan) sections. However, five stages may commonly be noted, namely: introduction, description, management objectives, prescriptions and implementation. Typically, the last of these sections comprises proformas setting out a phased sequence of tasks, the financial year in which these are to be undertaken, type of labour to be used, materials involved and costs. The starting points of management plan production are the definition of aims and the collection of survey material. The former entails a broad statement of the policies which are to underlie management of the land, indicating the desired balance between the various land uses and interests. The latter involves making a comprehensive record of what is present on the land and how it is managed, forming a baseline for analysis and the production of objectives. Next an analysis is undertaken of the options for land management, and the inter-relationships between existing and potential land uses. At this stage, potential problems and conflicts are identified and the various interests are weighed against each other. This enables the setting of 'management objectives', which serve as statements on the pursuit of aims for the longer-term and, more specific, short-term topics. A 'management prescription' is then needed to give an overview of future works and the resources necessary to achieve the management objectives. The 'implementation' section furnishes details of an integrated action programme, demonstrating detailed actions, normally tabulated on a yearly basis. Finally, monitoring and review programmes entail retention of a record and an assessment of the success of management achievements.

Similar plans are often produced as a basis for farm conservation, and environmental land management grants are now often only payable after the production of a 'whole-farm conservation plan', to ensure that environmental benefits on one part of a farm are not offset by damage and intensification elsewhere. Ideally, a comprehensive conservation management plan should be produced for a farm, and a number of guidelines are now available (e.g. Tait *et al.*, 1988). The first stage involves a survey and analysis of the farming system, the farm landscape and the extant wildlife habitats. Initially, therefore, there is a need to gain a broad understanding of the farm regime with respect to the implications it might have for conservation. Discussions with the farmer should cover the basic physical conditions (soil quality, growing seasons, etc.); past, present and future data on livestock numbers and arable cropping; and information on rotations and grazing regimes. An appraisal of the overall landscape, as seen by the passer-by and from within by the farmer, should be undertaken, including topography, vegetation, landform, buildings and artifacts. An examination of early maps and documents will help identify forgotten features of the farm. Both the landscape and the wildlife should be evaluated in terms of national, regional and local significance. Following this, consultations are held with the farmer as a prelude to the production of conservation proposals. Finally, feedback is obtained from the farmer on the acceptability and practicality of the proposals before the final production of a farm plan (Figure 5.6).

Biodiversity in the wider countryside is now frequently being addressed through the production of county-level local biodiversity action plans (LBAPs), following on from the national biodiversity strategy. The objectives for conserving biodiversity which arise from national strategy are to conserve and, where practicable, to enhance:

- the overall populations and natural ranges of native species and the quality and range of wildlife habitats and ecosystems;
- internationally important and threatened species, habitats and ecosystems;
- species, habitats and natural and managed ecosystems that are characteristic of local areas; and
- the biodiversity of natural and semi-natural habitats where this has been diminished over recent decades (LGMB, 1996).

Four key components have been established as the main elements of a BAP approach, namely:

- developing costed targets for our most threatened and declining species and habitats;
- establishing an effective system for handling the necessary biological data at both local and national level;
- promoting increased public awareness of the importance of biodiversity, and broadening public involvement; and
- promoting LBAPs as a means of implementing the national plan.

Perhaps the most important feature of the new approach is the setting of quantifiable targets for both species and habitats and the production of costed action plans for top-priority species and habitats.

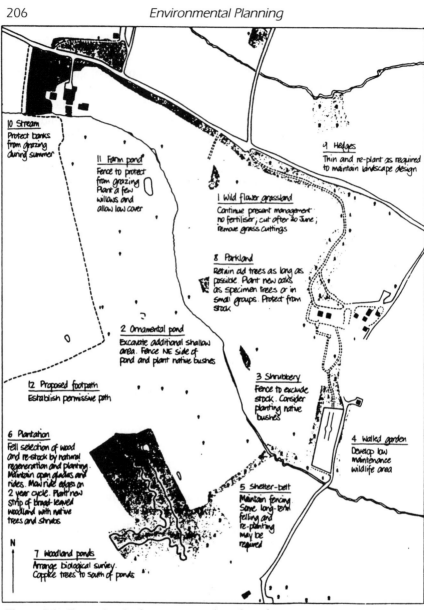

10 Stream
Protect banks from grazing during summer

11 Farm pond
Fence to protect from grazing. Plant a few willows and allow low cover

9 Hedges
Thin and re-plant as required to maintain landscape design

1 Wild flower grassland
Continue present management: no fertiliser, cut after 20 June; remove grass cuttings

8 Parkland
Retain old trees as long as possible. Plant new oaks as specimen trees or in small groups. Protect from stock

2 Ornamental pond
Excavate additional shallow area. Fence NE side of pond and plant native bushes

12 Proposed footpath
Establish permissive path

3 Shrubbery
Fence to exclude stock. Consider planting native bushes

4 Walled garden
Develop low maintenance wildlife area

6 Plantation
Fell selection of wood and re-stock by natural regeneration and planting. Maintain open glades and rides. Mow ride edges on 2 year cycle. Plant new strip of broad-leaved woodland with native trees and shrubs

5 Shelter-belt
Maintain fencing. Some long-term felling and re-planting may be required

N

7 Woodland ponds
Arrange biological survey. Coppice trees to south of ponds

Figure 5.6 Example of a farm conservation plan: site map
Source: Tait *et al.*, 1988

LBAPs are seen as a means whereby national strategy can be translated into effective action at the local level, and to ensure that national targets for species and habitats are attained in a consistent manner throughout the UK (LGMB, 1996). However, they have a broader purpose, the essence of which is 'to focus resources to conserve and enhance biodiversity by means of local partnerships, taking account of both national and local priorities' (*ibid.*, p. 5). The functions of LBAPs are thus to:

- ensure that national targets for species and habitats, as specified in the UK Action Plan, are translated into effective action at the local level;
- identify targets for species and habitats appropriate to the local area (which incorporate the viewpoints both of scientists and local people);
- develop effective local partnerships to ensure that programmes for biodiversity conservation are maintained in the long term;
- raise awareness about biodiversity conservation and of the entire biodiversity resource; and
- provide a basis for monitoring progress in biodiversity conservation, at both local and national level.

This process involves several distinct elements (Figure 5.7). Analysis and evaluation of the nature conservation resource are clearly essential, leading to detailed proposals for action within a specified period of time. In parallel with this is the development of an effective partnership with key players, particularly land managers, to identify appropriate delivery and funding mechanisms. A third component is the programme for monitoring the effectiveness of the overall plan including the extent to which both national and local targets are being achieved. Underlying all this is the need for adequate local databases, ideally integrated with the national biodiversity database. The local action plan is therefore an ongoing process comprising a sequence of steps which form a long-term strategy (Box 5.9).

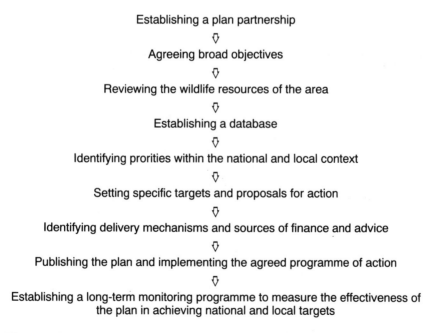

Establishing a plan partnership

⇩

Agreeing broad objectives

⇩

Reviewing the wildlife resources of the area

⇩

Establishing a database

⇩

Identifying prorities within the national and local context

⇩

Setting specific targets and proposals for action

⇩

Identifying delivery mechanisms and sources of finance and advice

⇩

Publishing the plan and implementing the agreed programme of action

⇩

Establishing a long-term monitoring programme to measure the effectiveness of the plan in achieving national and local targets

Figure 5.7 The processes involved in the formation of LBAPs (UKLIAG, 1996)

Box 5.9　Suggested content of an LBAP

1) Vision statement with broad objectives of the action plan partnership;
2) review of the wildlife resource of the plan area identifying national and local priorities for habitats and species;
3) review of priority habitats and species in terms of current status and factors causing loss or decline in the local context; also action already under way to meet their conservation requirements;
4) detailed action plans for priority habitats and species, covering for example site safeguard, habitat management, habitat creation, reintroductions, policy requirements, data needs, research needs and advisory work;
5) a geographical analysis of biodiversity within the plan area identifying issues specific to particular geographical areas and indicating how implementation of habitat and species action plans relate to areas of different ecological character. This should include biodiversity maps showing locations of key areas for action;
6) review of generic issues affecting biodiversity within the plan area with recommended action;
7) proposals for raising public awareness and involvement;
8) communication and publicity regarding the work of the partnership; and
9) proposals for monitoring progress of the overall action plan.

Source: Based on LGMB, 1996.

One of the earliest effective approaches to nonstatutory 'wider countryside' planning was the 'catchment management plan' (CMP) introduced by the former National Rivers Authority. The CMP provided a basis both for the authority's own investment programme and for liaison with planning authorities over development proposals within the catchments of particular rivers. Thus, in theory at least, it became possible to move away from incremental approaches which had characterised management of river channels, and location of new development on floodplains or critical aquifer recharge zones. Some past approaches to river management, it could be argued, have tended to take a very restrictive view of the overall catchment and have ended up by shifting problems up or down stream, rather than seeking an integrated solution. Since the creation of the Environment Agency, CMPs have been replaced by Local Environment Agency Plans (LEAPs), which incorporate the land-use aspects of all the Environment Agency's responsibilities and not only their water supply and regulation functions. They are, however, still based on river catchments, and have increasingly been implemented for our principal rivers. In addition, the Environment Agency has produced a policy approach to floodplain development, based on a number of key objectives, namely that:

- development should not take place which has an unacceptable risk of flooding, leading to danger to life, damage to property and wasteful expenditure on remedial works;
- development should not create or exacerbate flooding elsewhere;

- development should not take place which prejudices possible works to reduce flood risk;
- development should not cause unacceptable detriment to the environment; and
- natural floodplain areas are retained and where practicable restored in order to fulfil their natural functions.

These objectives are constrained by further environmental and sustainability considerations, aimed at respecting the 'regulatory', wildlife, aesthetic and other qualities of rivers. The basic principles are also intended to be translated into policy statements within local authorities' development plans (Environment Agency, 1998).

A key element of the LEAP process is the establishment of effective partnerships, though in practice the LEAP is 'owned' by the Environment Agency and is essentially an aid to help it meet its objectives, in consultation with other parties. LEAPs thus deal with strategic planning of agency interests, covering flood defence, water resources, navigation, conservation, fisheries, pollution, waste regulation, integrated pollution control, contaminated land and water, and air quality issues. Their purpose is to assess the problems and opportunities resulting from catchment pressures, activities and uses and to propose action to optimise the overall future well-being of the environment.

Earlier, we noted the highly complex environmental dynamics and administrative situations at the coastline. Some modest attempts have been made to achieve integrated planning solutions which reflect the partnerships advocated in current codes of practice and planning guidelines. The development of 'Shoreline Management Plans' is starting to provide sustainable coastal defence policies for virtually the whole length of the English coast and to set objectives for its future management. Prepared by Coastal Defence Authorities (the EA or local authority) acting individually or as part of a group, these plans take account of natural coastal processes and human and other environmental influences and are subject to extensive consultation with interested parties. Their extent and boundaries are often based on natural features such as sediment cells. The key bodies involved are:

- *the Environment Agency* – as the primary executive agency normally with ultimate responsibility;
- *MAFF* for giving guidance on coastal management and flood protection aspects, for determining project justification and for grant-aiding implementation;
- *DETR* for providing planning guidance for local authorities for managing the coast, and for being responsible for wildlife and amenity at a policy level;
- *the county councils* for determining structure plans and for co-financing recreation and access schemes on coastal zones;

- *the local district councils* for protecting the coast from erosion, and for controlling development in inundation-threatened zones, grant aided by DETR;
- *the Flood Defence Committees* (FDCs), consisting of county council member appointees plus MAFF selected nonpolitical representatives; the local FDCs have executive responsibility for programming flood defence works and for supplying the county council levies to match grant aid whilst regional FDCs' role is orientated towards strategy and policy guidance;
- *English Nature and the Countryside Agency*, the former as executive agency for managing key wildlife sites, and both as providers of grant aid for conservation and access; and
- *landowners*, including the main conservation bodies who own and manage 'critical' conservation habitats.

The key functions of Shoreline Management Plans are summarised in Box 5.10, and are complemented by English Nature's Estuary Management Plans, previously discussed.

Box 5.10 Key functions of Shoreline Management Plans

- To steer organisations away from local authority boundaries and towards the natural 'cells' of largely self-contained sediment movement;
- to improve understanding of the coastal processes operating within a sediment cell;
- to predict the likely future evolution of the coasts;
- to identify all human-made and natural features likely to be affected by coastal change;
- to pin-point zones for special investigation; and
- to facilitate consultation amongst organisations with an interest in a shoreline.

After a plan is completed, it should:

- help the assessment of strategic defence options;
- clarify the responsibility for monitoring change;
- inform the planning authorities of any areas where development would be undesirable;
- identify opportunities for maintaining or enhancing the natural coastal environment; and
- ensure continuing arrangements for consultation amongst interested parties.

Another major rural resource which has in places been subject to quasi-statutory planning is forestry. The only real influence which planners have over new afforestation is their status as consultee on some WGS applications. In order to clarify where planners would like to encourage and discourage new planting, therefore, some local authorities have prepared an Indicative Forestry Strategy (IFS), which sets out their likely response to such consultations in different parts of their area, and the rationale for this response. Importantly, it also provides an opportunity for elected members to become involved in the process of representing local voters' views on forestry proposals. IFSs identify suitable land for planting, first, by selecting land with desirable features such as land capability and, secondly, by eliminating land which is highly desired for other purposes such as nature conservation or hill farming. This procedure is undertaken on the basis of land-use attributes and consultations with other interests, including the forestry industry. The outcome is a zonal map indicating:

- existing forestry;
- preferred areas, with suitable land and few constraints;
- potential areas, where one or two specific constraints to afforestation exist but which could be resolved depending on the scale and design of proposals; and
- sensitive areas where there is a strong presumption (by the planning authority) against large-scale planting of commercial softwoods.

The maps may also show areas which are either physically unsuitable for forestry or economically impractical (e.g. mainstream agricultural areas with high land values). Once the IFS has been agreed, it is accorded some degree of official validity by being referenced in statutory development plan policies.

Conclusion

The landscape is a complex entity, which is both naturally and socially constructed, and which has multiple scientific and cultural meanings. The landscape must be understood, not simply as a resource of fine scenery and recreational opportunity, but also as a dynamic living system driven by fundamental environmental processes, and an archive of history and cultural treasure. Our traditional approach to landscape planning has been to fragment the countryside for administrative purposes, and to plan or manage different elements separately. Although this approach now seems less than satisfactory, it has nevertheless provided significant levels of protection to areas of critical capital which could otherwise have been debased for ever.

During the 1980s, countryside legislation started to include 'balance' clauses, requiring government departments and agencies to seek, when discharging their duties, a balance between conservation, agricultural, forestry and rural development activities. In line with the shift away from the 'balancing' philosophy of land-use planning, rural land management has now started to emphasise a holistic and capacity-based approach. In particular, this has entailed an appreciation of the 'wider countryside', rather than protected areas alone, and to seek to change the aims and purpose of rural land uses so that they are intrinsically more complementary to each other. The assumptions that only certain tracts of the countryside possess special qualities have given way to the recognition of a more universal existence of countryside 'character'.

Many of the ecological (and, indeed, cultural) assumptions which underlie the momentum towards 'wider countryside' planning remain controversial. Moreover, they are still weakly tested in practice, such as the targeted approaches which are now being taken to biodiversity action planning. Once more, however, all these emergent methods and concepts point to a greatly increased awareness by society of the 'free services' supplied by the total landscape, and suggest ways in which we can move towards a more sustainable future.

6 Urban Ecological Planning

Introduction

In an ecosystem, life is driven by energy flows and mineral cycles. Species in nature live within the carrying capacity of their biophysical environment, or else their population collapses as a result of factors such as starvation, intra-specific competition and disease. It is revealing to depict human societies in this fashion, by studying their energy throughput, materials cycles and waste production (Figure 6.1). Primitive human societies would be held in check once their populations exceeded local carrying capacity, and the subject of historical ecology has demonstrated how this has occurred in past civilisations (Crumley, 1993). However, sophisticated modern society rises far above the constraints imposed by local place and the stresses of seasonality by subsidising its ecological economy with large-scale energy and material transfers. Some of these transfers are 'temporal' in nature as, for instance, when we plunder fossil fuels from the geological past or dispose of chemical wastes which will persist for decades and even centuries into the future. Other transfers are 'spatial', for example, when we import exotic foodstuffs or minerals from different regions or countries, when we purchase commodities which have caused gross pollution elsewhere or when our air pollution is transported long distances. This wider impact may be thought of as our 'ecological footprint' (Wackernagel and Rees, 1996).

Humans, like any other animals, conform to the basic laws of ecology. All creatures require energy, materials/minerals, shelter and mobility. For much of history, humans have lived close to nature, gathering their food requirements from their immediate surroundings, moving mainly on foot and creating shelter from simple materials. Industrial communities, by contrast, have subsidised their natural economy to a vast extent; they have entailed massive amounts of movement, consumed huge quantities of minerals and created buildings from materials containing extravagant levels of 'embodied' energy. The sources of these subsidies (fossil fuels and minerals) are often distant, so that their impact is concealed from the end-users.

Similarly, cities observe fundamentally similar 'systems' rules to natural environmental processes. For example, 'negative feedback' can occur where a system reacts to change in such a way as to limit or contain it. An example might be the way a local authority will respond to increasing development pressures on urban greenspace by strengthening its protection in the development plan. Positive feedback arises where the system reacts to change in such a way as to reinforce it. An example might be the

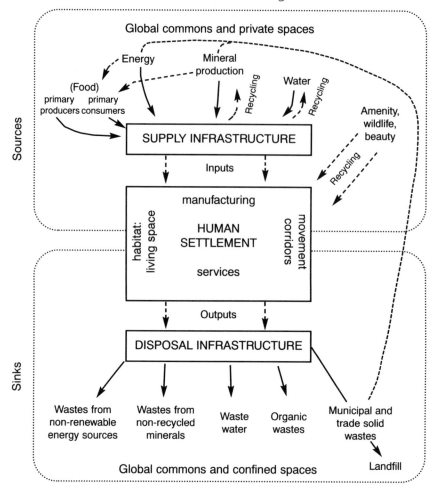

Figure 6.1 A human ecosystem

way the choice by some well-off households to move from an inner-city area prompts others to follow, contributing to a cycle of urban decline and suburban sprawl. Perhaps most significantly, 'state transition' can affect the urban system when a fundamental and irreversible shift occurs in its overall state. One example might be the change, which in many European cities probably occurred between 1950 and 1970, from homeostasis in travel patterns to positive feedback encouraging car use (EUEGUE, 1996).

Urban Growth and its Management

The first urban cultures began to develop about 5,000 years ago, but the world is now experiencing an unprecedented rate of increase of urban growth. In 1970 there were four world cities with a population over 10 million people; now there are 24 and almost half of the world's population is living in urban areas. The populations of cities in the developed world have now been relatively stable for a generation, but the megacities of the industrialising world show phenomenal recent growth. Cities are the major loci of production, consumption and civilised creativity as well as the source and site of much environmental damage. The key environmental effects of urban activities have been summarised as:

- the use of fossil energy in urban buildings, economic activities and transport, and corresponding emissions of 'greenhouse gases' and other pollutants;
- consumption of physical resources and production of wastes (which should be seen as two sides of the same coin); and
- releases of globally damaging pollutants such as ozone depleters and heavy metals (Middleton, 1995).

Similarly, cities are large entities in their regional ecosystems, and have substantial impacts on regional carrying-capacity thresholds through their concentration of activity. In particular, their high levels of resource consumption result in massive waste generation, so that urban environmental systems are invariably under constant pressure from human-generated wastes. Some 'headline' observations about cities have been made by Haughton and Hunter (1994), notably that:

- some 43% of the world's population was urban in 1990, varying from 72.6% in developed countries to 33.6% in developing countries; 60% of the world's GNP is produced in cities; 65% of cities with over 2.5 million people are coastal, several already at or below sea level;
- on average, each city of 1 million people daily consumes 625 kt of water, 2 kt of food and 9.5 kt of fuel, and generates 500 kt of waste water, 2 kt of waste solids and 950 t of air pollutants;
- urban populations in different countries vary hugely in their environmental demands – urban residents in developed countries generate on average .7–1.8 kg of domestic waste daily compared with .4–.9 kg daily in developing countries;
- in the USA almost one-third of urban land is now devoted to the needs of the car; and
- in many west European countries, 2% of agricultural land is being lost to cities each decade.

However, it has been authoritatively suggested that cities can offer people the option of living sustainably (CEC, 1990), and that cities have great potential to reconcile the various dimensions of sustainability. The population density which characterises cities can bring immense variety and choice of work, goods, services, recreation opportunities and social interactions within easy reach; the same density also enables a large proportion of travel to be carried out by environmentally efficient public transport, supports more efficient environmental services (such as reuse and recycling of wastes) and promotes more energy-efficient built forms. Thus, the positive aspects of cities should be borne in mind, notwithstanding people's alleged preferences in the developed world (where the privations of rurality and hazards of nature have largely been overcome) to live in the countryside. As well as being the major loci of production and consumption, cities are also home to civilised creativity.

Thus, cities represent four key 'dynamics', namely: the *economic dynamic*, driven by proximity, multiple contacts and activities, ability to assemble the economic factors involved in all stages of production, research and consumption; the *social dynamic*, which brings together a wide variety of social facilities to cope with deprivation, maximises social relations and delivers education, services and work; the *cultural dynamic*, enabling cities to be lively, vibrant and creative places, associated with factors such as density, proximity and choice; and the *political dynamic*, reflecting the city as the locus of government and home of citizenship (Haughton and Hunter, 1994).

Certain principles underlie the practice of sustainable urban management (EUEGUE, 1996). One of these is the notion of *environmental limits*, namely, that there is an upper threshold to the level of 'cramming' which is possible or desirable in an urban fabric. This typically reveals itself in terms of infill development, high-density building and road networks. A second is the concept of *demand management*, whereby planners abandon their habitual policy of 'predict and provide' to meet anticipated future levels of demands. Instead, it is assumed that not all needs will be supplied at the desired level, but sometimes that supply of environmentally based goods will be limited at a threshold commensurate with acceptable quality of life. The third is the principle of *environmental efficiency*, aimed at achieving the maximum benefit for each unit of resources used and wastes produced. It has been argued that environmental efficiency can be increased in several ways (European Sustainble Cities and Towns Campaign, 1994), notably by increasing durability of goods, increasing the technical efficiency of resource conversion, avoiding the excessive consumption of renewable natural resources and energy, and closing resource loops, for example by increasing reuse, recycling and salvage (and avoiding pollution). Overall, it is often best to pursue simple solutions which avoid the use of resources in the first place, and it may be possible in this respect to apply a 'principle of elegance' (Brugmann, 1992) based on reducing the economic cost of each unit of economic utility (Box 6.1).

Managing Demand

As we have just noted, the historic way of managing urban and regional systems has been that of 'predict and provide', whereby forecasts of

Box 6.1   Possible means of reducing the environmental cost of each unit of
economic utility

- Increasing the durability and reparability of products so that the resource 'costs' are spread over a longer useful life;
- increasing the efficiency with which resources (especially energy) are used, in both manufacture and use;
- simplifying production processes, avoiding overspecification, redundant elaboration and levels or dimensions of 'quality' which do not increase a product's usefulness or fitness for its purpose;
- minimising packaging and transport;
- using reclaimed and recycled materials in production, and in turn making products reusable and recyclable; and
- using renewable instead of finite resources, and producing wastes in bio-degradable forms.

---

demand have led planners to supply the anticipated level of infrastructure and developed land. Prominent examples have been additional land for residential purposes (to meet inexorable rates of peri-urban expansion), new and improved roads to match the growth in vehicular traffic, and additional minerals (especially aggregates) to satisfy the consequences of this growth. Even in long-industrialised countries, where rates of urban population growth are more or less static, expansion of the urbanic area continues due to reductions in the density of living and the decline in household size. However, this traditional approach has increasingly been challenged, especially in the name of sustainable development. Catering for this type of constant growth is probably unsustainable in terms of the supply of basic resources (such as land, water and minerals), the earth's assimilative capacity, the cultural capacity of humans to live in degraded conditions, and the physical and perceptual capacity of cities to accommodate change in the urban fabric or in traffic levels. Consequently, attention has been turning to restraining demand, for instance, in relation to the amount which people travel or the rate at which households are formed.

Guy and Marvin (1996) argue that emergent methods of demand side management (DSM) can help reconnect urban policy to infrastructure management. In particular, they suggest that adopting either supply-led or more demand-responsive modes of infrastructure provision critically shapes the intensity, and so the relative environmental impact, of networked services in cities. DSM can therefore help avoid environmentally and economically expensive supply investment by managing both the level and timing of demand placed on networks through the implementation of energy and water efficiency measures and the reshaping of transport patterns. There appear to be two main elements to this type of DSM: one entails introducing fairly broad-spectrum deterrents to avoidable consumption, such as water metering or taxing car parking; the other comprises far more focused targeting of specific consumers. In relation to this latter option, Guy and Marvin suggest that utility operators need to develop more intimate relationships with the most demanding users in order to modify demand on 'stressed' parts of their network. In this way, DSM initiatives can help infrastructure providers and users tailor the pattern of supply and demand, providing a powerful impetus to energy and water efficiency measures and more integrated transport packages.

Broadly, the transition to more effective DSM appears to require a strengthening of regulatory frameworks which encourage infrastructure providers to develop strategies which minimise resource use. These are generally the domain of central government, as they often involve the introduction of incentives and disincentives through the tax system: traffic generation, parking, use of leaded petrol, solid waste disposal and production of primary aggregates could all be influenced in this manner. At the more local policy level, DSM entails liaising with infrastructure providers to identify 'hot' and 'cold' elements of their networks, and promoting patterns of economic development which generally help to 'cool' demand. An indication of the kind of signals which might be sent to producers and consumers is summarised in Box 6.2.

Box 6.2   Signals to producers and consumers

To producers:

- setting a framework of long-term (20–30 years) objectives in negotiations between government and business;
- negotiated agreements with business sectors covering more detailed programmes of measures to encourage sustainability;
- identifying and disseminating best practice, including adoption of good-quality environmental management systems;
- improving consumer information on product characteristics, e.g. through labelling;
- encouraging responsible care and producer responsibility initiatives which ensure that sustainable development considerations are designed in from the outset (including more durability), and that once a product has finished its useful life it is reused, recycled or otherwise turned to productive use;
- encouraging improved communication and dialogue with stakeholders;
- 'sticks and carrots' to encourage cleaner technologies; and
- improved company reporting on the environment.

To consumers:

- improvement of information to purchasers (reliable ecolabelling);
- raising general levels of awareness and consumer education;
- more efficient use of domestic appliances, lighting and heating, and cars;
- minimal packaging and careful disposal and reuse of wastes; and
- shifting to longer-lasting, higher-quality and more satisfying products.

A notable policy shift towards DSM occurred with the 1998 white paper on transport (DETR, 1998b), which proposed the establishment of an integrated transport network, the reduction of congestion and the encouragement of public transport use. A key proposal was for introduction of local transport plans, involving consultation by local authorities with residents, businesses and transport operators, and covering all forms of transport. In addition, the plans are to include targets for improving air quality, road safety, public transport and traffic reduction. A principal cause of the declining use of public transport has been the inconvenience and delay associated with changing between different modes of public transport, and so the white paper sought (very ambitiously) to create the 'seamless' journey, facilitating easy movement between different modes of transport. In the future, an integrated strategy may include charging to drive in town centres (and possibly also in rural hotspots), as well as workplace parking charges, and differential taxing of cars according to the pollution which they cause.

Shelter and Urban Form

It is often argued that large urban settlements provide the best option for sustainability as the global impacts of many kinds of human activity can be reduced by locating them in densely settled areas. Indeed, settlement density is the most important single factor in determining the average distance people travel, and it is also generally the case that higher density facilitates more efficient provision of infrastructure services because of shorter distances. Equally, heat distribution (such as in district heating systems), waste reuse and recycling can be assisted by the concentration of economic activities, as there are sufficient businesses to support specialist activities which could not otherwise achieve economies of scale. The close links between urban form and patterns of energy consumption (especially for transport and residential purposes) are now starting to exert a strong influence on planning rationales (Owens, 1992). Consequently, some sustainability strategies support 'densification', even though this may be politically unpopular in cities which have an ingrained cultural dependency on low-density residential areas and high levels of car ownership. Conversely, it should be borne in mind that high urban densities may be associated with adverse effects such as social and welfare disadvantage, and pollution 'hotspots'. It has thus been noted that cities present a paradox: economic activities may cause less damage in cities than elsewhere per capita or per unit of production, but they may appear more damaging because of their greater concentration per hectare (EUEGUE, 1996).

The future of the city has led to a great deal of debate amongst environmentalists. Not long ago, it was fashionable to pronounce the death of the city, believing that urban megasystems could at best only be regenerated by vast public subsidies. However, many cities have experienced a remarkable degree of spontaneous renewal, based principally on their role as information nodes. It is commonly supposed that this can be associated with far less environmental damage than the 'old' industrial city. Brotchie *et al.* (1991) note that in the late twentieth century there has been a massive convergence of information and communication technologies which have hitherto been quite separate, and have had profound impacts on the physical form of cities. The traditional pattern of urban land use, in which residential activity was clustered around nodes of employment which maximise accessibility, has been breaking down for much of the twentieth century. These changes, combined with increasingly rapid and cheap forms of telecommunications, have tended to enmesh cities with international economic trading patterns, giving genuine expression to the concept of the global village. It is also likely that the internal structure of cities will become increasingly multicentred around information-based industries, so that the development of local centres may help to shorten travel generally and commuting distances in particular. However, the 'dream scenario' of information-rich economies enabling employees to work at home or in local clusters should be treated with caution. For example, Gibbs and Healey (1997) estimate, in relation to teleworking, that for every 1% of the UK workforce who take up telecommuting there is only a national energy saving of 0.06%; equally it has been argued that the higher levels of interconnection permitted by information and communication technologies will stimulate demand for travel rather than substitute for it. Further, information and communication technologies, through permitting the growth of just-in-time production and delivery systems, act to increase the amount of interaction, usually by road traffic.

Demand-side management considerations are now starting to lead to a revision of forecasts of future housing need, which have proved highly contentious, and to a deflection of much of this demand on to 'brownfield' sites. Greenfield sites are still subjected to a great deal of pressure, however, leading to the perpetual accretion of new development at the margins of existing settlements. Also, many brownfield sites are valuable for aesthetic, recreational and other environmental reasons, and often present difficult conditions for construction and engineering. Consequently, there has been renewed interest in the development of new settlements, especially where these can be designed as exemplars of sustainability. Moreover, we have noted that changes to the form of existing settlements, usually in terms of densification or co-location of complementary land uses, are difficult to accomplish as it is intractable to alter the inherited pattern of settlement. Thus, the scope for achieving more efficient urban form exists mainly in relation to new settlement options, which may include:

- centralised town/city regions;
- a 'string of beads' of semi-rural settlements;
- polycentric cities covering large areas; and
- linear development, typically along high-speed public transit systems.

The various options can be summarised as (UWE/LGMB, 1995): *cluster* approaches, based on the expansion of satellite towns, possessing greater or lesser degrees of autonomy; *corridor* strategies, where linear development is located between growth poles along public transport routes; and *compact* forms, where urban development is concentrated, and heavy reliance is placed on renewal and the regeneration of brownfield sites.

A study for the former Department of Environment (1993a) evaluated five alternative urban layouts (urban infill, urban extension, key villages, multiple villages and new settlements) against a range of economic, social and environmental criteria (Box 6.3). These evaluations led to a number of general findings. The *urban infill* option performed well against certain economic criteria, but badly in terms of public costs and potential for planning gain. It generally produced positive social results, though had a mixed environmental showing, scoring well on retention of land and habitats and on energy consumption, but less well on 'greening' and 'town cramming' issues. Peripheral *urban extensions* had a consistently good economic performance because of moderate costs and good planning gain prospects, and a reasonable attainment of social and environmental criteria; the weakest scores were on loss of land. However, towns and cities cannot expand indefinitely without placing stress on social and environmental factors, so one must be cautious in applying general conclusions to specific situations of urban expansion. *Key villages* proved to be a rather unpromising option, with poor prospects for planning gain and affordable housing, but with relatively favourable economic costs and social provision.

Box 6.3   Criteria for evaluating the sustainability of settlement patterns

Economic criteria:

- cost of end product;
- economy in provision and use of infrastructure;
- maintenance costs; and
- access to employment.

Social criteria:

- access to social facilities, potential sense of community;
- breadth of social mix;
- potential affordable housing contribution; and
- local acceptability.

Environmental criteria:

- loss of land;
- loss of natural habitats;
- energy consumption (transport);
- energy consumption (space heating);
- pollution levels;
- contribution to 'greening' of the urban environment; and
- town cramming effect.

Source: DoE, 1993a.

However, they tend to induce high car dependency. *Multiple village extensions* performed most weakly overall, with high costs and few compensating public, social or environmental gains. *New settlements* tended to perform to extremes, with relatively low costs of housing and many opportunities for planning gain (which reduce public sector costs); nevertheless, energy consumption from travel is likely to be high in smaller new settlements as is land and habitat loss. Within larger new settlements, though, there is reasonable scope for district heating and power schemes. How these different options compare depends on the weights which are applied to the principal criteria, and so it is difficult to generalise. Even the 'urban extension' option, which tends to be favoured by any weighting system, fails to emerge as a clear leader because its benefits vary relative to the size of and conditions in the 'anchor' settlement.

Moving down to the scale of the individual development, design guides have been produced which optimise the relationships between site development and environmental conditions. These consider, for instance, the development context, infrastructure, topography and ground conditions, water, microclimate (including daylight and solar heating), noise, flora and fauna, archaeology and use of local materials or existing buildings. Influential design guidance (UWE/LGMB, 1995) proposes the evaluation of the merits of new settlements in terms of their effect on:

- maintaining global ecology (atmosphere and climate, biodiversity);
- husbanding natural resources (air, water, land, minerals); and
- improving the quality of the local human environment (buildings, infrastructure, open space, aesthetic quality, cultural heritage).

Potential development sites may then be subjected to a 'Capacity Assessment Framework' which reflects a site's ability to perform in relation to these criteria. Another popular perspective is to focus on the people-friendliness of urban designs, and various principles which optimise this have been promulgated (Box 6.4).

Much of the scope for diminishing the impact of urban areas lies in changing the nature of buildings themselves. Energy is wasted in most buildings – whether domestic, commercial or industrial – and it is widely acknowledged that this could be changed by simple technical measures, such as

- low-energy light bulbs
- thermostatic radiator valves
- energy-efficient applicances
- loft insulation
- double glazing
- insulating hot-water tanks and pipes
- more efficient boilers
- heating controls, and
- efficient taps.

Box 6.4   Urban design: people-friendliness

Haughton and Hunter (1994), drawing upon Goodey (1993), show how diversity can be encouraged through good design:

- *variety* – multifunctional use, varied building styles;
- *concentration* – sufficient density of people to maintain variety and activity, preferably including people who are resident; short street blocks as this seems to encourage increased diversity in the ways streets and buildings are used;
- *democracy* – involvement of local people in design decisions;
- *permeability* – easy access to each other and to general urban facilities, whilst respecting needs for privacy and security;
- *security* – open and connecting spaces should be designed to enhance personal safety;
- *appropriate scale* – build on local context, reflect local conditions;
- *organic design* – locally distinctive, building on local context, respecting historical narrative within the urban fabric;
- *economy of means* – design areas *with* nature, using the most environmentally efficient means of achieving an objective;
- *creative relationships* – between buildings, routeways and open spaces;
- *flexibility* – buildings and spaces adaptable over time; and
- *consultation and participation* – including running of local resources.

In a more sophisticated sense, there is much 'embodied energy' in buildings, related to the processes involved in the manufacture and transport of their materials. Overall, it has been estimated that buildings account for around 40% of UK delivered energy and 50% of CO_2 emissions. In addition, many of the materials used in buildings are associated with ecological impacts, such as PVCu and unsustainably produced timbers.

Between the small site and the large settlement, some attention has more recently been focused on the 'mid-range' (Barton, 1998), especially the idea of the 'eco-neighbourhood', reflecting concern both with dispersal strategies and overemphasis on brownfield development. Key aspects of 'eco-communities' are that they are relatively small neighbourhoods with a heterogeneous social composition, and contain a diversity of land uses so that traffic-generating land uses are not unduly separated. They also require the adoption of 'ecologically responsible development principles', possess distinctive townscape and local identity, and involve high levels of public participation in local decisions. A number of variants on the eco-neighbourhood concept have been piloted, including: the *rural eco-village*, containing ostensibly 'green' dwellings and land uses; *tele-villages* based on low-impact, flexible communications technology and permitting locally-based but highly connected working; *urban demonstration projects*, containing 'showpiece' green buildings and energy supplies (for example, the Milton Keynes 'Energy Park'); *new urbanism* style developments, generally comprising under 1,000 dwellings, with compact pedestrian-scaled layouts (such as Poundbury in Dorset); and *ecological townships*, where urban neighbourhoods evolve towards sustainable living, measured across a range of variables.

Energy

The environmental impact of the use of traditional energy sources, mostly based on fossil fuels, has been widely documented. Broadly speaking, impacts are fourfold: fossil fuel extraction depletes natural capital which has accumulated over geological timescales; the extraction of geological resources typically transforms environments, for instance by destruction of scarce ecological habitats or the creation of subsidence, landscape scars and local pollution; there is a general dissemination of risk, hazard and pollution during the generation of electricity from geological sources; and conventional energy generation normally requires large production units which both concentrate environmental impact and require intrusive energy distribution networks, from which there are substantial losses *en route*, to channel electricity and hydrocarbons to widely dispersed end-users. Different conventional sources of energy production cause these impacts in differing proportions. For example, the use of natural gas for electricity generation is clean and efficient, but depletes finite carbon resources and thus is ultimately unsustainable. Nuclear power production is clean and efficient in certain respects, but poses discernible risks to general public safety; it has also been associated with some worrying evidence about standards of good practice at individual sites and creates problems of waste disposal and decommissioning. Coal production has benefits in terms of local economy and community, but disadvantages in terms of site impact and global air quality (from coal-derived energy). The great advantage of conventional energy production is its sheer scale, which presently dwarfs 'alternative' energy sources.

Alternative energy supplies draw upon renewable sources arising from energy flows that occur naturally and repeatedly in the environment. These include energy from the sun, the wind and the oceans, and the fall of water; the heat from within the earth itself (geothermal energy) which is extractable from certain underground hotspots; and combustible plant materials and agricultural and industrial wastes. These forms of energy are strongly promulgated by many advocates and, despite their small contribution at present, could become more significant as generation technologies gain in efficiency (Box 6.5). It is also possible that renewable sources can be generated at a neighbourhood scale, thereby reducing the need for widespread electricity transmission networks. However, they may be unreliable, particularly in terms of the variability of sun and wind at particular locations, and tend to produce a lower energy output for an equivalent area of land used in comparison with conventional sources. A boost to producers of renewable energy was provided by placing a Non-Fossil Fuel Obligation (NFFO) upon electricity supply companies, requiring them to purchase given levels of 'alternative' supplies for a specified period. Moreover, the relatively newly privatised markets of electricity provide greater choice for consumers to select their suppliers, and this creates opportunities for 'green purchasing' strategies favourable to nonfossil fuel sources. Very significantly, local authorities (which are major consumers of energy) now have the option to adopt 'energy procurement' strategies which favour green sources. However, no mode of energy production is without its environmental impact, and the most that can be hoped is that impacts are kept to a minimum, preferably kept local and not conveyed to future generations. For example, key planning considerations in respect of wind energy impacts include:

● the need to site turbines in exposed locations often in rural areas which may also be in attractive landscapes;
● the nature of noise emissions from the turbines;
● the movement of the blades; and
● considerations relating to safety and electromagnetic interference (DoE, 1993b; 1994e).

Box 6.5 Renewable energy

> Middleton (1995) has summarised some key issues associated with the principal types of renewable energy as:
>
> ● *hydropower* – already used on a large scale, though there is a substantial legacy of environmental impact from dam construction; there is considerable potential for small-scale hydroschemes designed to serve local needs;
> ● *wind energy* – e.g. Denmark aims for 10% from wind by 2005; environmental impacts are noise (though modern engines are reducing this), bird kills, interference to television and sensitive electronics, and land requirements;
> ● *solar power* – which has alleged great potential long term though supplies little as yet; photovoltaic cells produce no pollution or noise;
> ● *biomass* – may be derived from pulp and paper waste as in Scandinavia; also significant progress has been made in Brazil by substituting sugar-cane-derived fuel for petrol;
> ● *tidal* – the amount of energy depends on the range of tides and area of enclosed bay; however, it has controversial environmental effects including hydrodynamics, surface movement and indirect effects of these on fauna; and
> ● *geothermal* – limited use because of restriction to certain localities.

The general conclusion is that energy use must be minimised and we must buy time by making the most efficient use of resources. Certainly, a great deal of energy is currently wasted, in transporting fuels to power stations, in the construction and decommissioning of power stations, at the point of generation, during the distribution of energy, and in the wasteful use of appliances. A great deal of fuel is also consumed unnecessarily in the transport system. Consequently, many adverse impacts of energy production could be deferred if conservation measures were maximised. *Planning Policy Guidance Note 12* (DoE, 1992d) suggests that, whilst the planning system cannot compel people to live near their work, or to use public transport when it is available, or to walk or cycle, it can encourage development patterns that provide the choice. Various elements can, therefore, be incorporated into the energy-conscious plan, such as:

- development that makes full and effective use of land within existing urban areas without amounting to town cramming;
- development that is closely related to public transport networks;
- location of new development types that attract trips at points such as town centres which are capable of acting as nodes for public transport networks and where there may be advantages in enabling one journey to serve several purposes;
- housing that is located in such a way as to minimise car journeys to work, school and other local facilities;
- limitations (by capacity or price) on town centre parking, yet being careful that this does not encourage development in more energy-inefficient locations elsewhere;
- appropriate interchange opportunities between major public transport networks; and
- encouragement of facilities to assist walking and cycling.

More ambitiously, planners can also promote land uses which assist combined heating and power schemes, district heating networks and housing type/orientation/location. Such beneficial possibilities must, however, be tempered with actual experience of inexorably increasing average journey lengths for commuting, shopping and education, and the massive growth in out-of-town retailing which occurred in the 1980s and 1990s (from roughly half a million square metres in the mid-1970s to nearly 6 million square miles by the turn of the century).

Given the high-energy intensity of cities, a primary concern is to influence the behaviour of citizens as well as industry towards a more energy efficient and 'cleaner' operation of the urban system. The efficiency rate is dependent upon various issues, including available technology, resource management and land-use patterns. A good example of promoting energy efficiency at municipal level is the Cities for Climate Protection programme of the International Council for Local Environmental Initiatives (ICLEI, 1997), which requires of its participants a number of 'milestones'. These comprise:

- conducting an energy and emissions inventory – profiling energy and emissions for municipal operations and facilities, and for the wider community (in terms of residential, commercial and industrial buildings and the transport network);
- estimating future energy and emissions, based on population and floorspace growth;
- establishing a target for reducing emissions, which will guide planning and implementation of measures;
- developing and obtaining approval for a 'local action plan' to reduce CO_2 emissions, involving public awareness and education campaigns; and

- implementing policies and measures to reduce CO_2 emissions, requiring a rigorous political and analytical commitment.

Many lessons have been learned from the cities which have piloted these initiatives, but perhaps the most generally applicable observation is that partnerships between state, provincial and national governments, and with financial institutions, have been the key to successful implementation based both on retrofitting and new technologies.

Managing Wastes and Recycling Land

Environmental pollution may be of various types, and affects the media of air, water and soil. Most pollutants arise from waste materials, but waste energy can also lead to pollution from heat, noise and vibration. All these are, to a greater or lesser extent, capable of being influenced by planning actions, though this is generally indirect and the administrative lead will be taken by another authority.

A particular impact on land use is the production of solid wastes, for which landfill is still generally the principal option. Practice on this is now regulated in line with the European Framework Directive on Waste (CEC, 1975), which requires competent authorities to draw up plans relating to:

- the type, quantity and origin of waste to be recovered or disposed of;
- general technical requirements;
- any special arrangements for particular wastes; and
- suitable disposal sites or installations.

This is largely addressed via the production of Waste Disposal Plans, drawn up by Waste Regulation Authorities (i.e. upper-tier/unitary local authorities) under section 50 of the EPA 1990, and Waste Local Plans (often combined with mineral plans) produced under the TCP Act 1990. Collectively, these plans consider: the quantities and types of waste likely to be produced or imported, disposal site operating standards and licensing criteria; the waste management facilities likely to be required; means of achieving regional self-sufficiency in waste management facilities, land-use and transportation requirements of treatment facilities; and broad areas of search for future facilities which have suitable environmental, geological, hydrogeological and access features.

All forms of refuse disposal prove controversial, especially landfill (because of its land consumption, possible habitat loss, traffic, localised nuisance and potential health hazards from certain wastes), and incineration (which some see as providing efficient heat recovery and others criticise as expensive and polluting). Thus, the key principle of future strategies must be to minimise waste production in the first instance. As an overarching principle, waste is now managed according to a *hierarchy* commencing with *reduction* of wastes produced, maximum *reuse* of waste materials, followed by *recovery* (including material recycling, energy recovery and composting) and, when all other economically realistic options have been exhausted, *safe disposal*. In other words, landfill (the last of these) should increasingly be seen as a last resort, rather than standard practice. Whilst the viability of each option will vary according to the waste stream and local considerations, it is generally presumed that emphasis will increasingly be placed on the top of the hierarchy. A further axiom is that of the 'proximity principle' under which waste is managed close to the point at which it is generated. Ambitious reduction targets for landfill disposal, likely to result in significant increases in incineration, are now being set by the EC Landfill Directive.

The main categories of solid waste comprise domestic and commercial refuse, hospital waste, industrial by-products, reused industrial products, rubble produced by construction activities and fully or partly biodegradable products. Nonbiodegradable materials, e.g. all plastics but especially PVC, are the biggest problem because of their continuous accumulation and the dioxins that are emitted if PVC mixed with other waste is incinerated at a low temperature. The shortage of space for wastes is also aggravated by the rapid increase in wastes that are biodegradable, but which take a long time to decompose (e.g. metals). Simple landfill or even sanitary burial cannot provide the answer to the problem of solid waste management. The wisdom of incineration has been called into question, backed by scientific arguments, since it contributes to the greenhouse effect and releases toxic substances such as dioxins into atmosphere. The solution to this dilemma would appear to lie in mixed systems which advocate multiple use, and the reuse and recycling of materials in conjunction with the sanitary burial of rapidly biodegradable materials. However, progress to this optimum condition is still slow. A survey of recycling plans in English local authorities, conducted in the early 1990s (Coopers & Lybrand, 1993), showed that, for the four materials most frequently collected for recycling (glass, paper, textiles and cans), the recycling rate was only 3.4%. Policies now aim for recycling of 25% of household waste and a growing number of local authorities are achieving this target. However, there are concerns about the future availability of markets for collected materials due to decreases in the level of demand and reductions in the level of income from these materials, and the tightening of materials specifications which have to be met.

Land which has been disturbed by urban and industrial activity may be recycled for subsequent use. An important aspect of contemporary policy is the reuse of 'brownfield' land for urban and industrial development (present systems of grant-aid tend to favour end-uses such as industry, commerce and housing which result in the highest land values, rather than, for example, forestry or amenity uses). Reclamation work of this kind needs to restabilise land so that it has a high bearing capacity, and this may pose problems where underground mining has taken place and residual settlement and subsidence are a possibility. More serious problems arise where sites have been used for manufacturing industries which have produced long-term chemical contamination. This can cause persistent dangers to subsequent users – for example, where heavy metals are taken up in vegetables grown in gardens – and expensive remedial action may be necessary on affected sites. The problem is further compounded by widespread ignorance of the characteristics and incidence of affected sites, despite the production of registers of contaminated land. In such situations, planners are recommended to adopt a 'suitable for use' principle (DoE, 1994c), which involves determining the probability of contamination and the effects that this may have on suitable end-uses of a site. This is often a rather imprecise science, but may include actions within potentially affected areas such as informal discussions with potential developers at an early stage, setting out specific proposals for sites where contamination is known so that prospective purchasers and developers are aware of the issue, inclusion of standard questions on development applications, and close liaison with responsible authorities. However, where it is strongly suspected that a site is seriously contaminated, detailed investigation of the hazards by the developer and any necessary remedial measures will normally be necessary, and these are invariably expensive. Grant aid is sometimes available to assist with the regeneration of contaminated land.

A further situation in which the reversion of urban/industrial land occurs is where the land has been 'borrowed', especially following temporary disturbance such as mineral extraction. This requires both a full restoration programme and a period of 'after-care', during which land is managed according to carefully prescribed guidelines which seek to restore soil fertility. Agriculture has normally been assumed to be the desired after-use, though overproduction of foodstuffs and new policies for timber production have led to a growing emphasis on forestry as a preferred after-use.

Finally, some former urban/industrial land may simply be treated on a remedial basis, where relatively rapid and inexpensive methods are used simply to heal landscape scars cosmetically, and produce amenity land of some use for low-intensity recreation. Thus, whilst many reclamation schemes have involved massively expensive civil engineering solutions (which are entirely necessary in some circumstances), simpler options are often available. Indeed, in some cases it may be positively desirable to retain distinctive landforms such as conical shale tips as visual focal points, or to use flooded mineral workings as wetlands. A synopsis of the various methods for creating and modifying habitats on reclaimed sites is given in Table 6.1. Remedial treatment requires a detailed understanding of the environmental attributes of a site, and a discipline of *landscape science* has emerged which seeks to develop 'ecological' solutions to restoration, based on an understanding of the physical characteristics, exposure, stability, nutrient status and acidity of sites (Gemmell and Connell, 1984; Burt and Bradshaw, 1985). Ecological approaches to site restoration are still subject to fairly high rates of failure, or at least only partial success, due to the complex and intractable conditions associated with many old industrial sites (Box 6.6).

Table 6.1 Matching derelict land treatments with after-uses

Possible activities	Possible solutions					
	Close-cut sward	Medium-cut sward	Rough sward	Hard land-scapes	Shrubs/ trees/ wild-flowers/ wilder-ness	Allot-ments and urban farms
Intensive recreation	*	*		*		
Extensive recreation		*	*		*	
Passive recreation/casual open space		*	*	*	*	
Visual amenity		*	*		*	
Horticulture						*
Nature conservation and education			*		*	*

Source: Adapted from Burt and Bradshaw, 1985.

Box 6.6 Problems associated with the restoration of reclaimed sites

- Degree of compaction (resulting from buildings and passage of heavy machinery);
- excessive exposure to sun (on steep slopes, which maximise the angle of incidence of solar radiation);
- low levels of organic matter;
- instability (requiring regrading to reduce slippage and gully erosion);
- low nutrient levels (which may be partially remediated by fertiliser application and nitrogen-fixing species);
- rapid leaching due to high porosity or waterlogging due to compaction;
- acidity and alkalinity (urban sites are usually lime rich because of cement and other building rubble, but industrial wastes show extremes of both high and low pH) – colliery spoil in particular can often be associated with the mineral pyrites (FeS_2), whose transformation processes are complex, but broadly centre on the release of sulphuric acid by weathering and the production of very low pH levels. Where pH drops below 4.5, toxic concentrations of certain common metal ions may occur and the metabolic activity of most plants is inhibited. Where levels of pH greater than 8 occur, phosphorus and many important trace elements may be made unavailable;
- toxicity may be a problem, especially where certain residues of mining and smelting wastes remain (notably lead, copper, zinc and cadmium).

Industrial dereliction has led to a sizeable conservation resource, both actual and potential, in urban areas. Partnerships with industry, which seek to extract conservation gain from economic activity, have been promoted by English Nature through its Conservation and Industry Advisory Group, which provides brokerage of sponsorship for environmental action. Also, Industrial Nature Conservation Areas (INCAs) have hitherto been established within the curtilage of several major utilities and factories. Some sites have obtained their value over the passage of time by natural processes; others may have been consciously reclaimed, often by using 'ecological' techniques for vegetation establishment. Some derelict areas are, of course, inimical to amenity – such as highly acidic colliery spoil and toxic chemical waste – and there is a clear need for reclamation. Others, such as sand and gravel excavations below the water table, may convert readily to attractive sites, whilst some regionally rare plants may colonise particular types of industrial waste (Table 6.2).

Table 6.2 Types of industrial wildlife habitat in Britain

Industrial origin	Principal type of habitat
Solvay/Leblanc wastes, blast-furnace slag, lime wastes, pulverised fuel ash	Calcareous, species-rich grasslands and scrub
Clay, marl, sand and gravel pits	Species-rich damp grassland, marsh and aquatic habitats
Limestone quarries and chalk-pits	Calcareous, species-rich grasslands and scrub; woodland in old quarries; ferns and rock plants on cliffs; damp and aquatic habitats on quarry floors
Gritstone and other acidic, hard-rock quarries	Heath, acidic grassland, scrub and woodland; cliffs and quarry floors as above
Railway land including trackway, railway sidings, cuttings and embankments	Lime flora on ballast and on cuttings through calcareous rocks. Damp grassland and marsh in cuttings. Grassland, acidic flora, scrub and woodland on embankments
Mill ponds, reservoirs, canals and subsidence flashes	Damp grassland, marsh, scrub and aquatic habitats with rich emergent and submerged vegetation

Source: Adapted from Gemmell and Connell, 1984.

A most important initiative in remediating derelict land has been that of the Groundwork Trusts, comprising project-orientated locally based independent trusts with charitable status, whose boards are drawn from local authorities, landowners, industrialists and the local community. They offer a broad and flexible approach to environmental improvement and land management, ensuring wide public involvement, solutions tailored to local areas and designs which emphasise access and landscape benefits (Jones, 1989). A brief but important contribution, which operated during the 1980s and early 1990s, was provided by 'garden festivals', which were introduced to Britain from continental Europe. These transformed (at great expense) the appearance of a few of the most intractable sites of urban decay. Their benefits have been cited as a catalytic effect on investment and a city's image, attraction of private funding alongside public sector support and generating a variety of after-uses. However, critics rounded on their relatively transient impact, the limited extent to which the designed landscapes have persisted in subsequent site use and their minimal effect on city-wide greenspace (unlike some European counterparts) (Holden, 1989).

Other waste products in urban ecosystems are associated with noise, air and water. Noise is the result of waste energy, and the resultant sound waves are characterised both by the volume of noise and by the combination of frequencies. Those noises with a preponderance of higher frequencies (such as, for example, a circular saw) cause the greatest annoyance to the human ear. Consequently, measures of noise reflect particular frequencies, and thus its human impact; the most common weighting is known as the A-weighting [dB(A)] which approximates to sound as experienced by the human ear. As the incidence of noise varies throughout the day, sound levels must be reflected by a variety of measures, the commonest ones being the weighted mean (L_{eq}) and the level exceeded for a given number of occurrences, usually L_{90} (the level exceeded 90% of the time) and L_{10} (the level exceeded 10% of the time). These reflect, respectively, average levels, background levels and peak levels. Planners need to be aware of the way in which these different measures are applied and how they can be used to frame planning conditions about hours of site/plant operation. Particular care is necessary in rural areas (where excesses over very low background levels may be significant), in locations where particular atmospheric conditions occur (temperature inversions, in particular, make noise carry further and seem amplified), and close to sensitive noise receptors (such as hospitals).

The main goal of sustainable urban management in relation to air is to ensure quality, which can be influenced by the reduction of pollution sources and quantities, and the regeneration and filtering of atmosphere. The principal sources of urban atmospheric pollution are from the combustion of fossil fuels for domestic heating, power generation, in motor vehicles, in industrial processes and in the disposal of solid wastes by incineration. These sources emit a variety of pollutants, especially SO_2, NO_x, CO, SPM, and Pb. Ozone (O_3) is the main constituent of photochemical smog, formed photochemically in the lower atmosphere from NO_x and VOCs (e.g. from road traffic) in the presence of sunlight. Pollution reduction is largely the domain of statutory agencies, as we have previously noted, but the location of new plant and the concentration of industrial premises can be significantly influenced by planners. The capacity for air regeneration and filtering is reflected in the provision of green elements in the city and selecting suitable plant species that maximise the transformation of CO_2 into oxygen. This helps to counter urban emissions, particularly from road traffic. Also, apart from cleaning the air, green elements serve to reduce noise pollution and assist the formation of suitable microclimatic conditions by reducing the impact of wind and moderating temperature variations. In Germany, it is common for planners to consider channels that provide city centres with fresh air from the urban fringes. The fresh air channels are essentially green corridors, planted with vegetation that does not obstruct the wind and therefore the air flow into the city. These channels can, of course, serve simultaneous roles, such as routes for footpaths and cycleways.

Issues of water use and quality pose some of the most pressing municipal engineering problems. Middleton (1995) has summarised the principal hydrological impacts of urbanisation as being associated with *surface-water* problems, such that the poor water quality often associated with pollution and run-off from construction sites means that many rivers that flow through urban areas are biologically dead, and *groundwater*, which often suffers seepage from heavy metals, synthetic chemicals and other hazardous wastes such as sewage. In respect of this latter point, aquifers do not have the self-cleansing capacity of rivers and, once polluted, are difficult and costly to clean. Also, overusing groundwater leads to lowering of water table levels and consequent ground subsidence; equally, though, the abandonment of pumping leads to problems, so that rising groundwater levels have become a critical problem for many 'postindustrial' cities as manufacturing industries have given way to service industries which are much less demanding of water.

Sometimes, water may not always need to be improved, as grey waters are acceptable for a variety of uses and may be recycled within plants. It has therefore been suggested that cities can promote the use of two-water supply systems through the spatial planning system, and by incorporating such requirements into building regulations; where the installation of complete two-water supply systems is considered infeasible, the option of installing double sewerage networks may provide a compromise (separation of water into two categories – washing water and toilet waste). Washing water could be treated separately (to remove harmful substances such as phosphorus) and then reused, for example, for agricultural purposes. Urban flooding has been an increasing concern, as the accumulation of relatively modest human interventions in river basins – such as cutting down small areas of woodland, urban expansion, straightening the courses of streams, laying drains for agricultural field improvement, waste disposal from factories and the use of herbicides by local authorities – all contribute to floods and their environmental consequences. Thus, the principles of sustainable water management require consideration of the complete cycle of urban water systems including water extraction, treatment, distribution and consumption, sewerage, waste water treatment and waste water disposal. It has been argued that sustainable water management should have four-fold aims, namely: to secure the availability of a basic level of service; to ensure the protection of habitats, flora and fauna; to preserve the water's capacity to dilute and remove pollutants; and to safeguard the aesthetic and recreational values of water elements in the landscape (Guy and Marvin, 1996).

Traffic and Transport

Over the last 40 years, increased personal mobility, due primarily to growth in car ownership, has been a central factor in the dispersal of homes and jobs into suburban and rural areas. Major new road building has enabled people to commute long distances, and the highly concentrated settlement patterns deriving from the historical anomaly of the Industrial Revolution are progressively being replaced by a continuation of preindustrial population distributions, similar to the even distribution of hierarchical settlements predicted by early economic geographers. More recently, evidence suggests that a 'principal component of growth in urban traffic during the 1990s has been the rapid expansion of numbers of women in the waged labour force; this is likely to be compounded by the effects of increased longevity, as an ageing population becomes heavily reliant on the accessibility provided by the car. Moreover, car dependency has been exacerbated by the creation of 'edge cities' comprising major new office complexes, high-technology factories, warehouses and retail parks. All this is leading to an increasingly complicated pattern of movements and a large increase in the average length of journey to work trips: the prospect for simple interurban movements is reduced; even journey-to-work travel is likely to be complex and multipurpose; and households are likely to create rising journey-to-work distances as two partners search for rewarding work within commuting distance of home. Car-based patterns of land use have developed, and these are not well served by public transport. Indeed, it is difficult to see how mass transit systems can cope with the increasingly complex travel patterns being created by new settlements and modern lifestyles.

In many cities in the EU the car accounts for over 80% of urban mechanised transport, and in some (e.g. Milan) the car's share is over 90%. Despite the rhetorical claims made by government about 'greening' the urban environment, forecasts in fact suggest that very substantial increases over present levels are likely. Road traffic, as well as being the main cause of noise pollution, is also responsible for most of the summer smog formation in Europe, and WHO guidelines for ozone, NO_x and CO_2 emissions are regularly breached. Roads also consume much of the urban land budget and typically account for 10–15% (rising to 30% in some cases) of the area of large cities in Europe. In general, transport can be seen to have four major areas of environmental impact:

- *land* – use and wastage of land and its associated ecosystems; excavation and use of minerals for road construction; generation of solid wastes as vehicles are withdrawn from use;
- *biospheric* – introduction of migrant species to new environments or, alternatively, barriers to species migration;
- *atmospheric* – emissions of greenhouse gases, particulates, fuel and fuel additives; noise and vibration; and
- *hydrologic* – contamination of surface and groundwater from surface runoff and spillages of petrol and oil and transported substances; modi-

fications of hydrological regimes during construction of transport facilities (EUEGUE, 1996).

Distance travelled to work is strongly linked to the mode of transport used by commuters, as people travelling by car and rail tend to have longer journeys to work. This has the effect of extending the working day, so that commuting to and from work takes up a large part of many people's lives. How people commute is largely determined by what they do for a living, where they live, the modes of travel that are available to them and their income and personal preference. The level of investment into public transport influences people's transport decisions, and efficient, clean public transport systems with relatively low fares will undoubtedly attract more commuters. The revival of public transport, and to a lesser extent cycling and walking, for travel to work journeys will require a substantial diversion of funds from road-based transport investment. Economy, efficiency, reliability, quality and safety are major factors which influence people in their selection of transport mode, and these are all significantly affected by investment levels. In response to these needs, local authorities are now being required to assemble packages of 'balanced' transport and demand management measures designed to promote public transport, cycling and walking. However, the sums allocated through this method thus far have been too low to achieve a significant shift in the distribution of transport investment.

The other side of the transport problem is to focus on reducing people's need and propensity to travel. Thus, there are proposals to restrain the amount of traffic entering major city centres, including tolls or meters. Techniques of traffic calming have some effect in diverting heavy and fast flows of traffic away from residential areas. More ambitiously, substantial investment in high occupancy vehicle (HOV) routes and light rail transit may be necessary to create fast, reliable and safe systems. The OECD study *Urban Travel and Sustainable Development* (1995) sets out three strands of an integrated policy aimed at moving towards sustainability:

1) best practice in urban policy;
2) using innovative land-use and transport measures to reduce the need to travel; and
3) application of progressive increases in fuel taxation to reduce car kilometres and CO_2 emissions.

This study estimated that it would take two to three decades to see the full effects of this integrated policy, but that benefits would start to flow as soon as it was implemented.

Nature in the City

The subject of landscape ecology has been addressed mainly in the context of the countryside, but very similar arguments can be raised in favour of retaining green networks in cities. Urban networks, however, are less likely to contain outstanding examples of natural capital, though they may more readily be justified on multiple use grounds. In particular, urban greenspace frequently has a greater appeal to nonspecialists than do key rural sites, and may thus encourage public participation and support. Urban networks typically comprise formal parks, footpaths and cycleways, old railway tracks, areas of derelict land, patches of waste ground and some specially designed sites such as urban nature reserves and community gardens.

Social aspects of greenspace are those which enable people to make contact with nature in the course of their normal daily lives (Barker, 1997). For example, English Nature (1996) advocate that people living in towns and cities should have:

- an accessible natural greenspace less than 300 m from home;
- statutory Local Nature Reserves provided at a minimum level of 1 ha per thousand population; and
- at least one accessible 20 ha site within 2 km of home, one accessible 100 ha site within 5 km of home and one accessible 500 ha site within 10 km of home.

This would be even further improved if these sites were nodes on a green network. Certain urban sites, such as woodlots, allotments and city farms, may also give city dwellers the opportunity to engage with rural pursuits such as forestry and agriculture. However, there are a number of social concerns about public safety associated with urban greenspace, and linear links of various sorts may be perceived by residents as 'crime corridors'. These dangers may be more apparent than real, but this scarcely matters if adverse perceptions alienate people from such ideas. Various solutions have been proposed, including information programmes, improved site design, involving local people in site proposals and management, providing a ranger service and clearing litter.

Dawson (1994) has suggested that landscape ecological dynamics can be promoted by applying some general principles, namely:

- corridors should be preserved, enhanced and provided, as they do permit certain species to thrive where they otherwise would not;
- corridors should be as wide and continuous as possible; and
- their habitat should match the requirements of the target species.

However, Barker (1997) notes that many species able to live in urban areas do not need to have seamless continuity in their preferred habitat to let them move from place to place, but that a close mosaic of habitats is sufficient. Nevertheless, this requires detailed knowledge of the kinds of fine detail which permit the diffusion of particular species, facilitating the retention of microhabitats such as old tree stumps, underpasses and ponds. Within these networks, core sites are also essential, requiring the implementation of habitat protection, creation and expansion programmes.

Wider environmental benefits from urban greenspace include water management, air hygiene, landscape and leisure. Careful design of urban river corridors can assist flood prevention, aquatic wildlife and groundwater recharge; sometimes, features may assist more than one of these, such as reed beds being established to remove pollutants from sewage whilst also providing a valuable ecological habitat. Also, wide green corridors or major open spaces within the urban fabric can help to mitigate the urban heat island effect, especially where located on ridges or along river corridors. Air flows in these spaces can help flush pollutants from the urban system, whilst vegetation can help trap and remove air-borne particulates. It has also been contended (Spirn, 1984) that the establishment of a green network can impart a macrostructure to the urban landscape, whose usually gentle gradients are in turn often compatible with the provision of human movement corridors (e.g. walkways and cycleways).

Barker advocates that the most effective way of achieving urban green networks is by preparing long-term, environmentally led strategies for settlements. This would need to address such aspects as:

- the grafting of new network elements on to extant natural features and transport corridors;
- retention of natural elements as viable, integrated systems, which then serve as a constraint on the scale, configuration, design and management of future development;
- the promotion of urban green networks which serve multiple needs, creating open space which is 'multifunctional and pluricultural' (EFILWC, 1996);
- protection of some gap sites and brownfield land from redevelopment, where they serve a more positive use as an environmental resource; and
- using approaches to civic design which are landscape led, rather than including the planting of ecologically simple communities as an afterthought.

Perhaps the most important outcome of this approach to planning could be that the urban-rural divide would be broken down, and that town and country would become more connected both in people's minds and in the functioning of environmental systems.

Industrial Ecology

An emergent possibility for future city planning is offered by the concept of 'industrial ecology'. This explicitly represents urban-industrial systems as ecosystems, which cycle materials and conserve energy. In the same way that elements of natural ecosystems are crucially interdependent, so industries can become intimately connected by supplying each other with

material and energy needs; thus one factory's wastes may become another's raw material or heat supply. The first step in such a scheme is to
prepare an ecosystem approach to industry, involving:

- conscious mapping of resource flows at a city and regional level;
- co-ordinated development of industry sectors, technologies and individual companies to maximise resource synergies;
- minimisation of imports of materials, exports of wastes and long-
  distance movement of part-finished goods; and
- encouraging businesses to locate near their workforces, suppliers, cus-
  tomers and other businesses with synergies (Socolow *et al.*, 1994).

One feature of industrial ecology is that it may involve some revision of
traditional town planning principles: for instance, even greater sympathy
towards mixed uses, high-density development and urban (rather than
rural) locations may be desirable. The historic problems associated with
close juxtaposition of industrial and other land uses may be reduced by
encouraging safe, clean, quiet and benign industries, so that it may be
necessary to target specifically this type of inward investment, for example
by providing specialised infrastructure.

In a sense, industrial ecology can be seen as promoting 'symbiotic' links
between industries so that environmentally conscious practices can also be
profitable. Ecological systems are characterised by energy capture and
material cycling, but urban systems are typically linear and less efficient in
their use of matter and energy. Thus, advocates of industrial ecology argue
that human production systems should aim to apply the dynamics of natural systems to increase efficiency and to minimise demands on environmental sources and sinks (Cosgriff Dunn and Steinemann, 1998). Conventional
approaches to environmental management in industry have tended to focus
on waste reduction strategies within individual service and manufacturing
units; conversely, the essence of industrial ecology is that a 'systems view' is
taken of production activities. As in nature, industries may evolve to occupy specific 'niches' where they utilise or transform particular industrial
by-products into useful resources. Thus, it is claimed that there is a variety
of ways in which industrial ecology can create symbiotic links between
environmental quality and economic development:

- waste products from one industry can provide inputs for another (reducing costs and protecting earth resources);
- waste disposal costs can be reduced because waste products serve as
  inputs for other industrial or municipal processes;
- industry profits can be increased because by-products previously considered waste now have economic value (niche industries have the
  opportunity to evolve and fill in the gaps between two other entities to
  exchange energy and material flows); and
- the local economy can be improved by encouraging a larger mix of
  environmentally conscious businesses.

Beatley (1995) has also claimed that industrial ecology can assist social sustainability, in so far as it counters the traditional trend towards separation of land uses and sprawling development patterns, which tend to separate groups by income and race, with an accompanying inequitable distribution of economic and life opportunities. Although there is still little evidence that industrial ecology is yet a universally realistic model, there has been some progress towards establishing it in practice. Most notably, the town of Kalundborg in Denmark has been widely studied by advocates of this concept. This small industrial town is home to an intricate web of energy and material flows between the city and its local industries, and it does appear that 'symbiosis' between industrial partners is being achieved through a series of bilateral agreements.

Conclusion

Urban areas – in the developed world especially – are notorious for the scale of their ecological footprint. Whilst planning controls have helped to contain the land-use impact of urbanisation, the often unseen impact of cities spreads far beyond their physical boundaries. Air and water pollution, solid waste generation, derelict land, congested roads, waste energy and disruption of hydrological and biological processes, all reflect the dysfunctional current state of urban ecology.

Urban environmental management is, however, an area of intense activity both at the research and policy levels. After decades of indecision, societies are now starting to accept the need for much more rigorous demand management, and are turning to a variety of regulatory and market-based instruments to shape our use of natural resources more efficiently. More speculatively, there is growing interest in the potential of industrial ecology, which could provide a basis for a more equal relationship between environmental systems and economic activity.

In practice, a major constraint to improved design and use of urban ecological capital is the inherited nature of most of our urban stock. Planners rarely have the possibility of starting afresh – indeed, when they have had opportunities for comprehensive redevelopment of large areas, their results have often been questionable. Consequently, most measures must involve a degree of retrofitting solutions into an ageing stock of houses, roads, infrastructure and open space. Occasionally, new settlements or substantial urban extensions may be developed and, whilst these are typically controversial in terms of land take, they offer huge scope for model sustainability planning. Perhaps most problematic of all will be the ingrained attitudes of city dwellers who have unwittingly become accustomed to lifestyles which depend on unacceptable environmental damage: planners' skills and knowledge will be to little avail unless efforts are made to build a new social consensus about sustainable living.

7 Information and Decisions in Environmental Planning

This chapter covers two areas which are essential for effective environmental planning in practice: the types of information which are available to assist planners in reaching decisions, and types of technique which can assist planners in reaching decisions based on the best available information. Historically, these areas have been associated with reductionist scientific approaches, assuming a very high quality and fitness of information, a highly rational decision-making process, and a seamless link between policy generation and implementation. It is premature to sound the death knell of this 'rationalistic' approach to planning, as no planner would want to abandon the tenets of well informed and carefully reasoned decisions. However, it is true to say that there is now far greater awareness of the patchy nature of 'scientific' data, of the need to combine 'expert' and 'lay' sources of information, and of the weaknesses of decision-support and implementation methods. The following discussion centres on information regarding natural resources and environmental quality, both of which increasingly seek to integrate lay knowledge, and on methods for optimising environmental investment decisions.

Information for Environmental Planning: Interpreting the Natural Resource Base

Interpretive mapping of environmental resources has a long history, and serves two main purposes. One is to tell us the extent and quality of resources which are present, so that the best can be safeguarded as critical natural capital. The other is to identify the potential of land, so that it may be used for new purposes to which it might be better suited. Thus, some classifications are 'neutral', simply presenting as objective an account as possible of land resource properties, whilst others are more purposive, suggesting potentials and capabilities of land.

Land Classification

Researchers in both the natural and social sciences often wish to investigate phenomena on the basis of a sample survey and, in order to enhance the representativeness of a sample, often seek to draw it upon a stratified basis. In respect of land, however, there may be no obvious sampling frame to adopt as a basis for stratification, despite the obvious benefits this would have for the conduct of surveys or the selection of particular land types for closer scrutiny. Thus, a method developed by the Institute of Terrestrial Ecology (ITE) (Bunce *et al.*, 1981; Smith and Budd, 1982) aims to classify land, according to readily available information, into types which are reasonably internally homogeneous.

In identifying classes on the basis of a large quantity of attributes, a choice must be made whether to cluster together these attributes into associations on the basis of statistical linkages (from the bottom up) or whether to subdivide (from the top down) in order to identify characteristics which discriminate diagnostically between different groups. The ITE method adopts the latter approach, using a technique known as 'indicator species analysis' (TWINSPAN), and this produces land classes which are described in terms of a set of diagnostic criteria. More specifically, a large suite of geologic, topographic and land use attributes are recorded from Ordnance Survey maps, and it is left to TWINSPAN to select those which are most important in determining the land classes. It successively divides the dataset into 2, 4, 8, 16 and finally 32 classes, at which point the exercise is terminated. This provides a natural grouping of samples based on presence/absence data for selected attributes. The result is an efficient key to classification enabling grid squares not surveyed in the field to be allocated to their most likely land classes on the basis of desk studies (Box 7.1).

Resource Evaluation – Agriculture and Forestry

High-quality farmland is a valuable and essentially irreplaceable resource. Historically, it has been strongly safeguarded and has been a major constraint on development in the countryside. The emphasis on protecting prime agricultural land has somewhat abated in recent years as food production has moved into surplus, though this situation could well be reversed if we return to a style of agriculture which is more reliant on intrinsic soil quality and less on chemical and energy subsidies. As agriculture operates in a climate of policy variability and changing demands, the land which is suitable for the widest range of production purposes is most highly valued. The major consideration in respect of evaluating farmland is thus the presence of *limiting factors* which affect its flexibility, the principal ones being soils, slope, climate, wetness and soil erodibility. An evaluation of these factors underpins the two principal farmland classifications in the UK.

Thus, in Britain, the Land Use Capability Classification (LUCC) (drawing heavily on the US Department of Agriculture approach) recognises seven classes of farmland. In addition to the main gradation from 1 (excellent, unconstrained and highly flexible land) to 7 (of little or no use even for very low-intensity grazing), there is a further subdivision of classes 3

Box 7.1   An example of statistical land classification

In a study of the land characteristics of Northern Ireland, Cooper and Murray (1992) based a multivariate classification on information derived from map grid squares. The analysis initially produced the conventional 32 ITE land classes but these results were simplified into 10 land groups. The key qualities of these groups were:

- *Group A* – flat land below 50 m (typically coast, estuaries and river valley bottoms), dense road network, Mesozoic rocks and alluvial soils.
- *Group B* – relatively flat, 51–150 m, frequent small villages, mainly schists and quartzites with overlying brown podzolic soils.
- *Group C* – mostly below 150 m, many small buildings, ubiquitous streams, basalt and Mesozoic rocks, soils mainly alluvial and gleyed.
- *Group D* – sloping land between 51 and 150 m, at lowland–upland transition, many farm buildings, schists and brown podzolic soils.
- *Group E* – mainly 151–250 m, mainly middle mountain slopes, reasonable density of buildings, mainly schists with brown podzols/peaty podzols.
- *Group F* – rolling hill land, mainly 151–250 m, frequent streams, loughs and rough grazing, varied rocks with brown podzolic and peaty soils.
- *Group G* – upland group centred on 151–250 m contours, mainly steeper and higher land of glen sides and escarpments, mainly schists but some basalts and other rocks.
- *Group H* – mountain terrain and uplands, mainly schists and basalt, with peaty gleys and brown podzolic soils.
- *Group I* – mainly 251–500 m elevation, less steeply sloping, mainly schists and basalt, peaty gleys and blanket peat.
- *Group J* – highest mountain terrain, mainly schists, usually with peaty gleys or blanket peat.

and 4 according to intrinsic quality and enterprise suitability. There is also a complementary notation denoting the nature of constraints, if any, and these comprise soil deficiencies (s), wetness and drainage problems (w), adverse climate (c), liability to erosion (e) and excessive gradient (g). This detailed assessment of land capability is of greatest use to those involved in the agricultural industry, as it forms a basis for technical advice and decisions relating to investments such as enterprise type, field drainage and use of machinery. Perhaps of more interest to planners are classifications which provide a straightforward indication of the relative quality of farmland, so that development plan policies can be written which give appropriate levels of safeguard. Thus, in England and Wales a simpler five-grade Agricultural Land Classification (ALC) has also been compiled, whose top four grades are broadly comparable with the best four classes of the LUCC, whilst grade 5 is generally thought of as 'rough grazing'. A complementary upland land classification has also been produced, focusing on the poorest grades of land and distinguishing between hill and upland sites which can potentially sustain some cropping, improved pasture, nutritious unimproved pasture or which are inherently very poor. Problems of interpreting the ALC have arisen because of the broadness of grade 3, so that this is now subdivided to enable more refined policy interpretation. Certain types of land, even though not nationally important on the ALC, may still, however, have a strong importance in the local agricultural economy. The most important example of this is in-bye land in the uplands, where small areas of good-quality valley land or improved lower slopes may have a disproportionate effect in maintaining stocking densities. Loss of in-bye land through infrastructural developments in valley bottoms may thus have a severe impact.

The use of land capability assessment in planning situations is illustrated by the kinds of assistance given in relation to transport schemes. At 'stage 1' the appraisal is based on ALC maps, whilst at 'stage 2' the Regional Land Use Planning Unit (LUPU) of MAFF will commission, free of charge, a broad appraisal, based on existing documentation, of the agricultural implications of route options and the extent to which each might be expected to affect the national interests. LUPU, however, charge for evaluating the impacts of single route options. If more than 20 ha of grade 1, 2 or 3a is likely to be lost, MAFF must be consulted, whereas in Scotland consultation must take place if over 2 ha of prime land or over 10 ha of nonprime land is affected.

For developing countries, an approach based on land potential – i.e. 'positive' rather than 'limiting' factors – is more relevant. Here the method of the Food and Agriculture Organisation (FAO) is often used, in which 'land mapping units' are identified and a land-use recommendation is allocated to each of these. The steps in the FAO approach are:

- initial consultations, concerned with identifying the objectives of the evaluation, and the data and assumptions on which it is based;
- descriptions of the kinds of land use to be considered and the establishment of their land and social requirements;
- description of land mapping units and derivation of land qualities;
- comparison of land-use requirements with types of land present;
- economic and social analysis;
- land suitability classification; and
- presentation of recommendations from the evaluation study.

It will be noted that socioeconomic appraisal forms an integral part of this process, for such considerations may be paramount in developing countries. Similarly, economic criteria – for instance, gross margins – are becoming increasingly popular as a component of land evaluation, as these represent actual rather than hypothetical farm enterprises and yields.

With regard to land evaluation for forestry, three approaches are commonly used (Booth, 1985): those based on site indices, volume or indirect criteria. The site index method involves measuring the height of a specified number of the tallest trees within a defined area, and this enables comparisons to be made between different sites. The assessment of volume in forests is usually based on detailed mensuration of a large sample of trees. Once these volumetric measurements have been made it becomes possible to predict standing timber and final yield from a simple measurement of the girth of trees at a height of about 1.3 m above ground, referred to as the 'diameter breast height' (DBH). From measurements such as these can be calculated the yield class of a forest, namely, the increment in cubic metres per hectare per year. Indirect productivity assessment, which is essentially similar to agricultural capability mapping, is based on measurable site attributes such as climate (especially exposure), soil (for instance, fertility and rooting-depth limitations), vegetation and topography. This approach is often necessary where new plantations are to be established, and there is no pre-existing forest stand on which to base predictions of future yield. The approach developed on this basis for Scotland (Bibby *et al.*, 1989) is summarised in Box 7.2.

Box 7.2   Classes in the Macaulay Land Capability for Forestry Classification

---

- *Class F1* – land with excellent flexibility for the growth and management of tree crops; deep soils well supplied with moisture; no serious adverse climate or site factors; wide range of broadleaved and coniferous trees possible.
- *Class F2* – land with very good flexibility for the growth and management of tree crops; soils have only limited waterlogging; minor restrictions of soil depth, slopes and climatic restraints; both broadleaved and coniferous species suitable, but choice is more restricted than in F1.
- *Class F3* – land with good flexibility for the growth and management of tree crops; land management is primarily concerned with limitations imposed by drainage, sloping land or patterns of variable soils; suitable for a wide range of conifers or restricted range of broadleaved species.
- *Class F4* – land with moderate flexibility for the growth and management of tree crops; local ploughing difficulties may be encountered due to stony or shallow soils; small areas at risk from windthrow; suitable for many coniferous species and in places for the less demanding broadleaves.
- *Class F5* – land with limited flexibility for the growth and management of tree crops; ploughing possible but may be difficult; some sites at risk from windthrow, sufficient to affect management/thinning practices; mainly suitable for spruces, larches, pines; possibly birch, alder or other hardy broadleaves.
- *Class F6* – land with very limited flexibility for the growth and management of tree crops; adverse climate and poor soil; windthrow may effectively prevent thinning and seriously curtail rotation length; surface terrain may impose great difficulty on ploughing; species choice mainly restricted to lodgepole pine and Sitka spruce.
- *Class F7* – land unsuitable for producing tree crops; sites subject to extremes of climate, wetness, rockiness and slope.

Source: Adapted from Bibby *et al.*, 1989.

Resource Evaluation – Wildlife

The evaluation of wildlife has typically taken place in relation to habitats. This is because it has long been assumed that habitats are suitable proxies for individual species, which would be much more time-consuming to map and analyse. Consequently, various approaches have been taken to identifying key sites. The most systematic mapping in Britain has been undertaken according to protocols developed by the former Nature Conservancy Council. These drew upon a three-phase system of mapping ecological resources, comprising:

- *phase 1 habitat survey* – providing a general description of habitat/ vegetation types within a study area and to fit these to a standard classification;
- *phase 2 survey* – assembling further information, usually on selected sites. In phase 1 habitat survey information on the *species composition* of communities is normally restricted to species lists. Although these are useful, they give no indication of *species importance* in a community. Phase 2 survey involves the collection of quantitative vegetation data, again with the aim of applying a standard classification to facilitate comparative evaluation, and of abundance data on selected animal species and/or groups; and
- *phase 3 survey* – comprising intensive sampling to provide detailed quantitative information on species populations and/or communities.

Typical categories of habitat which are used in ecological mapping are listed in Box 7.3.

Box 7.3   Typical categories of habitat used in ecological mapping

- deciduous forest
- coniferous forest
- mixed forest
- acid grassland
- basic grassland
- dwarf shrub heath
- fen
- carr
- saltmarsh
- raised bog
- lowland bog
- marsh
- unimproved grassland
- improved grassland
- arable
- boundary habitats
- standing water
- flowing water
- other soft coastline.

The main classification employed now in Britain is the National Vegetation Classification (NVC) (Rodwell, 1991; 1993), which is an elaborate key for the identification of a wide range of communities. The notion of discrete, distinct vegetation communities is debatable and some scientists emphasise, instead, the continuum of conditions which exist in nature (Sanderson *et al.*, 1995). Nevertheless, the recognition of NVC categories as noda of vegetation within continuously varying species composition has intuitive appeal to plant ecologists. The NVC is based on a series of multivariate classifications of data derived from quadrat surveys throughout Britain, and this classification has been published as a series of manuals that include maps of the distribution of sampled quadrats of each vegetation community, the community species composition, and summary environmental and management regimes deemed to be responsible for its development. The utility of this classification has been ensured by the inclusion of phytosociological keys and the development of computer programs that can be used to assign membership of new stands of vegetation within the classification. These programs do not identify a stand of vegetation as belonging to one single plant community type, but instead provide measures of how close the observed species composition of the vegetation stand is to the species composition of the plant communities described in the NVC. Some geological and geomorphological features such as 'natural rock exposures' and 'limestone pavements' are also recognised as important habitats. Box 7.4 refers to a number of the key types of association found in the NVC.

In addition to mapping vegetation types, ecologists sometimes need to evaluate sites – i.e. place relative values on them which reflect their conservation priority – which is a far more fraught task. Approaches to ecological evaluation have been heavily influenced by the principles adopted in the *Nature Conservation Review* (Ratcliffe, 1977). This and other approaches have indicated the major evaluative criteria to be – in descending order of popularity – diversity, naturalness, rarity, area, threat of human interference, amenity/education value, representativeness of habitat type, scientific value, recorded history, population size, typicalness, as well as a variety of minor criteria including fragility, potential value and management considerations. A clearer indication of what is meant by the more important of these is given in Table 7.1. Usher (1986) has, however, noted that not all these are equally effective in distinguishing between sites of differing value: on the basis of a test exercise, he identified area, diversity, rarity and typicalness as the most efficient discriminators. Whilst it would be difficult to generalise from this particular example, it is worth noting the problems of basing assessments on particular criteria, and of combining the results of diverse criteria into a single index of relative value.

Approaches to conservation evaluation often tend to be either use-specific or ecosystem-specific. An example of the former is that used by the statutory nature conservation organisations in their selection of SSSIs. Since each of the main ecosystem types shows important variations, especially those related to climatic gradients, the aim of conservation has been to select and consolidate a countrywide network of each habitat type. Here, site selection has involved the careful sifting out, through a comparison of similar types, of those sites which represent the best examples of

Box 7.4   Key types of association found in the NVC

Woodland and scrub is typically first divided into broadleaf and coniferous communities, though the NVC includes many woodland communities and subcommunities, and English Nature's 'semi-natural' category can be divided in relation to woodland histories. It is almost impossible to distinguish primary woodland from very old secondary woodland, so for practical purposes a distinction is drawn between *ancient woodland* (before AD1600) and *recent woodland*. Scrub is a seral community which will tend to colonise grassland, though it often contains significant ecological value.

Grassland and marsh are typically, in western Europe, anthropogenic climax communities maintained by mowing and grazing, and will tend to convert to scrub and then woodland if these controlling factors are removed. The NVC contains a detailed classification of grassland plant communities, which may broadly be equated with the three main EN habitats, namely, acid, neutral and calcareous grasslands. Most acid grasslands are species-poor hill pasture on thin, acid soils, and may intergrade with moorlands. Changing practices, including overgrazing, have led to the spread of bracken and gorse scrub. Neutral grasslands are mainly found on lowland loams and clays – in general, the most valuable are 'meadows' which are species-rich in both plants and invertebrates. Wet meadows intergrade with marsh. Calcareous grasslands occur on chalk or limestone substrates, and are typified by English downlands; traditionally, they were maintained by low-intensity sheep grazing which, together with rabbit grazing, encouraged a species-rich sward and associated animal communities.

Heathland is, like grassland, an anthropogenic climax community that was created by forest clearance and maintained by grazing, fire and the use of materials for thatching and fuel. The substratum is always acid and oligotrophic – 'dry dwarf shrub heath' is dominated by ericoids and dwarf gorses, whilst 'wet dwarf shrub heath' is similar but contains high proportions of hydrophilous plant species and intergrades with 'mire'.

Mire is normally represented by bogs or fens. Bogs are confined to high rainfall areas, especially 'blanket bog' (i.e. forming a continuous cover, except on steep slopes or exposed ridges), which develops only where there is a meteorological water surplus for most of the year. Most of the 'rich fens' have been lost, due to agricultural improvement; 'poor fen valley mires' are quite common in heathland areas where they intergrade with wet dwarf shrub heath; 'basin mires' are often found over deep depressions that were evidently created by glacial activity. Because they are 'geogenous', fens are highly susceptible to changes in the quantity or quality of groundwater supply.

Source: Based on information in Morris *et al.*, 1995.

reference points within the British field of variation. At the outset, survey teams are given some discretion to decide what degree of variation around each ecosystem type should be represented. When more than one possible candidate site for each regional variant is found, comparative evaluation is needed to identify the best. At the site level the criteria previously described (extent, diversity, etc.) come more prominently into play. More recently, a semi-quantitative scoring system for the selection of SSSIs has been elaborated but, whilst greater standardisation of evaluation procedures is desirable, it is limited by inherent subjectivity in the values

Table 7.1　Major evaluative criteria for nature conservation sites

Criterion	Major evaluation principles
Diversity	Community diversity is based on a count of communities present at a site (presumes existence of a suitable classification). Species diversity is measured by one of: 1) species richness (count of number of species known to occur); 2) functional analysis of the number of different species and their relative abundance; and 3) number of different trophic levels and the interactions within and between these
Area	Assumes the larger the area/extent of the site, the greater its nature conservation value; related to the area requirements of species in the site (e.g. large predatory species require large hunting grounds)
Rarity	Considers the distribution and range of a species, isolation and size of populations
Naturalness	Implies minimum disturbance by human influence, or that species have not been introduced by a human agency
Representativeness	A classificatory criterion, based on a list of the kind of ecosystems occurring within a geographical area; can only properly be assessed in relation to surveys of total biogeographical area in which site is located
Typicalness	Quality of a site as an example of a particular ecosystem
Fragility	Degree of sensitivity of habitats, communities and species to environmental change

Source: Ratcliffe, 1977; Usher, 1986.

themselves. Ecosystem-specific evaluations are generally based on phytosociological approaches, for instance those which have been devised to assess the conservation potential of woodlands (e.g. Kirby, 1986).

Resource Evaluation – Landscape

Assessments of 'landscape' have to satisfy more subjective criteria than purely 'scientific' resource evaluations, and selection of the appropriate attributes and ascription of value judgments provoke continuing controversy. With respect to landscape assessment, the major problem is determining aesthetic merit, so that one vista can be ranked relative to another. This is clearly a contentious matter, influenced by such factors as education and contemporary taste (in respect of the observer), season and lighting conditions (in respect of the landscape) and, where it is impractical to take groups of surveyors out into the field and prints/ transparencies must be used, the quality and framing of the photographs. Many researchers have sought to reduce the intrinsic variation in the process by adopting quantitative approaches to the description and evaluation of landscapes. In these, simple numerical indices based on measurable landscape features have progressively been replaced by multivariate statistical measures based either on clusters of landscape elements or on observers' consensus. These may have particular applicability in planning decisions, where the landscape calibre of an area may require to be judged in terms of its regional/ national importance. However, their use in comprehensive national survey has not been a success.

Thus, for many practical purposes purely numerical approaches are considered to be unreliable, and not appropriate to the public objectives associated with landscape designation, protection and management. Latterly, assessments based on a combination of expert subjective opinion and local sources have gained increasing favour, though these need not exclude a degree of quantification. The method adopted by the Countryside Agency is based on the assumption that people value their surroundings according to cultural and historical associations as well as the basic character of the landscape (Countryside Commission, 1993). Associations may include links with writers, composers, painters and famous people, as well as historical patterns of landscape development. Intrinsic landscape character incorporates physical features, form and colour, length and breadth of views, amount of clutter or detail, and features which focus the eye. Together they produce a unique sense of place. The approach accommodates objective and subjective qualities of landscape, by combining verbal descriptions with structured checklists and sketches. The survey then follows the sequence:

1) define the purpose of the survey – set objectives and criteria for judgement;
2) desk survey;
3) field survey – visit, travel around/through, complete systematic survey sheets, make sketches, record impressions;
4) analysis – break landscape down so that the observer understands what makes it special;
5) evaluation – judge landscape against criteria established at the outset, decide whether it measures up; if there are different character zones, are all of equal value in relation to the criteria?; and
6) reach decision.

This process is expanded upon in Box 7.5.

An important subset of landscape assessment has been associated with the description, classification and evaluation of river valleys; indeed, the principles associated with this are transferable to more general situations of landscape planning. For example, the former National Rivers Authority (NRA) produced guidelines for landscape assessment based upon three elements (NRA, 1993):

- *inventory/description:* a factual documentation of the landscape including a description of character, the elements which contribute to this and their interactions. In order to convey the essential character of landscape, or its sense of place, aesthetic and perceptual factors are included in the description, and this involves subjective judgements;

Box 7.5 Stages in landscape assessment

Landscape assessment comprises a number of steps, including 1) definition of purpose, 2) desk study, 3) field survey and 4) analysis and reporting. The process follows:

- *general familiarisation* with the survey area, through preliminary visits and desk study; and
- *structured survey*, undertaken by formal observations at sample points.

Recording and presenting information from structured field survey work are important. Methods include:

- *map annotation*, to record where different features are in relation to each other and their significance, and aspects of visual analysis such as view lines, eye-catching features, edges, dead ground and boundaries between areas of different character;
- *checklists*, to record the presence or absence of a variety of landscape elements, their conspicuousness and whether they make a positive or negative contribution to the landscape. More subjective, judgmental factors can be recorded, for example, by a checklist of descriptive adjectives of aesthetic characteristics;
- *written descriptions*, to provide an overall impression of the landscape – to paint a word picture. These are difficult to do well because capturing the essence of place in words is not easy. Descriptions should incorporate information about the landscape elements present, the contribution they make, the aesthetic characteristics of the landscape and the way the surveyor perceives the landscape as a whole;
- *annotated sketches*, to convey information about the way different aspects of a landscape interact when viewed at ground level. Annotation ensures that a record is made of the way the surveyor perceives the landscape; and
- *photographs*, to provide a supplementary record. They are a quick way of recording a landscape and, while they do not interpret the scene, they provide a basis for subsequent drawings to illustrate the character of the landscape.

Source: Countryside Commission, 1993.

- *classification:* a division of the study sites into landscape areas which have distinct and recognisable character, and grouping together areas of the same type (involving a degree of professional judgement); and
- *evaluation:* a judgement of the relative value of different areas of landscape or of different features within them. Where resources allow, it may be helpful for the professional judgements of the assessor to be complemented by independent perceptions and valuation of the landscape. This can involve review of published material about the landscape and, where necessary, assessing local public perception.

A particular aspect of river landscape evaluation is the distinction between strategic and detailed assessment, the former analysing the macro- and micro- river landscapes (e.g. the valley's visual envelope and environmental characteristics, and the channel itself together with the profile and vegetation of its banks). The detailed assessment concentrates on a smaller area defined by a specific purpose of work (e.g. a flood defence scheme) and considers important features and areas of special sensitivity. Evaluation and management strategies are addressed by seeking to place river segments in particular 'value classes' and proposing management strategies in respect of conservation, restoration or enhancement. The former NCC also considered the management of drainage channels, the aim being to encourage those practices which benefit and in many cases enhance wildlife within drainage channels and their banks, whilst achieving a standard of maintenance appropriate to the flood defence and land drainage needs of the area (Newbold *et al.*, 1989). Various factors of surface and groundwater hydrology, and associated fauna and flora, are drawn to the attention of engineers (Box 7.6).

Box 7.6   Ecological assessment of rivers

Types of drainage channel comprise:

- lowland valley river upstream of coastal plain;
- lower reaches of lowland rivers;
- internal drainage channels in urban and arable areas; and
- internal drainage channels in grassland areas (which keep a higher water table to act as a barrier to stock and prevent further land shrinkage in peaty areas).

General principles for channel maintenance, in respect of sustaining wildlife interest, include:

- the range of aquatic wildlife is dependent upon the depth of water held in the drainage channel;
- most aquatic wildlife is dependent on a stable water level over any year;
- the quality of water is crucial to both wildlife and drainage;
- conservation management within the channel and on the banks is essential; and
- habitat creation, reinstatement or restoration can provide more varied habitats for wildlife, increasing the number of species in or along any drainage channel.

Various options may be pursued for achieving these in channels, off-channel areas, channel margins and banks.

Resource Evaluation – geology

Environmental geology deals with the interaction between people and their physical environment, and is primarily concerned with the sustainability and physical capability of land and of natural resources for different types of development. Guidance produced for the DETR (Thompson *et al.*, 1998) draws attention to the role of resource evaluations in alerting planners to:

- the natural characteristics of the ground, such as natural underground cavities and mineral resources;

- the legacy of previous land use, including industrial contamination and mining-related subsidence; and
- the characteristics of natural physical processes such as flooding, land-slips and coastal sediment movement.

These attributes raise issues of potential hazards, resources and opportunities for conservation. Thus, planners need to understand them in terms of the constraints which they impose on, or create for, development, and the effects which development can have on the geological system.

Sound environmental planning of geological resources can, the guidelines suggest, be based on:

- high levels of awareness (notably consulting with expert sources);
- acquiring information from various sources (BGS, 1998);
- implementing systems for storage, handling and retrieval of geological information;
- implementing procedural frameworks for taking into account environmental geological issues in plan preparation and planning decisions;
- using expert advice; and
- allocating resources to improving the ways in which earth science information can be accessed and integrated into planning decisions.

Various sources of government advice are available to assist planners in interpreting environmental geology (Table 7.2).

Table 7.2 Government advice for interpreting environmental geology

England	Wales	Scotland
PPG 14 (Development on Unstable Land)	TAN(W) 13 (Coastal Planning)	NPPG 7 (Planning and Flooding)
PPG 20 (Coastal Planning)	TAN(W) 19 (Development on Unstable Land)	NPPG 10 (Planning and Waste Management)
PPG 23 (Planning and Pollution Control)	TAN(W) 20 (Development of Contaminated Land)	PAN 33 (Development of Contaminated Land)
MPGs on mineral resources generally, and on mined ground	TAN(W) 21 (Development and Flood Risk)	PAN 51 (Planning and Environmental Protection)
DoE Circular 30/92 (Development and Flood Risk)	TAN(W) 13 (Planning and Pollution Control)	NPPG 4 (Land for Mineral Working)
National Rivers Authority (1992) Policies and Practice for the Protection of Groundwater		Scottish Environmental Protection Agency Groundwater Protection Strategies for Scotland

Notes:
MPG = Mineral Planning Guidance; NPPG = National Planning Policy Guidance; PAN = Planning Advice Note; PPG = Planning Policy Guidance; TAN(W) = Technical Advice Note (Wales).

Rapid and Participatory Rural Appraisal

During the 1970s, a suite of methods commonly referred to as 'rapid rural appraisal' (RRA) started to emerge. Primarily introduced for developing countries, where indigenous technical knowledge was sought, they have more recently found application in developed countries where detailed information on local natural resources is scarce. Essentially, it embraces a range of qualitative and quantitative methods of gathering information from local people on land resource and community potentials. Chambers (1994) notes that it provides a better way for 'outsiders' to learn, especially where lay knowledge is considered to be at least as important as expert knowledge. The approach seeks to enable outsiders to gain information and insight from lay people and about local conditions, and to do this in a cost-effective and timely manner. The physical survey element typically draws upon methods of agroecosystems analysis, which seek to ascertain:

- systems properties, such as productivity, stability, sustainability and equity;
- pattern analysis of space, comprising maps and transects;
- time, including seasonal calendars and long-term trends;
- flows and relationships, represented by various diagrams (e.g. flow, causal, Venn);
- relative values, relating to matters such as relative sources of income; and
- decisions (mapped as decision trees/diagrams).

Social topics are recorded through conventional survey techniques, and increasingly through methods of applied anthropology. Thus, it is less one-sided than questionnaire surveys, where much of respondents' time is taken by the outsider, and little or nothing is given back. However, it still entails outsiders collecting data, which they then have to take away and analyse elsewhere. In response to this criticism, therefore, methods of 'participatory rural appraisal' (PRA) have been developed.

PRA entails a more active role by the host community, not only to increase their sense of 'ownership' of the product, but also to engage their analytical capabilities. Both types of appraisal have moved closer together, though it is still broadly accurate to claim that RRA methods are more verbal, with outsiders more active, while PRA methods are more visual, with local people more active. The principal techniques upon which both approaches draw are summarised in Box 7.7.

Economic Evaluation of Rural Resource Use

Wildlife and landscape values are literally invaluable, immeasurable. However, there are various ways in which we can ascribe value functions to

Box 7.7 Principal techniques of PRA and RRA

- Secondary sources (e.g. files, reports, maps, aerial photos);
- semi-structured interviews;
- key informants (perhaps identified through participatory social mapping);
- groups (e.g. focus groups) and group interviews;
- do-it-yourself (outsider asks to participate in community activities);
- they do it (villagers as investigators and researchers);
- participatory analysis of secondary sources (e.g. aerial photographs);
- participatory mapping and modelling (similar to planning-for-real);
- transect walks (walking with or by local people through an area and observing/listening);
- time lines and trend and change analysis (e.g. major remembered events in a village);
- oral histories and ethnobiographies;
- seasonal calendars (e.g. to track seasonal weather changes, labour patterns, patterns of borrowing);
- daily time use analysis (e.g. tasks with time demands, drudgery);
- livelihood analysis (stability, crises, coping mechanisms, credit and debt, etc.);
- participatory linkage diagramming;
- institutional diagrams (identifying individuals or institutions important in and for a community);
- well-being and wealth grouping and ranking;
- analysis of difference (e.g. by gender, age, social group, occupation, wealth/poverty);
- matrix scoring and ranking (to score perceived performances of different seeds, trees, soil conservation methods);
- estimates and quantification (to explore 'what might happen if . . . ');
- key probes (questions which can lead to key issues);
- stories, portraits and case studies;
- team contracts and interactions;
- presentation and analysis;
- sequences (use of a combination of several methods in a given sequence);
- participatory planning, budgeting, implementation and monitoring;
- group discussions and brainstorming;
- short standard schedules or protocols (for short and quick questionnaires or to record data); and
- report writing (without delay, so that feedback is instant).

Source: Based on Chambers, 1994.

environmental assets, as an aid to making logical and reasoned planning decisions. They may also have particular relevance in the determination of compensatory arrangements and in environmental impact assessment. Broadly, it is possible to distinguish 'nonuser benefits' from 'user benefits'. The former comprise option values (associated with amenity and recreation potential, genetic material, ecological stability and research potential), vicarious consumption and landscape (research and amenity) values. User benefits comprise landscape quality, education, research (both pure and applied) and recreation (specialist and nonspecialist activities). (The productive value of land is also a user benefit, but is normally calculated separately.) All these can be costed with varying degrees of accuracy, and four approaches are typically adopted:

1) direct cost methods, in which only those costs resulting from the conservation practices are considered;
2) opportunity cost of the land involved, i.e. the economic return from alternative land uses which is forgone as a result of conservation;
3) cost of alternative means of achieving similar objectives, where these have directly attributable costs; and
4) willingness to pay (WTP) or willingness to accept compensation (WTA) by various user groups.

This last, which is perhaps the most widely used approach, can itself be subdivided according to the questions posed to the potential user. These can crudely be summarised as: 'what would you pay?' (compensation variation), 'how much would you pay in a hypothetical market?' (contingent valuation) or 'what would I have to pay you?' (equivalent variation). Where an individual has no property rights over the resource in question, WTP is the appropriate measure rather than WTA, though the notion of 'rights' may itself be rather complex in relation to environmental resources. Similar values can also be reached on the basis of a community decision-making question and a judicial award question. The choice between these methods is vexed and should be left to the experienced economist, although contingent valuation (CV), backed up by a carefully designed questionnaire, has gained a marked popularity.

CV relies on people making 'bids' which reflect the values which they place on intangible assets such as visual and ecological resources, including recreational resources for which there is no entry charge. Cobbing and Slee (1993) note that the first challenge in CV is to identify the beneficiaries of the environmental values. These are typically people within the catchment of the area (this itself is problematic to define), who might value the environmental resources either for their own use (*use values*), or for use by others either now or in the future: they may also value it for its existence without the rights of use (*nonuse values*). Obviously, obtaining reliable replies from interviewees in response to these hypothetical market decisions requires a very delicate approach, and careful questionnaire design is essential. Expert assistance is needed to design questions which preclude certain types of tactical response from the interviewees, or which do not give respondents the opportunity to 'embed' their more general preferences for environmental protection.

This type of approach has been illustrated in relation to the Pevensey Levels *Wildlife Enhancement Scheme*, where English Nature pay farmers to undertake various measures aimed at retaining the low-lying, wet grazing meadows and their associated system of drainage ditches (Willis *et al.*, 1996). At the time of the survey, total costs for running the scheme were just under £150,000, but bids from interviewees indicated a total 'passive use' value of the site of over £2.5 million/year. Such results must be treated with caution for various reasons, but they do suggest that the environmental values of key conservation sites easily justify the amount of public money spent upon them.

Capacity Analysis

The previously discussed concept of carrying capacity can be extended beyond its ecological and recreational applications to the wider issue of sustainable development. Thus, it can be assumed that towns, areas of countryside, stretches of coastline and so forth, have inherent capacities which must not be exceeded if they are to be managed sustainably. However, even in the relatively straightforward instance of recreation planning, the concept of capacity has proved intractable to determine, and there are many difficulties in applying it to much more complex human ecosystems.

The concept has been articulated in general terms at the policy level (e.g. CPRE, 1997; DETR, 1997), where it is seen to be a useful tool of 'environment-led' planning. In one of the few practical examples the West Sussex Environmental Capacity Study (WSCC, 1996) examined the county's environmental resources and the effects of development pressures on them. The study also attempted a practical methodology to identify least unsustainable locations for development, including the highly contentious issue of substantial additional housing provision.

Remote Sensing

Most information on environmental resources has traditionally been obtained by field surveyors, but for a long time now certain information has been obtained from the air. Aerial photographs have been widely used for the interpretation of landform, land resources and environmental change, and careful interpretation can yield high-resolution information on surface features. Photographs are available in various formats, such as monochrome, colour, infrared (heat emissions) and stereoscopic pairs; skilled interpretation is necessary, especially when making precise estimates of distance, area and height.

Since the advent of earth orbiting and geostationary satellites, huge quantities of information about the earth's surface, oceans and weather systems have been obtained by remote sensing. Instead of relying on photographic images, satellite antennae possess the capability to sense spectral signatures which are emitted by various different types of surface. These can be transmitted back to earth in various wavebands, and unscrambled to give reasonably accurate information on ground cover. This can then be plotted as images which may look entirely similar to photographs, but are in fact 'false colour' plots, where the colours have been carefully selected to approximate to real life. However, colours are often allocated to emphasise specific features or to draw out fine contrasts, in which case the composite nature of the images is more obvious. When coupled with 'ground truthing' – i.e. detailed assessments from complementary sources which corroborate the estimates of ground-cover types derived from the remotely sensed image – this can produce results which are acceptably accurate for many purposes. Not only does ground truthing provide a check on actual types of land cover, it can be used to produce more accurate calibrations of the spectral range for each type of land cover under examination, and permit an estimate of the level of accuracy of the classification of the whole image.

Sensors in a satellite may be used to detect and measure three types of energy. Two of these – the reflected solar energy from a surface (e.g. water, vegetation, a building) and energy radiated from the surface (principally thermal infrared radiation) – are measured by passive sensors, which are the most commonly used. The third – active sensors – involves an energy source on a spacecraft which is directed at the earth and the reflected energy is then measured back at the satellite (radar). Satellites are typically located in one of two positions:

- around 700 km above the earth, which is the commonest altitude for earth observing satellites, following a sun-synchronous path; and
- 'geostationary' satellites, orbiting at 35,000 km and more, which are placed in the plane of the equator and with a direction of rotation in the same direction as the rotation of the earth.

By keeping at least three geostationary satellites in position, global coverage is achieved. These have mostly been used for weather monitoring, but increasingly sophisticated equipment has allowed them to be used for a wider range of purposes, though their distance does limit the degree of resolution. Observations made from these platforms are necessarily limited to wavelengths which are readily transmitted by the atmosphere, and these are referred to as spectral windows, which can be divided into 'bands'; the most recent satellites have a greater number of (narrower) bands and can thus more effectively discriminate between types of surface (Clayton, 1994).

The major problem with remote sensing is inaccurate information about particular sites. This may arise because of the coarse level of resolution, which groups several land covers in one pixel (picture cell). It may also be because of errors in the statistical analysis of wavebands. Water vapour (sometimes as impenetrable cloud or fog) can also distort the signal, or even restrict the availability of information to periods of clear skies. Remote sensing has thus tended to be most popular in developing countries where large land masses are often sparsely settled or poorly mapped, and rapid appraisal is needed. However, as images improve, remote sensing is becoming a relatively inexpensive and more widely used facility in many situations. Even higher quality will be available when satellites currently only in military use become available for civil use.

Geographic Information Systems and Expert Systems

Such is the complexity of information when trying to make decisions on environmental resources that management information systems are now often required to integrate and interpret it. In recent years computerised systems for handling spatial data on computer have been developed, known as 'geographical information systems' (GIS). These can allow the operator to combine map and statistical data, to measure values within boundaries of various types (political or physical) and to study spatial patterns or the degree of spatial correspondence between different variables within the GIS.

The heart of a GIS is a mapping facility, into which data are input laboriously by a digitiser or, perhaps, through a scanner. Different layers of data are usually input, and they can then be combined or manipulated. The data are stored in one of two principal forms. The first is raster, in which the mapped area is divided into a grid of pixels and stored as a bitmap. This is a relatively simple and inexpensive approach to storing spatial data, but it has the disadvantage of fairly crude representation of continuous data, especially in relation to the precise positioning of linear features. This may be a serious problem, especially for superimposing different layers. A refinement on this basic approach is the quad tree, in which cells can be subdivided according to the complexity of data in a particular layer. Thus, where there is little variation in information across a surface, a fairly coarse grid may be used, but where many types of area and boundary are present information can be stored in much smaller cells. The alternative way of storing data is by vectors, which describe the precise position and trajectory of a line (either a boundary of an area, or a linear feature). Provided lines are digitised at very small intervals, this is an extremely accurate way of recording data, but its processing and storage demands are high.

A GIS is much more than an automated mapping facility, however, and its real power lies in its relational database and modelling capabilities. The former enables factual information (e.g. land quality, census data, species records) to be stored in relation to each geographical location. Thus, to each spatial entry corresponds a set of management information which can be called up by the user. The latter enables a predictive computer model to be linked to the database information, so that forecasts and simulations can be derived and their results displayed on the GIS. Thus, decision-makers can study the simulated results of alternative planning scenarios in 'real time'. Traditionally, GIS modelling has required a great degree of knowledge and skill on behalf of the operator, and use of this type of software has been limited to fairly specialised situations. Increasingly, though, user-friendly versions are being developed which allow more general use, and some GIS models are even enabling lay users to participate in future choices about development of their areas.

It has been noted that two general types of application exist for GIS. One is very focused upon specific and tightly specified objectives, supporting inventory-type questions asked by administrators (e.g. 'who owns this land parcel?') or specific analytical queries asked by investigators (e.g. 'which viewpoints are contained within the visual envelope of this feature?'). The second main application is designed to address issues of a more strategic nature, such as land-use changes flowing from a proposed policy innovation. In this type of application, policy issues cannot usually be anticipated far in advance, and output is needed quickly by policy users who wish to explore patterns and test their working hypotheses (Haines-Young *et al.*, 1994).

An example of the latter approach is the Countryside Information System (CIS) developed for the former DoE during the early 1990s, to provide accessible information about the countryside which could assist in national-scale policy development (Haines-Young *et al.*, 1994). The database for the CIS drew principally on the 1990 Countryside Survey (CS90), although it has subsequently been possible to add in other land-use and census information. Given the impracticability of obtaining land-use data for every grid square in Britain, the database included a modelling facility capable of predicting the likely composition for each square based on the ITE land classification scheme. One concern about basing a GIS on policy requirements is that these can change, leading to redundancy of the system. Consequently, it was necessary to design the CIS around three generic questions, namely:

1) What is the state of an environmental parameter in a given region?
2) Where do areas with a given environmental characteristic occur?
3) What are the ecological properties of a given landscape type or region and how will they be affected by a given set of changes or impacts?

In addition, it was necessary to educate the users of the system both to provide technical competence and to create awareness of the inherent limitations to the data and embedded models.

Going beyond even the policy-support role of a strategic GIS, is the facility of an 'expert system'. In this, a GIS with its related database is coupled with a predictive capability, typically based on ecological, physical environmental and economic models. Decision-support systems such as these are defined as computer-based information systems that combine models and data in an attempt to solve poorly structured problems with extensive user involvement. An impressive example of this was the Natural Environmental Research Council/Economic and Social Research Council Land Use Programme (NELUP), for which Newcastle University developed an expert system based on the River Tyne catchment in northeast England (O'Callaghan, 1995). The model accommodates the fact that decisions about the allocation of land between competing uses are influenced to a large extent by market forces, often short-term in nature, but that these may have long-term implications for the landscape, for the quantity and quality of the water reaching rivers and reservoirs, and for terrestrial and aquatic ecology. These elements are linked by a suite of models and an extensive database. The economic model interprets economic and social trends in order to produce changes in land-use policy and evaluate the likely economic outcome of a potential decision; the hydrological model simulates surface and groundwater flows, and changes to their quality, arising from land-use change; whilst the ecological model predicts the assemblages of plant, vertebrate and invertebrate species that result from a particular land use.

Websites

Enormous quantities of environmental information are now available through the World Wide Web (WWW) which is a segment of the Internet developed for the purpose of disseminating information to a wide audience. To explore the web, it is necessary to go through a 'browser' application designed to display web pages and provide an interface for navigating the net. Many environmental organisations now publicise their website addresses (identified by a URL – uniform resource locator), but alternatively these can usually be discovered by using an appropriate 'search engine' which permits keyword searches, via the browser, on given themes.

Such is the rate of recruitment of new websites to the Internet that any *résumé* here would be out of date long before it was published. However, four general types of website are likely to be of particular interest to environmental users: government departments and agencies, academic departments, environmental organisations of various kinds and custom-designed environmental/ecological directories and indexes. Even whole books and journals of interest are now being placed on the Internet, though these generally require payment of a subscription for access. Some discussion groups on the Internet are also dedicated to environmental issues.

Publishers on the Internet typically enter a 'homepage' which gives basic details of their organisation or service, and this may also be accompanied by a 'hot site' of current newsworthy information. Typically, the homepage will serve as a directory to more detailed sources of information on particular topics, which are accessed by clicking on highlighted words and icons. This linking of documents (through 'hotlinks') is achieved through the use of Hypertext, which enables the user to click on a highlighted word/phrase/icon and be transported instantly to the referenced document. Thus, the web page is an encoded data document which contains a set of Hypertext Markup Language (HTML) instructions. (The idea of hypertext is that, unlike conventional books with sequential page numbering, there is no physical adjacency of pages, and the only sequence is that determined by the user.) Alternatively, these more deeply embedded pages can be accessed directly if their full URL is known to the user. Some of these pages are interactive, and the user can send messages to the host. A useful summary is provided by Briggs-Erickson and Murphy (1997), whose book can (or in 1998 could) itself be found on the net at: http://www.si.umich.edu/~cbriggs/environ—murphybriggs2.html (see review by Miles, 1998). Also, at the time of writing, an excellent starting point for environmental planners is at the UK site: //www.naturenet.net/

Green surfing on the net carries with it various pitfalls, and the user must be discriminating about the information encountered. Some websites are of spurious value, viruses can be contracted and discussion groups in particular can be sources of unedited and unreliable information. However, the best sites – especially those of government departments and leading organisations – are often exceptionally high quality and up to date, and represent an essential first port of call when making inquiries. Indeed, some carry copious data which would otherwise be slow and expensive to obtain.

State of Environment Reports

Ideally, environmental information should be presented in an accessible, comprehensible format, regularly updated. As far as possible, common timescales should be used so that trends can be compared. Also, given the increased emphasis placed on lay expertise and citizen concerns, datasets should incorporate local knowledge. This ideal situation is generally far from the case, but increasing efforts are being made to produce State of Environment (SoE) reports which contain a suite of information reflecting environmental conditions in varied and reliable terms. This may be presented in book form and/or on CD-Rom.

Thus, an SoE report represents an attempt to provide accurate and comprehensive profiles of the status of environmental components at regular intervals. SoE reports analyse and interpret trends and conditions in the environment, and identify developmental pressures likely to alter environmental quality and resource availability. Typically, the benefits of SoE reporting are threefold. First, it increases public awareness about the condition of the environment, the effects of human activities on it and the implications of environmental change. Secondly, it enables decision-makers to define problems, establish priorities for action and identify yardsticks for measuring progress. Thirdly, it can promote sustainable development through the better management of environmental resources.

Although the monitoring of environmental conditions may seem a logical and self-evident activity, it is in practice very difficult to undertake, principally because of the problems of obtaining adequate information. Acceptable standards of sustainability are difficult to define, and the yardsticks against which these can be tested are still controversial. It is thus essential that agreement is reached rapidly on an appropriate set of indicator variables which can be measured at regular intervals so that changes in the status of critical environmental components can be detected. A valuable example has been set by Environment Canada (1991) in attempting to select a comprehensive set of indicators which are published not only in a one-off report, but also in regular updates. These measures had to satisfy a number of criteria, namely that they should: be scientifically valid; be supported by a sufficient run of data to show trends over time; be responsive to changes in the environment; be representative and understandable; be relevant to stated goals, objectives and issues of concern; and (ideally) have a target or threshold level above which concern is triggered. The preliminary indicators selected by Environment Canada are summarised in Box 7.8, although it is recognised that these will require further refinement and consolidation into standardised management information systems in the future. Very comprehensive sets of indicators have now been produced at central government level in Britain and by a number of local authorities. Some of the latter have made considerable progress towards 'quality of life' indicator series, which stress socioeconomic aspects as well as environmental ones, and which balance quantitative measures with qualitative ones. These indicator sets are still relatively untested in practice, and the extent to which they assist improvements in environmental quality or raise general public awareness remains to be seen.

Box 7.8   Set of SoE indicators used by Environment Canada for briefing reports

---

Principal themes and issues in Canada's National Environmental Indicator Series:

1) *Ecological Life-Support Systems*
   stratospheric ozone depletion
   climate change
   toxic contaminants in the environment – persistent organochlorines
   acid rain

2) *Human Health and Well-Being*
   urban air quality
   urban water – municipal water use and waste-water treatment

3) *Natural Resources Sustainability*
   sustaining Canada's forests – timber harvesting, forest biodiversity
   sustaining marine resources – Pacific herring fish stocks

4) *Pervasive Influencing Factors*
   Canadian passenger transportation
   energy consumption

Source: Environment Canada, 1994.

In an attempt to summarise trends in the state of the environment, the UK government consulted on the use of 13 key indicators as a means of monitoring changes to citizens' quality of life (DETR, 1998a). These included:

- populations of wild birds representative of farmland and woodland;
- emissions of greenhouse gases;

- days of air pollution;
- road traffic motor accidents;
- river quality;
- average life expectancy;
- educational qualifications;
- new homes built on 'brownfield' sites;
- waste and waste disposal;
- homes judged unfit to live in;
- economic growth;
- social investment (spending on public assets); and
- percentage of working-age population in employment.

The choice of this range of indicators reflects the strong shift away from the purely 'green' environmental agenda of the 1980s and early 1990s, to a more broadly based sustainability agenda at the close of the century.

Where sufficient data are available, the changing state of the environment and the effectiveness of policy may be evaluated more critically by means of an *environmental audit*. Audits may be of various kinds, all of which aim to facilitate evaluation of various types of environmental performance, namely:

- compliance audits, which ensure that regulations are not being breached;
- site audits, comprising spot checks of known problem areas;
- corporate audits, examining the performance of an entire business or agency, and more positively, ensuring that technical and advisory support on environmental matters is available throughout the corporation;
- issues audits, responding to specific environmental issues (such as a company's policy towards rainforests);
- associate audits, in which vetting of environmental action is extended to an organisation's contractors, agents and suppliers ('chaining'); and
- activity audits, evaluating policy in activities crossing business boundaries, notably over distribution and transport networks.

The legislature surrounding pollution control and environmental protection is very complex, and industries need to be sure that they are at least complying with this legislation, whilst many companies are also evaluating their performance in relation to a more broadly assumed environmental responsibility.

In Britain, the two principal 'green auditing' frameworks are ISO 14001 (which effectively replaced the British Standard BS7750 in 1996) and the European Eco-Management and Audit Scheme (EMAS). These methods enable organisations to address the environmental effects of their activities, taking into account regulatory requirements, the development of environmental policy statements, and management programmes and procedures. Both ISO 14001 and EMAS require an organisation to have:

- a detailed understanding of the range of environmental effects attributed to its practices and processes, either at the corporate level or with respect to one or more of its sites;

- an environmental policy stating the intentions and principles of the organisation in relation to its overall environmental performance;
- objectives and targets which define broad environmental goals and more detailed performance requirements;
- an environmental management system (EMS) setting out the necessary actions and responsibilities required to achieve the set objectives and targets;
- appropriate procedures in place to control activities; and
- internal audits of the organisation's environmental management system.

Methods for Deliberation and Participation

In relation to community participation, there are a variety of approaches which seem capable of improving people's involvement in environmental decisions. Beetham (1991) suggests that legitimation of power can be based on:

- conforming to established and agreed upon rules;
- obeying rules supported by beliefs shared by both the dominant and the subordinate parties; and
- operating through consent by those subordinate to the particular power relationship.

A situation which commonly arises is that of NIMBY, in which alienation between parties arises at an early stage. This can be represented as a failure to recognise the significance of the locally valued assets. Responding constructively to local resistance or misperceptions can be handled in two ways: through conflict mediation or consensus-building. Conflict mediation draws upon methods of industrial dispute resolution, and requires placing a skilled mediator between conflicting parties to reconcile opposed viewpoints. Consensus-building has focused on a range of situations, from major industrial facilities to parish-level wildlife areas. These typically entail positive methods of involving people in imagining future possibilities and agreeing consensual courses of action (Selman, 1996).

In part, these methods reflect a fundamental concern to attempt to rectify the failings of contemporary western democracy, where there may be a lack of mutual trust and poor rates of participation. However, at a more pragmatic level it can be used to improve the quality, and sense of community 'ownership', of decisions. The process of consultation is itself an opportunity for collective deliberation, and thus for different parties to communicate their viewpoints and to engage in a learning process. It is important during the process to ensure that relatively inarticulate or under-represented groups are 'facilitated' so that their views can be expressed effectively. A number of inclusionary and deliberative techniques are becoming fairly common (Box 7.9), although their more widespread use is

Box 7.9   Methods used for community deliberation, participation and consensus-building

Conflict mediation/resolution – a suite of methods aimed at negotiating agreement between parties with polarised views, particularly in relation to a specific development. A skilled arbitrator is able to hold meetings on various terms with stakeholders until a mutually agreed solution is forged. Techniques are similar to those used in industrial and diplomatic arbitration and reconciliation. There is growing interest in 'alternative dispute resolution' approaches which can, in some circumstances, dispense with normal 'adversarial' appeal situations.

Consensus-building – a range of approaches in which stakeholders focus on 'brainstorming' and areas of common ground, before starting to home in on solutions; useful in breaking away from 'mindsets'. A typical technique involves participants jotting ideas on adhesive labels which are progressively sorted and distilled until common themes emerge.

Round tables – a mechanism used in situations where it is appropriate to have a standing committee of stakeholders. During meetings, representatives are organised at a 'round table' where there is no 'head' and thus no power hierarchy, so that all members' views are treated with equal weight.

Focus groups – small groups of invitees selected to represent the views of particular sectional interests (e.g. Asian women, unemployed youths). The groups need to be skilfully facilitated, and the discussions carefully summarised or transcribed. A seemingly biased, unrepresentative and subjective approach, but one which is, in practice, extremely revealing about what people really think.

Citizens' juries – a jury of lay individuals who listen to key evidence on a particular issue and have the opportunity to cross-examine expert witnesses. At the end of the hearing, they reach their verdict.

Planning/sustainability for real – given the difficulty which many people have in interpreting maps or policy documents, this approach relies on visualisations and interactive displays which help people to design their neighbourhood in a very hands-on way. Much use is made of physical models, with movable pieces (to represent buildings, etc.). Although this is most widely used in relation to built development, it has been applied to community forests, national parks, etc.

Future search – an attempt to 'envision' future environments, typically on a more strategic scale than planning-for-real. It may be used to speculate on options for sustainability or renewal of whole municipalities. Essentially, it entails trained facilitators involving participants in imaginary time travel so that they leave their normal sets of assumptions and daily experiences behind. The concluding phase focuses on 'how do we get from here to there?'

(NB   These techniques are not mutually exclusive and there is in practice significant overlap between them.)

limited by cost, lack of expertise and, it must be admitted, continuing apprehensions about their representativeness or effectiveness. None of these approaches is a complete substitute for the normal processes of representative democracy, but they can be useful in helping reach informed decisions. As an example of a deliberative process, a *citizens' jury* was held to feed views into possibilities of reversion of fenland areas in eastern

England to wetlands. Over a four-day period a 16-strong jury of citizens from the Fens were able to cross-examine invited expert witnesses on proposals for enhancing the conservation and recreational interest of the area. Their report, which favoured incremental development of new areas of wetland, was used to inform a 'Wet Fens' initiative led by various countryside organisations (CSEC, 1998).

Decision-Making in Environmental Planning

Environmental planning requires anticipatory action in order to direct land-use change to the most suitable locations. In terms of rational planning theory, this presupposes high-quality information and reliable decision-making methods, supported by effective legislative powers. In practice, this is an ideal to which we merely approximate, and much planning is inevitably reactive: thus, assessments of developers' proposals are often *ex post* rather than *ex ante*. Planners constantly try to improve on this situation, however, and produce statutory plans and other strategic frameworks to ensure that development proposals are sympathetic to predefined priorities. This tension is well illustrated by methods of environmental impact assessment, which have now become well established as a method of *ex post* project appraisal, but which gradually are also starting to form the basis of a more anticipatory and strategic environmental planning regime.

Environmental Impact Assessment

Environmental impact assessment (EIA) or, as it is now more commonly called in Europe 'environmental assessment' (EA), has been defined as a systematic procedure for considering the effects of certain proposed projects on the environment prior to a decision being made by the competent authority on whether the project should be given permission to proceed (Jones *et al.*, 1998). An essential output of an EIA is a report containing its findings, normally referred to as an *environmental impact statement* (EIS) or *environmental statement* (ES).

Weston (1997), though, has alluded to the paradox of how EA, which is a rational and systematic environmental management procedure, is integrated into a sometimes irrational planning system which is in essence a political process. In this context, the 'science' of predicting environmental impacts within EA is largely secondary to the politics of the decision-making process. Moreover, EA has frequently been bureaucratic and adversarial in practice, often being applied late in the project planning cycle. This may result in its inability to prevent or modify undesirable projects because they have gained too much momentum before being scrutinised for environmental effects. Similarly, environmental safeguards may be incorporated at too late a stage, resulting in costly and potentially ineffi-cient retrofitting of abatement measures rather than optimising resource use in the initial design. EA should ideally ensure that thorough examina-tion of a proposal takes place at the earliest possible opportunity, prefera-bly to coincide with the earliest stages of project planning, so that design and performance standards can be influenced from the outset. In this con-text, EA should potentially be able to make an important contribution to the realisation of *best practicable environmental options.* Criticisms and defences of EA in practice abound, but a number of sanguine observations on UK practice can now be made on the basis of a growing body of evidence, namely:

- ES quality has tended to improve over time;
- EA has resulted in environmentally positive project modifications in most cases;
- environmental issues have been examined earlier and in greater depth as a result of EA;
- EA has led to improved communication of information to interested parties;
- there is still considerable scope for improvements in ES quality;
- EA is expensive and can result in project delay;
- guidance on screening and content needs improvement; and
- monitoring of ESs is rarely carried out.

EA is used in two main contexts: in countries with rudimentary planning systems where EA can be used to short circuit the longer-term develop-ment of a comprehensive land-use management framework, and in mature legislatures, where it is used to complement existing planning procedures. Thus, the purposes and statutes for EIA vary markedly for developed and developing countries. In advanced economies, studies tend to focus on the risk of adverse consequences on environmental quality, as befits countries with severe air and water pollution problems. In developing countries, ESs tend to have a broader concern for the balanced development of the natu-ral resource base and to provide much-needed environmental baseline information. They also often display a greater awareness of the high poten-tial costs of risks and errors, as projects are frequently of a revolutionary nature, whereas proposals in developed countries tend to be superimposed on an extensive existing industrial infrastructure, and therefore are more evolutionary in nature (Ahmad and Sammy, 1985; United Nations En-vironment Program, 1988). Whilst there are fundamental similarities in international practice, numerous differences of detail occur between indi-vidual legislatures (Wood, 1995).

The Sequence of EIA Activities

Despite the marked differences in EA procedures in different parts of the world, on close examination they tend to contain a large amount of very similar elements (Box 7.10). The first stage is generally taken to be the description of the proposed development provided by the applicant. This entails identifying aspects of the project – at both the constructional and operational stages – for which information is sought. It also requires an estimate of the resources to be used and the wastes to be created during construction, initial commissioning and subsequent operation. A description of the existing and projected environmental conditions is then required entailing collation of existing environmental data and identification and supplementation of any information gaps. The core of the assessment is a forecast of the probable impact of development, which involves predicting the magnitude of project-induced changes and gauging their importance by reference to scientific benchmarks and the responses of affected parties. Next, a check must be made for compliance with other environmental plans, policies and controls. It is also desirable, and under some legislatures mandatory, to review alternatives to the proposed development in terms of their likely environmental impacts and socioeconomic benefits. Finally, a nontechnical summary of the EIA should be prepared, using the most effective means of presentation to draw attention to the salient features of the assessment.

EIA procedures normally contain a requirement for consultation and public participation. This is a two-way process, ensuring that the public is kept fully aware of project proposals and their likely impact, and that the assessor is alerted to perceived key issues and locally important factors. It is thus better to make constructive use of the public concern aroused by proposals rather than seek to minimise its extent. The process of consultation and participation may provide factual information and analysis, assist in the early identification of key issues and potential alternatives to the project, provide expert advice (especially from statutory consultees) on the prediction of particular impacts and provide an indication of the importance which ought to be attached to individual issues in the final analysis. A fuller discussion of this is given in the section on 'deliberative and communicative methods'.

At the outset, comprehensive assessment must clearly entail collation of adequate datasets on relevant aspects of both the project and the affected environment. From the developer's point of view the first step may be to evaluate potential sites for development suitability. Although it has typically been based purely on engineering and commercial criteria, this step now more regularly involves a consideration of environmental and socioeconomic factors, enabling EIA to form part of the early design process. Thus, physical constraints would be established, such as land too steep or liable to flooding, sites with high foundation or reclamation costs, and relative costs of servicing or road and rail access. For certain types of industrial installation, hazard and safety matters would be considered, including the feasibility of achieving adequate separation from existing housing, airport safety zones and land where escapes of explosive gases might flow into dangerous pockets. Areas would also be screened for ecological,

Box 7.10   EA procedures

Preliminary procedures include:

- defining the scope of the assessment:

  –key environmental issues
  –use of standard checklists and matrices
  –preliminary consultations
  –consideration of alternatives, need and demand
  –deciding on the role of public consultation;

- preparing a project specification;
- defining requirements for environmental baseline surveys;
- establishing environmental trends; and
- considering existing plans and policies.

An *initial checklist* of matters to be considered for inclusion typically comprises:

- information describing the project:

  –purpose and physical characteristics
  –land-use requirements and physical features
  –production processes and operational features
  –alternative sites and processes;

- information describing the site and its environment:

  –physical features
  –policy framework;

- assessment of effects:

  –on humans, buildings and human-made features
  –on flora, fauna and ecology
  –on land
  –on water
  –on air and climate
  –other indirect and secondary effects;

- mitigating measures; and
- risks of accidents and hazardous development.

Baseline studies should contain the data necessary to identify and assess the main effects which the development is likely to have on the environment; whilst the *project specification* should:

- describe what is known about the requirements of the project;
- identify the range of uncertainty applying to specific aspects of the design; and
- start to spell out potential environmental impacts.

It should also supply:

- a description of the likely significant effects, direct and indirect, on the environment of the development; and
- where significant adverse effects are identified, a description of the measures envisaged in order to avoid, reduce or remedy those effects.

(*continued over*)

The process of *predicting impacts and formulating mitigating measures* involves:

- identifying potential impacts;
- describing resources and receptors;
- cause and effect;
- predicting the nature and magnitude of impacts;
- selecting methods of prediction;
- uncertainty;
- mitigating adverse effects;
- evaluating impacts; and
- hazard and risk assessment.

Evaluation usually entails considering the significance of an impact under a number of criteria:

- extent and magnitude;
- short-term and long-term impacts;
- reversibility and irreversibility;
- performance against environmental quality standards;
- sensitivity of the site; and
- compatibility with environmental policies.

The typical *content and structure* of an ES comprises:

- the physical characteristics of the proposed development, and the land-use requirements during the constructional and operational phases;
- the main characteristics of the production processes proposed, including the nature and quality of the materials to be used;
- the estimated type and quantity of expected residues and emissions (including pollutants of water, air or soil, noise, vibration, light, heat and radiation) resulting from the proposed development when in operation;
- (in outline) the main alternatives (if any) studies by the applicant, appellant or authority and an indication of the main reasons for choosing the development proposed, taking into account environmental effects;
- the likely significant direct and indirect effects on the environment of the development proposed which may result from:
  –the use of natural resources
  –the emission of pollutants, the creation of nuisances and the elimination of waste;
- the forecasting methods used to assess any effects on the environment; and
- any difficulties, such as technical deficiencies or lack of know-how encountered in compiling any item of specified information.

visual, cultural and agricultural sensitivity. Problems of data availability and confidentiality may arise, although experience suggests that these tend not to be insuperable: indeed, developers may find that acquisition of the additional data required for an EIA is a useful aid to project design and management. Information regarding the attributes of the proposed development is normally supplied in a 'project specification report' (Box 7.11).

Box 7.11   Project specification report – typical content

1) Purpose and physical characteristics of the project, including details of proposed access and transport arrangements, and of numbers and origins of employees.

2) Land-use requirements and other physical features of the project:

   a. during construction;
   b. when operational; and
   c. after use has ceased (where appropriate).

3) Production processes and operational features of the project:

   a. types and quantities of raw materials, energy and other resources consumed; and
   b. residues and emissions by type, quantity, composition and strength including discharges to water, emissions to air, noise, vibration, light heat radiation deposits/residues to land and soil.

4) Main alternative sites and processes considered, where appropriate, and reasons for final choice.

In addition to inventories of the development proposal, it is also essential to obtain particulars of the predevelopmental, or *baseline* conditions. These are likely to include a range of socioeconomic, technical and environmental factors. Sometimes data will be available from secondary sources, such as population censuses or resource surveys, and maximum use should be made of these as the time available within an EIA for original survey and experimental work is very limited. Within the EU, projects for EIA require a baseline description of the environment likely to be significantly affected, and this is expected to include: population, fauna, flora, soil, water, air, climatic factors, material assets (including architectural and archaeological heritage), recreation, landscape and the interrelationships between these factors. A summary of the considerations pertinent to the review of natural environmental baseline conditions is given in Box 7.12.

Having obtained basic data about the project and its environmental setting, assessors can then proceed to the *impact identification* stage, and identify the range of impacts likely to result from development. There are various standard techniques to assist with impact identification, most of the major ones having been devised within a few years of the passage of the US National Environmental Policy Act 1970 (NEPA), although they have since been subject to considerable refinement. It is important to distinguish between *first-order (or primary) impacts,* having a direct effect on the environment and arising from a cause immediately related to the project, and *higher-order (indirect) impacts* occurring incidentally, where the project affects the behaviour of biophysical or socioeconomic environmental systems. Indirect impacts are clearly more difficult to identify and are more likely to be missed by planners in the initial stages. A popular type of identification technique has been the presentational or mathematical

Box 7.12   Some key elements of a baseline survey

Climate and air quality:

- reference should be made to typical and unusual conditions – wind (direction and speeds) humidity (including fog) and precipitation air quality, including past trends.

Water:

- the hydrological balance (relationships between precipitation, flowing water, standing water, groundwater, etc.);
- the groundwater regime (storage, recharge, abstraction, etc.);
- drainage/channel patterns;
- sedimentation;
- flooding risk;
- water quality; and
- surface waters (volumes, flow rates, seasonal variations, ecological characteristics, water use).

Geology:

- unique/special features;
- tectonic/seismic/volcanic activity;
- mineral resources; and
- weathering/landslide/subsidence characteristics.

Soils and land capability:

- soil types in terms of erosion, slope stability, liquefaction, bearing capacity, soil structure; and
- capability of soils to sustain different land uses, especially agriculture and forestry.

Ecology:

- species checklists, including rare and indicator species;
- diversity (species and spatial);
- productivity levels and carrying capacities of different communities; and
- nutrient levels and budgets.

'Environmentally sensitive' areas such as:

- prime agricultural and forest land;
- wetlands/coastal zones/shorelines; and
- landfills (solid/toxic waste disposal sites).

Aesthetic attributes, such as:

- noise and vibration levels of existing environment; and
- visual quality of existing environment.

matrix. Although there is now less obsession with contingency classifications of developmental impacts than there was for the first two decades of EA practice, impact matrices still often occur in Environmental Statements. An archaeology of matrix-led approaches, which explains the numerous variants likely to be encountered, is summarised in Box 7.13.

Box 7.13 A historiography of 'matrix' methods of environmental impact identification

Presentational matrices – the most common and straightforward type of matrix – are grid diagrams in which two distinct lists are arranged along perpendicular axes. One axis typically presents items arising from the baseline survey and the other, aspects of the development gleaned from the project specification. Thus, the columns and rows running from the axes intersect as cells in which a mark can be placed to indicate potential impacts. For example, the generation of solid wastes would show up in a column marked 'landfill' in the project specification axis, and would have implications for the 'land use' row of the baseline axis; the cell where this row and column crossed could be marked in some way as a possible sphere of impact. Thus, there is assumed to be a potential for cause–effect relationships between opposing axes and, where one is identified as being likely to occur, its presence is marked in some manner in the cell common to both. Interactions in cells may be marked simply to show their existence (commonly by a cross), or as a brief description, or as a symbol (e.g. 'a' = direct, 'o' = indirect). In more sophisticated variants scores may be incorporated for impact magnitude (physical scale) and importance (relationship to the sensitivity of the environment). The Leopold Matrix (Leopold, 1971), one of the earliest presentational matrices, contained a possible 8,800 cell entries, each divided by a diagonal so that both magnitude and importance scores could be included. Although it was highly comprehensive with regard to information content, it was excessively laborious to operate and offered no rules for obtaining cell scores. A manual prepared for use within the British development control system (Clark *et al.*, 1981) contained a greatly simplified form of matrix, with far fewer cells, and interactions marked only with a cross. It does, however, allow for selective expansion of part of the grid for particular purposes and may be arranged in the form of several separate, sequential matrices to represent different stages of the project cycle, such as construction, operation and decommissioning. An alternative form of matrix was used by Sphere Consultants (1974) to assist in selecting oil platform construction sites on the west coast of Scotland. In this, the locational alternatives were listed along the horizontal axis, and each was scored according to its relative performance against a range of environmental and engineering criteria on the vertical axis. These scores could be combined to obtain an indication of the relative acceptability of each site, and they could be weighted to give precedence to either environmental or engineering considerations.

A particular problem for impact assessors is presented by higher-order impacts which, because they are indirect, may be overlooked. One way of approaching this is by taking advantage of the mathematical properties of matrices. The most common method is the 'component interaction matrix', which considers only the environmental setting of a project: thus, project actions are omitted, so that both columns and rows list the same baseline components. First-order impacts are identified by placing '1' in the appropriate cell, and these are used to establish higher-order relations by successive multiplication of the matrix. For instance, consider the interaction matrix (Table 7.3) for a (very much simplified!) estuarine ecosystem.

(*continued over*)

Table 7.3 Simplified interaction matrix for an imaginary estuarine ecosystem: initial matrix

	wt	DO	sal	cur	det	m-i	fish
Water temperature				1			
Dissolved oxygen	1			1			
Salinity		1		1			
Currents							
Detritus		1				1	1
Macro-invertebrates		1	1	1	1		
Fish		1	1		1		

Notes:
wt = water temperature; DO = dissolved oxygen; sal = salinity; cur = currents; det = detritus; m-i = macro-invertebrates.

To multiply this by itself, each new cell entry is obtained by summing the product of each quantity in its corresponding row and column. (For instance, the new top-left cell – reference 1,1 – will be calculated by obtaining the sum of the products of each cell in row 1 and its corresponding cell in column 1.) The rather trivial solution to the entry in the first cell in the new matrix is thus:

$$0x0 + 0x1 + 0x0 + 1x0 + 0x0 + 0x0 + 0x0 = 0$$

The new matrix obtained by this cross-multiplication is shown in Table 7.4.

Table 7.4 Simplified interaction matrix for an imaginary estuarine ecosystem: derived matrix of second-order interactions

	wt	DO	sal	cur	det	m-i	fish
wt	0	0	0	0	0	0	0
DO	0	0	0	1	0	0	0
sal	1	0	0	1	0	0	0
cur	0	0	0	0	0	0	0
det	1	2	2	2	1	1	0
m-i	1	2	0	2	0	1	1
fish	1	2	1	3	1	0	0

Notes:
As Table 7.3.

Thus, the first multiplication has produced a matrix showing second-level linkages in which, in this example, ecological components are linked by two-step paths (i.e. involving an intermediate component) in either none, one, two or three different ways. For instance, we can tell by referring back to the original matrix that fish can be related to currents via three intermediaries – dissolved oxygen, salinity and macroinvertebrates – since both have one-link chains with each of these. Successive repetition of this procedure will reveal three-link, four-link and five-link pathways respectively, at which point the exercise is usually terminated. An extension of this is the *minimum link matrix* in which the shortest number of steps connecting two components is shown: in this example, the original matrix could be combined with the second, so that both direct and two-link chains were entered (a direct link being entered in preference to a second-order interaction). A major advantage of this technique is that it can assist the fraught task of identifying indirect impacts within physical environmental systems and food chains. However, it is unable to consider project actions, only the potential interactions between baseline components; moreover its use of an ordinal scale means that there is no way of telling whether interactions are particularly significant.

Shopley *et al.* (1990) advocated the matrix multiplication procedure as a general means of tracing the nature of indirect impacts, or the 'secondary impact potential', within a particular system. They emphasised the advantages of:

- calculating measures which describe disruptive effects, showing the relative potential of components to initiate secondary impacts;
- describing 'dependency tracks', depicting how any one component is indirectly related to another;
- identifying 'critical components', whose removal from the system (for instance, obliteration by industrial development) would sever the shortest path of dependency between two indirectly linked components; and
- identifying 'cut components', whose loss would completely sever one part of a system from another.

A simpler, but popular, identification method is the *checklist,* which attempts simply to list either those features of a proposed development which could cause impacts ('project characteristic checklists'), or aspects of the environment which are potentially sensitive to disturbance ('environmental characterisic checklists'). In Britain, published guidelines appear to endorse the checklist approach (DoE, 1995a) requiring consideration of the following list of impacts and mitigation measures:

- effects on human beings, buildings and human-made features;
- effects on flora, fauna and geology;
- effects on land;
- effects on water;
- effects on air and climate;
- other indirect and secondary effects;
- measures taken to avoid, reduce or remedy impacts (such as site planning, technical measures, and aesthetic and ecological measures);
- assessment of likely effectiveness of mitigating measures; and
- risks of accidents and hazardous development.

Many official methods now employ a tailored checklist to ensure that assessments are conducted in a replicable fashion. For example, the design manual for roads and bridges (*Volume 11: Environmental Assessment*) (DoT, 1994) covers the impacts, effects, scheme appraisal, mitigation and assessment techniques of highways schemes. In this, it includes guidance on:

- vehicle exhaust emissions;
- cultural heritage (especially archaeology and monuments);

- ecology and nature conservation;
- landscape (landscape assessment and visual impact assessment);
- land use (demolition of property, loss of land used by the community, effects on developed land and effects on agricultural land);
- traffic noise;
- effects on pedestrians, cyclists, equestrians and community effects;
- vehicle travellers (including driver stress);
- water quality and drainage; and
- pollution prevention and mitigation.

Screening and Scoping

It is evident that not all development proposals will require EIA, and so a 'screen' must be operated in order to sieve out candidate projects for further analysis. The majority will fall short of the threshold necessary to trigger full-scale evaluation, whilst a few will clearly exceed them. Screening is now most commonly undertaken in relation to a government-prescribed set of thresholds and criteria which determine the applications of EIA. This is not an absolute benchmark, however, as some planning authorities have evidently had a poor awareness of how it should be applied in particular situations, whilst developers often volunteer ESs even where they are not strictly necessary.

Many early EIAs, especially in America, were excessively lengthy and exhaustive, and were written to be litigation-proof rather than to provide succinct analysis and information. Consequently, it is now usual to focus effort on those impacts which are expected to be of significance in making the final decision. This process is referred to as 'scoping' (i.e. limiting the scope of coverage) or 'significance testing'. In the USA, a review undertaken by the Council on Environmental Quality in 1978 required agencies to reduce their paperwork by making only brief reference to issues other than significant ones, and using scoping methods both to highlight important issues and to de-emphasise insignificant ones. This begs the question of how the assessor determines 'significance' without introducing bias, and Thompson (1990) identified no less than 24 methods for determining impact significance. UK guidance indicates that significance may be related to:

- impacts of international importance (e.g. loss of wetlands listed under the Ramsar convention);
- impacts of national importance (e.g. effects on a national park);
- impacts of regional/county-wide importance (e.g. having implications for 'regional planning guidance' or affecting a green belt);
- impacts of district-wide importance (those of relevance within the planning authority's boundary); and
- impacts of local importance (those affecting a limited area which are largely contained within the site itself).

Many ESs include summary sheets for those impacts which are identified as being of especial significance. These comprise succinct statements of the impact and its main attributes, for instance whether it is likely to be beneficial/adverse, long term/short term, direct/indirect, local/strategic and reversible/irreversible (Table 7.5).

Table 7.5 Example of summary sheet

Potential impact of proposed development	Classification of impact		Description of potential impact
Reduced shellfish yields resulting from reclamation of wetlands	B	= beneficial	50–65% of current shellfish yields will be lost. This will affect 30–35 local shellfishers who provide the principal or only source of income for 220–280 people. Up to 12 of the fishermen have skills which could be used in the new plant but 20–25 will have to switch to other types of fishing. This will impose considerable stress on existing inshore fishing and may result in overfishing and bankruptcies
	A	= adverse	
	St	= short term	
	Lt	= long term	
	R	= reversible	
	I	= irreversible	
	D	= direct	
	In	= indirect	
	L	= local	
	Sg	= Strategic	

Source: Based on information in UNEP, 1980.

Useful ESs are those which include realistic predictions of the scale and magnitude of the occurrence of specific impacts likely to arise from a proposed development. This creates significant problems of experimental design, as it lacks classical control situations, so that forecasts tend to be unreliable. Forecasts are required for a range of subject areas: the topics on which information is required under the British regulations are reviewed in Box 7.14. Before embarking on the forecasting stage, it is essential to clarify the data, staff, computing and laboratory facilities which will be necessary. A compromise will have to be struck between the resources and time available to the assessor and the reliability and sophistication of the methods which can be undertaken. It must be borne in mind that technical forecasts will probably not be well understood by most decision-makers and it is therefore important to display the results in a comprehensible format.

In essence, there are four key stages to impact prediction, namely:

- identifying potential impacts which may be harmful or beneficial to the environment;
- describing resources and receptors which are vulnerable to change;
- examining the chain of events or 'pathways' linking cause with effect; and
- predicting the likely nature, extent and magnitude of the anticipated changes or effects.

Box 7.14 Environmental effects for which forecasts are required under the British regulations

Effects on human beings, buildings and human-made features: change in population arising from the development, and consequential environmental effects; visual effects of the development on the surrounding area and land-scape; levels and effects of emissions from the development during normal operation; levels and effects of noise from the development; effects of the development on local roads and transport; effects of the development on buildings, the architectural and historic heritage, archaeological features and other human artifacts, e.g. through pollutants, visual intrusion, vibration.

Effects on flora, fauna and geology: loss of, and damage to, habitats and plant and animal species; loss of, and damage to, geological, palaeontological and physiographic features; other ecological consequences.

Effects on land: physical effects of the development, e.g. change in local topography, effect of earth-moving on stability, soil erosion, etc.; effects of chemical emissions and deposits on soil of site and surrounding land; land-use/resource effects (quality and quantity of agricultural land to be taken; sterilisation of mineral resources; other alternative uses of the site, including the 'do nothing' option; effect on surrounding land uses, including agriculture; waste disposal).

Effects on water: effects of development on drainage patterns in the area; changes to other hydrographic characteristics, e.g. groundwater level and flows, water courses; effects on coastal or estuarine hydrology; effects of pollutants, waste, etc., on water quality.

Effects on air and climate: level and concentrations of chemical emissions and their environmental effects; particulate matter; offensive odours; any other climatic effects.

Other indirect and secondary effects associated with the project: effects from traffic (road, rail, air, water) related to the development; effects arising from the extraction and consumption of materials, water, energy or other resources by the development; effects of other developments associated with the project, e.g. new roads, sewers, telecommunications; effects of association of the development with other existing or proposed development; secondary effects resulting from the interaction of separate direct effects listed above.

Mitigating measures: site planning measures; technical measures (e.g. process selection, recycling, pollution control, containment/bunding); aesthetic and ecological measures (e.g. mounding, colour, landscaping, site safeguard); indication of hazard preventative measures, assessment of the likely effectiveness of mitigating measures.

Source: Based on material in DoE, 1995a.

Clearly, there is a great deal of uncertainty surrounding the actual outcome of environmental impacts, no matter how careful the work of the impact assessors, and this might arise from a number of sources. These include the relative probability of an event occurring, the potential severity of that event if and when it does occur, the indirect consequences which may stem from a significant event, and the interactions which may occur between two or more events (i.e. the 'domino effect' which may be triggered by an isolated disturbance). Some key types of impact, and the aspects which need to be considered when forecasting, are summarised in Box 7.15.

Box 7.15   Key types of environmental impact

Human impacts
The main impact is likely to occur where major development is proposed in a relatively remote and lightly populated area. There may be benefits – outward migration may be stemmed, the local economy and services are sustained, more stable and better paid local jobs are created – but there may be adverse impacts, if inward migration leads to conflicts of interests between incomers and the host population, and if the development brings anxiety, loss of amenity, social upheaval and the destruction of traditional local jobs. A rapid increase in population may lead to an overloading of services, infrastructure or housing stock. The construction phase is relatively contained, but long-term effects are likely to occur at the development stage. Forecasts of these impacts are typically based on census information, but additional data on infrastructure and housing will be available from the local authority.

Major projects usually have direct impacts (e.g. local employment demand) which filter into the community and generate a range of indirect impacts (e.g. expenditure by local employees on goods and services). These impacts vary according to the life-cycle of the project, so that socioeconomic pressures may be very peaked and stressful at the construction stage, but may subside once the plant is operational. Labour demand will also vary according to the mix of skilled, semi-skilled and unskilled workers over time. Construction work is likely to make significant inroads into local unemployment levels, although many employees will commute long distances or live in temporary work camps.

Operational labour demand is often much smaller and, since it requires high skill levels, may recruit on a national market and have little effect on the local workforce; however, key workers will probably move into the area, and begin to make demands on housing and services. It is common to find that actual impacts and perceived impacts may differ: for instance a community may perceive considerable benefits from site development, whereas most of the expenditure may 'leak' outside the local economy, especially in remoter areas with little indigenous industrial base.

Noise and vibration
Principal sources of noise and vibration impacts are transport, industrial activities and, especially at the construction stage, blasting and pile-driving. Before sound levels can be calculated, the total sound power output of the plant or development should be calculated. Sound power levels should then be adjusted to allow for factors affecting the attenuation of sound (e.g. distance, wind direction, topography, humidity) and converted to dB(A). Noise contours can then be superimposed on a map showing lines of equal noise exposure. Prediction of vibration impacts is more complex, and is related to soil and geology conditions. Sometimes they are assessed by small test explosions being monitored in various sensitive locations. BS4142 on noise suggests that the 55 dB(A) L_{eq} daytime outdoor level should be the level below which noise need not normally be a material consideration in determining an application for planning permission. British Standards 6472 and 7385 respectively cover human exposure to vibration in buildings and the measurement and evaluation of vibration in buildings. Baseline surveys are expected to consider the location and nature of sensitive receptors and a range of mitigation measures (such as barriers and double glazing).

Noise nuisance is caused not only by increases in the average levels, but also by wide ranges of noise levels. Thus, residents in rural areas, accustomed to only a low level of background noise, may prove sensitive to levels which

(continued over)

would not disturb city dwellers, whose degree of acclimatisation is much greater. It is thus important to observe the amount by which new peak levels are likely to exceed background levels and pre-existing maxima. The range of noise events is often expressed in terms of the (background) level exceeded for 90 or 95% of the time (L_{90} or L_{95}) and the peak exceeded for only 10 or 5% of the time (L_{10} or L_5). Commonly, analysts first select locations likely to be sensitive to noise from a proposed development (such as hospitals and schools) and then calculate noise levels at source, correcting for factors such as cladding, surrounding walls, topography, wind direction and distance. It is also important to consider particular tonal qualities and impulsive/ intermittent noises (which can be particularly annoying), as well as ground-borne vibration resulting from blasting or pile-driving.

Traffic

Traffic impact may be associated with increased noise, vibration, CO_2 emissions, delays and congestion, especially where roads pass through or close to residential areas, schools or recreation areas. It may also be exacerbated where there are numerous junctions (because of 'stop-start' traffic patterns). The overall impact is related to both the 'whole journey' (its frequency, length, timing and mode of travel) and the effect on the local transport network at critical times. The volume and type of traffic generated by development is related to its accessibility, which in turn relates to its location, nature and size (for example, if it is not easily accessible by walking, cycling or public transport). The type of development is also influential: shops, offices and leisure facilities may attract many, but relatively evenly spaced, car journeys, whereas warehousing will be associated with freight trips, whilst single-use developments in general will tend to generate distinct peaks of traffic flow.

Land use

Principal 'greenfield' land uses likely to be affected by built development are agriculture and, to a lesser degree, forestry. The significance of land for both these uses can largely be determined from land capability assessments which are published in map form, often with an accompanying memoir; where they are not available, it is normal to commission a special survey. In addition to land capability assessments, the role of farming and forestry in the local economy should be considered and, if part of an enterprise is to be alienated by the development, the viability of the remaining portion of the unit should be evaluated.

Soil, geology and water

Estimates of impact should consider the loss or destruction of soils and any physical damage which they may experience (e.g. as a result of storage in soil heaps), and any chemical effects of land contamination. Geological impacts may include the damage of geological or geomorphological SSSIs, the alteration of groundwater properties and subsidence. Effects on water may be more diverse, and affect both its quantity and quality. The former can be expressed as hydrological impacts upon the catchment, including infiltration through soils to recharge groundwater, lateral flows through soils and surface flows to streams and drainage channels. New developments often contain large areas of impervious or semi-pervious surfaces and stormwater drains, and so can raise peak flows; increased abstraction from rivers may result in additional siltation or reduced flows. Water quality may suffer from:

- discharges from sewage works and industrial plants at identifiable point sources;
- intermittent discharges, such as those from storm flows and land runoff;

- continuous leaching from ground within which the water body is enclosed (nonpoint discharges), which may also affect groundwater;
- deposition of materials from the air;
- miscellaneous events, for example those arising from accidental spillage;
- releases from dead/decaying aquatic flora and fauna; and
- releases from construction activity.

In respect of water pollution the assessor will need to know:

- existing uses to which the water is put (domestic supply, recreation, etc.);
- existing sources of pollution including both point (i.e. industrial and municipal outfalls) and nonpoint (land runoff) sources; and
- background information on dependent habitats, current usage, plans for the water body, capacity of existing treatment plants, flow rates (including dry conditions) and quality of groundwater connected with the water body.

A normative approach is often taken in order to minimise the effects of 'mixing zones', where cumulative impacts could arise from the combination of effluents from several outfalls. Water pollution forecasting is best considered within the broader context of hydrological modelling, and many models are available to calculate flow rates in rivers over time. Water quality models are closely related to the flow field, and are based on mass conservation equations. Groundwater quality models are especially complex as they have to predict both transport through porous materials and flows along fractured rock. Hydrological impacts can arise from projects not directly associated with the manipulation or utilisation of the hydrological system. In particular, roads, urban/commercial development, industrial development, wastewater treatment works, landfill, quarrying and mining, forestry and deforestation, and intensive agriculture can lead to factors such as changes in drainage systems, groundwater flows, increased runoff velocities, changed groundwater recharge, chemical, biological, particulate and thermal pollution, changes in interception and increased soil erosion.

Air and climate
Prediction of meteorological impacts is one of the most complex aspects of an ES, as it may vary in relation to the range of different pollutants emitted, the combination of these to form secondary pollutants, the effects of local climate and topography, and the variety of scales at which the effects of air pollution may be experienced. A key consideration is whether the development complies with any guide values for various gases and particulates. Emission characteristics of the proposed development may be estimated in a number of ways, including use of data derived from similar developments elsewhere. Where mathematical models are used, these normally seek to predict dispersion of air masses from source; they are mainly useful for hazard determination during emergencies or measuring the compliance of new sources with air quality standards. Analysts will require background information on emission characteristics, ambient levels of pollution, meteorological conditions, past trends and local topography. Local conditions may sometimes be simulated by the use of wind-tunnel experiments, but these are expensive and only applicable in limited situations. Short and medium-range estimates (about 1–30 km from source) are often based on the Pasquill–Gifford equations, which yield illuminating answers although their reliability is limited by their simplified treatment of the horizontal and vertical distributions of pollutants. Refinements have been made to some models to allow for varying

(*continued over*)

meteorological conditions and topographic effects, whilst 'statistical' models have been developed to cope with the complexities of various boundary layer conditions. Prediction of long-range air pollution transport, potentially tracking the fate of pollutants hundreds or thousands of kilometres from source, may utilise models based on the composition of trajectories between sources and receptors. For instance, it is often assumed that such pollutants are contained in a hypothetical 'box', so that assumptions can be made about the behaviour of pollution concentrations in this box as it expands along its trajectory, and ultimately deposits its pollutants by wet or dry deposition.

Landscape
Direct effects on landscape resources can be estimated by considering local landscape character (i.e. landform and surface patterns), the nature and extent of landscape changes likely to take place, the status of the landscape in terms of national and regional designations, and the significance of impacts in terms of their recorded rarity, typicalness, etc. Public perceptions of landscape change are usually based on:

• views from public vantage points overlooking the development site, and the degree of visibility which exists at each location (the 'zone of visual influence');
• the visual characteristics of the proposal in relation to its surroundings (e.g. its scale, form, height, colour and materials);
• identification of the population most likely to be affected by changes in site appearance; and
• the magnitude and significance of perceived changes in landscape quality, including the degree of visual assimilation.

The evaluation of visual impact is then based on carefully selected vantage points from which the surveyor can take or create sketches, scale drawings, computer generation perspectives, colour photographs or film/video montages. Hitherto, simple but laborious analytical methods based on trigonometry were used to predict the zone of visual intrusion of a structure, for instance using a contoured map and specially constructed templates. However, visibility models which consider views from a wide range of critical points and allow comparison of alternative locations are obviously impractical to construct manually, and analysts now rely on computer packages (including those within GIS) which permit digital terrain modelling. These models are constructed from a grid of spot-height data, digitised from contour maps, from which perspective plots can be produced.

Cultural heritage
Much of the information on the cultural heritage can be found from the Sites and Monuments Record (SMR) for archaeological artifacts and on official 'lists' of historic buildings (buildings can be 'listed', at varying categories of excellence, on the grounds of architectural interest, historic interest, close historical associations or group value). Many historic landscapes are also now on official registers. Impact assessment should consider the likelihood of direct damage to such sites, as well as indirect impacts which might affect them, such as lowering of the water table. Very frequently, a developer will be required to pay for a full archaeological field survey in advance of construction works taking place.

Flora and fauna
Potential impacts to flora and fauna are usually interpreted from field survey of habitats and species, which may be available from previous records of both

official and amateur sources, but which will frequently have to be conducted afresh. Surveys often have to take place at short deadlines, and thus probably at inappropriate times of year to record key species. Thus, impact prediction is typically based on direct loss of habitat. Indirect impacts are, inevitably, far more difficult to gauge, and may rely on assumptions based on past experience, or on the use of ecological models to simulate the consequences of changing site conditions. Sources of indirect impact principally comprise:

- pollution of air, soil or water;
- microclimate modification;
- changes to river regimes;
- changes to public access and disturbance; and
- altered relationships between the balance of habitat types locally.

Ecological models pose many problems to the assessor, and most have been used for research rather than EIA purposes. Many difficulties arise, such as: the location-specific nature of most studies, whose results do not necessarily have a wider applicability; the prohibitive data requirements associated with many models; and a basic lack of scientific knowledge to model certain processes with confidence.

Source: Based mainly on material in DoE, 1995a.

In a private company, a 'financial appraisal' would be undertaken to estimate whether a project would be likely to yield a large enough profit to justify the necessary expenditure. Development of the environment clearly entails balancing financial benefits and costs, but also raises much wider issues of trading off economic gain against environmental loss. Essentially, this involves discounting future streams of benefits and seeing if their aggregate value exceeds the costs (financial, social and environmental) of the project. An argument against this approach is that economic assessments are concerned with trade, and that some environmental components – especially those which represent critical natural capital – are simply not tradable. If formal project appraisal of this nature is to be conducted, therefore, it needs to assess development proposals from the standpoint of society as a whole, and examine the total costs to society entailed in developing the project and comparing these with the benefits that the project provides to society. The principal technique in this respect is cost benefit analysis (CBA), which has been extensively applied and progressively refined. Despite its almost universal use, however, it continues to be subject to two main strands of criticism:

1) conventional CBA does not assess all items in the same manner. In particular, whereas most cost or benefit items are represented by their money value, the environmental impacts of a project are often given nonmoney descriptive evaluations which may, in turn, not be given equal weight by decision-makers; and

2) conventional CBA does not have a 'sustainability criterion', i.e. it does not have an inbuilt mechanism to ensure the preservation of environmental services between generations.

As we have noted previously, there now exists a wide range of valuation techniques to derive reasonable estimates of nonmarket goods. The sustainability problem is a relatively new one, which requires considerable further development, though basically it entails incorporating the constant natural assets rule which states that 'compensation requires passing on to future generations a stock of natural assets no smaller than the stock in the possession of current generations' (Bateman, 1994).

EIA Legislation

The term *environmental assessment* (EA) is generally taken to mean the whole process by which information about the environmental effects of a project is collected both by the developer and from other sources, and taken into account by the decision-taking authority in forming its judgement on whether the development should go ahead. This information is assembled in an *environmental statement* (ES), submitted with the application for planning permission. Presentation of the ES should be in a form which provides a focus for public scrutiny of the project and should also allow the importance of the predicted effects, and the scope for modifying or mitigating them, to be properly evaluated by the planning authority before a decision is taken.

To be systematic, the analysis should:

- examine the environmental character of the area likely to be affected by the development (through baseline studies);
- identify relevant natural and cultural processes which may already be changing the character of the area (which may alter baseline conditions by the time a development becomes operational – sometimes referred to as the 'no development' alternative – such as considering 'normal' increases in traffic on local roads);
- consider the possible interactions between the proposed development and both existing and future site conditions;
- predict the possible effects, both beneficial and adverse, of the development on the environment; and
- introduce design and operational modifications to minimise or avoid adverse effects and enhance positive effects.

The analysis may indicate ways in which the project can be modified to anticipate possible adverse effects (e.g. through identification of a better practicable environmental option) or by considering alternative processes.

Within Britain, EA practice has been developed in the context of European legislation (Directive 85/337), and has mainly been introduced through secondary legislation (i.e. regulations and circulars) which amends primary legislation (e.g. town planning, agriculture, forestry, pollution control acts) in appropriate ways. The majority of developments fall under the remit of the planning system, where EA is provided for under the Town and Country Planning (Assessment of Environmental Effects) Regulations 1988 (SI 1988/199), as amended in 1994 and subsequently. EA is provided for by separate regulation in Scotland and Northern Ireland. In addition, the Planning and Compensation Act 1991 allows the Secretaries of State to require EA for projects needing planning permission other than those listed in the directive. The Planning (Environmental Assessment) Regulations, in Schedules 1 and 2, list the types of project that require EA: for Schedule 1 projects, EA is mandatory in every case whilst, for Schedule 2 projects, EA is required if it is felt the project is likely to give rise to significant environmental impacts by virtue of its size or location. Guidance has been issued on indicative criteria and thresholds to help planning authorities reach their opinion as to whether or not EA is required for any particular Schedule 2 projects. Schedule 3 sets out the required content of the ES and Schedule 4 names the bodies who must statutorily be consulted. Parallel to these are regulations for electricity, roads and land drainage, forestry and salmonid farming in marine waters. The major requirements of the regulations may be summarised in the following terms:

- an applicant may, before applying for planning permission, request in writing the planning authority's opinion of whether the proposed project will be liable to EA (a brief plan and description of the project must accompany this request);
- if an application is submitted which in the view of the planning authority should be accompanied by an ES but is not, they should notify the applicant of this requirement within four weeks;
- an applicant may appeal to the Secretary of State (SofS) against a requirement by the planning authority to prepare an ES, and the SofS will issue a direction accordingly (the SofS may also issue a direction even where this has not been sought);
- the SofS may call in the application and receive the ES directly;
- a developer may volunteer an ES;
- an ES may be inspected in the planning office for a period of four weeks;
- as soon as it is established that an EA is to be prepared, the planning authority must notify the Schedule 4 (see above) consultees, and the developer must make a copy of the ES available to each of these free of charge;
- the planning authority should determine the application within 16 weeks (instead of the normal 8 weeks), and this period includes the submission of any supplementary information which the authority requires from the developer, but which was lacking from the original ES; and
- when the application has been determined, the planning authority must notify the applicant, the SofS and all consultees, and place a copy of the decision, including any conditions imposed, on the planning register.

Developers may wish to publish separately the nontechnical summaries provided in environmental statements, and make further publicity arrangements such as mounting an exhibition. It is also open to planning authorities to provide further publicity for particular proposals for which ESs have been supplied.

We observed that Schedules 1 and 2 stipulate those categories (or 'use classes') of development for which ESs are mandatory and those for which ESs may be required where they exceed a certain magnitude or significance. Further clarification of the interpretation of Schedule 2 is provided in accompanying circulars (i.e. 'indicative criteria and thresholds'), and these indicate that EA is required if projects are likely to have significant effects on the environment, by virtue of such factors as their nature, size or location. The basic test is thus significance, and not the amount of opposition to, or controversy surrounding, a proposal, except to the extent that the substance of opponents' arguments may expose significant environmental factors. In general terms, EA will be needed for Schedule 2 projects:

- which are of more than local importance;
- occasionally where, even though they are small scale, they are proposed for particularly sensitive or vulnerable locations; and
- occasionally where the proposal would have unusually complex and potentially adverse environmental effects, where expert and detailed analysis of those effects would be desirable and would be relevant to the issue of principle as to whether or not the development would be permitted.

Projects of 'more than local importance' include those where their sheer scale is likely to result in wide-ranging environmental effects. Examples include large mining operations, substantial new manufacturing plant and major infrastructure projects; most development applications which depart substantially from approved development plans may also require EA. Projects in 'sensitive locations' may often be identified in consultation with statutory bodies. Such locations may be expected to include national parks and other key landscape designations, Sites of Special Scientific Interest, Special Protection Areas, areas or monuments of major archaeological importance and green belts. Projects proposed for urban areas are likely to trigger an EA where the characteristics of development would be likely to have significant effects on heavy concentrations of population. However, the Secretary of State will not always deem that Schedule 2 projects affecting a sensitive area will require an EA, nor operate an automatic presumption against development within a 'designated' area. (The classes of projects covered by the two schedules are excessively lengthy to reproduce here, and may readily be found on DETR's website, http://www.detr.gov.uk. At the time of writing, the full URL of the page giving EA classes is http://www.planning.detr.gov.uk/eia/assess/.)

Subsequent to project approval and implementation, it is highly desirable that monitoring and auditing of new developments takes place, although it is generally difficult to enforce this practice and so it is widely neglected. Monitoring is concerned with the identification and measurement of impacts from development, and is a process of repetitive observation of one or more elements or indicators of the environment according to prearranged schedules in time or space. Auditing is the comparison of predicted impacts of development with those impacts which appear to have occurred; its purpose is to test the accuracy and coverage of the predictions made in the ES. Often, however, it is impossible to conduct an adequate audit because of the vague wording of predictions. Auditing of many impacts can only be undertaken when monitored data allow statistically valid interpretations of cause–effect relationships to be derived for projects with a long operational life. Ideally, an audit should cast light on the accuracy of forecasts, the performance of measures to mitigate impacts and the effectiveness of environmental management practices, and may comprise: 'baseline' monitoring of initial environmental conditions against which change can be measured; 'effects' monitoring during the construction and/or operation phases of a project, preferably including consideration of the performance of mitigation measures; 'compliance' monitoring of adherence to specific consent conditions; and 'comfort' monitoring, to ensure that unacceptable disamenities do not materialise (Secter and Wiebe, 1985).

Environmental damage from projects and policies has often arisen because environmental considerations were taken on board at too late a stage – generally as an afterthought – by which stage damage was already inherent in the proposals. One solution to this may be to subject plans, policies and programmes to environmental review at an early stage. Whilst the principle is attractive, however, the practice is very difficult because of the inherent uncertainties associated with the future. One approach is to evaluate the contents of a plan in terms of their intrinsic sustainability. For example, a common failure is for excellent 'green' proposals in one part of a plan to be negated by damaging industrial or transport measures in another. Thus, a plan may be subject to a compatibility analysis, in which a matrix of policies is compared to explore any potential contradictions. Policies may then be explored in relation to their contribution to attaining specific environmental targets.

More broadly speaking, project-based EIA by itself is inadequate as an environmental management system. In particular, options may be foreclosed because of decisions made at a higher level, so that EAs must often focus on matters of design detail and ignore controversial strategic issues. Equally, individual development consents may have cumulative impacts and so cannot be adjudicated in isolation. Thus, many people would argue for a broader system of policy review and plan evaluation. Whilst this occurs in relation to financial and efficiency/effectiveness criteria, it rarely happens for indirect environmental effects. Thus, there have been calls to introduce environmental and sustainability appraisals of entire policy and programme areas, as well as of land-use plans. Indeed, the possibility of introducing strategic environmental assessment (SEA) has been the subject of a draft European directive, which would (if adopted) require consideration of:

- the policies and plans, and their main objectives;
- environmental protection objectives and related measures established at the EU or member-state level, relevant to the proposed action;
- existing environmental problems, especially those related to protected and/or sensitive areas;
- likely significant environmental impacts of the proposed action, main alternatives and mitigation measures;
- monitoring arrangements, and an outline of projects or other measures expected as a result of implementing the action;
- an outline of difficulties encountered in compiling information; and
- a summary of the above.

The type and extent of public consultation would be decided by the member states. The main problems of SEA in practice have been those associated with the vague and uncertain nature of speculative proposals, future technology development and policy outcomes. These are formidable obstacles to the general acceptance and effective operation of SEA, but the concept is fundamentally attractive as environmental controversies are increasingly associated with justification of the need for particular developments rather than with the detail of the developments themselves.

Multicriterion Analysis

A common problem in environmental planning is the need to decide between alternative options for the use of a particular resource. One approach which has been advocated by many researchers (though relatively few practitioners) is multicriterion analysis. In this, alternative scenarios for future development or conservation are scored according to evaluative criteria; these criteria are also often weighted to reflect their relative importance. Aggregate scores can then be studied to determine the most favourable option.

Broadly speaking, there are two main approaches to multicriterion methods, namely, multiple attributes decision-making (MADM) and multiple objectives decision-making (MODM). The former is used for selecting an alternative from a relatively small, explicit list of alternatives, whilst the latter usually involves choice among a large set of choices implicitly defined by a set of constraints. Thus, MADM seeks a choice between a number of discrete alternatives, whilst MODM explores a continuous solution space, bounded by constraints, and there are many more feasible solutions. Various programming models are available to determine the mathematical optimum on the basis of a series of rational choices (Malczewski *et al.*, 1997).

In a study of rural development conflicts in northeast Scotland, van Huylenbrock and Coppens (1995), used a suite of seven criteria to evaluate five alternative scenarios, which ranged from no site development, through recreational options, to differing levels of residential development. Criteria related to:

- maintenance of plant species
- maintenance of bird species
- protection of landscape value
- suitability for informal recreation
- suitability for formal recreation
- adequacy of housing facilities for new residents, and
- adequacy of provision of social infrastructure.

Formal analysis of development options then entailed a pairwise comparison between each different scenario according to each different criterion. Local residents and experts were invited to score these pairwise comparisons on an ordinal scale. An interaction matrix was then constructed to summarise the comparisons and yield a range of summary statements.

From the basic matrix, it is possible to derive a 'conflict analysis multi-criteria framework', through four steps:

1) transformation of the differences in criterion scores between two alternatives into preference intensity on the basis of preference functions;
2) aggregation of the preference intensity scores for each pair of alternatives into two preference indicators by taking into account the relative importance of the criteria;
3) pairwise comparison of the preference indicators for each pair of alternatives; and
4) ranking the alternatives on the basis of pairwise comparisons.

This can be used to rank alternative outcomes, and to conduct a sensitivity analysis of the effects of modifying assumptions about weighting and aggregating. Equally, the analysis can be used to examine trade-offs between different issues, such as the balance between more ecologically or more economically orientated development. Although multicriterion analysis falls into the general category of rational decision-making methods, its particular interest is that it makes the multidimensional elements of environmental dilemmas and trade-offs explicit, whereas these are often conflated in techniques which reduce the 'bottom line' to a simple index or ratio.

Assessing Hazards and Risks

An aspect of project assessment which is complementary to EA is that of the analysis of industrial hazards and risks. Hazards are circumstances or events which can cause harm or damage to people and property, while risk is a measure of the likelihood that such an event may occur. Two main categories of hazard can be identified in relation to major projects. The first is largely the concern of developing countries and relates to major infrastructure projects such as irrigation, hydropower and water supply schemes, where there may be increased risk to human health through the spread of parasites and disease. The second category includes 'major' accidents or malfunction of plant, resulting in physical damage and/or the release of toxic materials.

An assessment of hazard and risk should include:

- the identification of inherent potential hazards;
- the description of unavoidable hazards;
- the quantification of levels of risk to operatives and the public;
- the submission of proposed modifications to plant or processes to reduce hazards; and
- the determination of whether or not the levels of risk are acceptable.

Identification of hazards relies upon a number of techniques including hazard and operability studies, technical audits and the application of past experience to draw up a potential 'failure list' for plant, equipment and operating regimes. For instance, failures may include rupturing of vessels, tanks and pipelines and shearing or malfunction of pipe connections and control valves resulting in spontaneous combustion or explosions of gases.

The risk of primary failures is then analysed to arrive at the statistical probability of any event occurring in a given time period. The possibility that one failure may trigger another (the 'domino effect') must also be considered. Fault tree analysis is used in this process along with historical data on failure frequencies. Description and quantification of hazards resulting from primary and secondary failures involve calculating the potential for escaping gases or liquids to mix with air/water, on the basis of flow rates, dispersal patterns and predictions of physical and chemical reactions. For most types of flammable, toxic or explosive releases, computer models have been developed which enable the operator to define the areas which might be affected by a particular type and size of release, and also to assess the severity of injury or damage likely to be caused in given locations. This information can be used to plot risk contours.

Risks of injury or damage are usually quantified separately for operatives of plant and the community at large, on the grounds that the workforce of hazardous installations are trained in emergency procedures and can be evacuated much more rapidly than the general public. The risks of injury or fatality to workers are generally related to fatal accident frequency rates calculated for various industries. The assessment of what constitutes an acceptable level of risk of injury to the general public is far more problematic and subjective. Human response to risk is a highly individual judgement which varies with age, personal circumstances and whether or not the individual feels in control of the activity giving rise to that risk. However, for the population at large any hazard which would increase the risk of injury or death significantly above that arising from natural events like flood or lightning would generally be regarded as unacceptable: against this benchmark, the level of acceptable risk would be about one in a million per year.

It is clear that there are many uncertainties in risk and hazard analysis and, indeed, these methods were developed specifically for situations in which experimental research would have been too expensive, too dangerous or have required too long a time to produce results. Uncertainty may arise in two, often interconnected, ways: namely, related to incomplete knowledge about the project or host environment, and due to the unpredictability of future events. Analysts must thus seek either to reduce the degree of uncertainty, by committing more resources to data collection and project design, or to evaluate impacts in the presence of uncertainty, through probability analysis.

Inevitably, however, there is a gap between the level of understanding of risk amongst professional analysts and the general public. With the knowledge of previous nuclear accidents, for example, most people find it difficult to accept the very low mathematical probabilities which are placed on such events. Social scientists have, therefore, argued that more humanistic methodologies now need to be integrated into risk assessments, in order to determine the trade-offs between impacts. Thus, whereas assessments typically used to focus purely on the analysis of morbidity and mortality, they now seek to incorporate the views of various interest and value groups. Risks then need to be managed so that their consequences can be mitigated, and awareness and training programmes have to be consciously implemented in affected localities. The growing public sensitivity to risks and hazards is strongly associated with the emergence of a 'risk society', which has previously been discussed.

For the purposes of practical decision-making, risk assessment may be thought of in terms of the 'precautionary principle'. Farmer (1997) notes that, if the precautionary principle is not to prevent all development, then it must be applied in a practical manner. Thus, it is necessary to distinguish between strict precaution and adaptive precaution. The former means prevention of development or pollution discharge, in cases where uncertainty is combined with high potential damage. The latter acknowledges uncertainty and, in allowing development or discharge, builds in procedures and monitoring arrangements which can assess whether adverse changes are occurring, and thus has the flexibility to respond to these. Thus, there are several reasons why uncertainty exists in dealing with the management of environmental pollution, namely:

- lack of time to collect sufficient information. Very often decisions need to be taken and it may be impossible to gather enough data to reduce sufficiently the uncertainty. This is true especially when examining longer-term impacts, e.g. chronic health problems, effects of low-level radiation or subtle changes to ecosystems;

- some environmental systems are so highly complex that our ability to predict their responses is very limited;
- some aspects of the environment are still poorly understood;
- it may be necessary to undertake experimental studies to obtain sufficient information, but this might mean manipulating the very aspects of the environment that one is concerned to protect;
- there may be significant practical problems in obtaining data – for example, one may need to deal with remote sites, deep seas or subsections of the human population that are difficult to identify; and
- the costs of obtaining sufficient information may outweigh the economic or environmental benefits associated with the study (Farmer, 1997).

Conclusion

During the last quarter of the twentieth century, there was a veritable explosion of environmental information, owing much to the advent of satellite imagery, geographic information systems and the Internet. This complemented the continued mapping and evaluation of rural resources by more traditional methods such as field survey, and the growth of statistical techniques to assist in the sampling and summary of data. Much of this information is highly technical, but considerable efforts have been made to communicate selected 'headline' indicators to wider audiences. Equally, many attempts have been made to incorporate lay knowledge into datasets and decisions.

Whilst this is allegedly something of a movement away from rationalistic styles of planning, there is nevertheless an evident need to weigh the arguments for and against development options very carefully. To assist with this, a range of environmental decision-making techniques have been refined over a substantial period. Some of these, such as environmental impact assessment, have become adopted as standard practice, whereas others, such as capacity planning, are still ill-defined and experimental. Ideally, environmental planning methods should have formal ways of incorporating public viewpoints, and not merely be treated as expert tools.

It is clearly naive to believe that improved information and decision-making techniques alone can deliver sustainability. Much more profound debates must take place at a social level in order to create the changes necessary for truly environmentally led planning to take place. However, they are an important hallmark of a sustainable society, and give a distinctive edge to the planner's expertise and professionalism. Although not perfect, correctly used they can help us navigate more surely the murky waters of the sustainability transition.

References

Ahmad, Y.J. and Sammy, G.K.(1985) *Guidelines for EIA in Developing Countries*, Hodder & Stoughton, London.

Aitken, R. (1986) *Scottish Mountain Footpaths: A Reconnaissance Review of their Condition*, Countryside Commission for Scotland, Perth.

Barker, G. (1997) *A Framework for the Future: Green Networks with Multiple Uses in and Around Cities and Towns*, English Nature, Peterborough.

Barr, C. and Fuller, R. (1997) Commentary on the rural results, in Walford, R. (ed.), *op. cit.*, pp. 44–50.

Barton, H. (1998) Econeighbourhoods: a review of projects, *Local Environment*, 3, pp. 159–178.

Bateman, I.J. (1994) Environmental and economic appraisal, in O'Riordan (ed.), *op. cit.*, 45–65.

Bateman, I.J., Diamand, E., Langford, I.H. and Jones, A. (1996) Household willingness to pay and farmers' willingness to accept compensation for establishing a recreation woodland, *Journal of Environmental Planning and Management*, 39, pp. 21–43.

Battersby, S. (1998) Climate change and Kyoto – just more hot air? *Croner: Environmental Briefing*, 22 January, p. 3.

Beatley, T. (1995) Planning and sustainability: the elements of a new (improved?) paradigm, *Journal of Planning Literature*, 9, pp. 383–95.

Beck, U. (1992) *Risk Society: Towards a New Modernity*, Sage, London.

Beck, U. (1996) *Environmental Politics in an Age of Risk*, Polity Press, Cambridge.

Beetham, D. (1991) *The Legitimation of Power*, Macmillan, London.

Bell, M. and Evans, D. (1998) The national forest and Local Agenda 21: an experiment in integrated landscape planning, *Journal of Environmental Planning and Management*, 41, pp. 237–53.

Benson, J.F. and Willis, K.G. (1992) *Valuing Informal Recreation on the Forestry Commission Estate*, Forest Bulletin no. 104, HMSO, London.

Bibby, C.J. (1987) Effects of management of commercial conifer plantations, in Good, J.E.G. (ed.) *Environmental Aspects of Plantation Forestry in Wales*, ITE/NERC, Merlewood, Cumbria, 70–5.

Bibby, J.S., Hyslop, R.E.F. and Hartnup, R. (1989) *Land Capability Classification for Forestry*, Macaulay Land Use Research Institute, Aberdeen.

Bishop, K. (1991) Community forests: implementing the concept, *The Planner*, 77, pp. 6–10.

Bishop, K. (1992) Assessing the benefits of community forests: an evaluation of the recreational use benefits of two urban fringe woodlands, *Journal of Environmental Planning and Management*, 35, pp. 63–76.

Bishop, K., Phillips, A. and Warren, L. (1995) Protected for ever? Factors shaping the future of protected areas policy, *Land Use Policy*, 12, pp.291–305.

Bishop, K., Phillips, A. and Warren, L.M. (1997) Protected areas for the future: models from the past, *Journal of Environmental Planning and Management*, 40, pp. 81–110.

Black, J.S. and Conway, E. (1996) Coastal superquarries in Scotland: planning issues for sustainable development, *Journal of Environmental Planning and Management*, 39, pp. 285–94.

Blair, A.M. (1987) Future landscapes of the urban-rural fringe, in Lockhart, D. and Ilbery, B. (eds) *The Future of British Rural Landscapes*, Geobooks, Norwich, 157–84.

Bond, A.J. and Brooks, D.J. (1997) A strategic framework to determine the best practicable environmental option (BPEO) for proposed transport schemes, *Journal of Environmental Management*, 51, pp. 305–21.

Boorman, L.A. (1987) Sand dunes, in Barnes, R.S.K. (ed.) *The Coastline*, Wiley, Chichester, 161–97.

Booth, T.H. (1985) Resource evaluation – forestry, in Basinski, J.J. and Cocks, K.D. (eds) *Environmental Planning and Management*, Proceedings of a Commonwealth Science Council Workshop held at CSIRO, Canberra, 77–88.

Briggs-Erickson, C. and Murphy, T. (1997) *Environmental Guide to the Internet* (3rd edn), Government Institute, Rockville, MD.

British Geological Survey (BGS) (1998) *Guide to Sources of Earth Science Information for Planning and Development*, BGS, Kenilworth.

Bromley, P. (1990) *Countryside Management*, Spon, London.

Bromley, P. (1994) *Countryside Recreation: A Handbook for Managers*, Spon, London.

Brotchie, J., Batty, M., Hall, P. and Newton, P. (eds) (1991) *Cities of the 21st Century: New Technologies and Spatial Systems*, Longman Cheshire, Melbourne.

Brugmann, J. (1992) *Managing Human Ecosystems: Principles for Ecological Municipal Management*, ICLEI, Toronto.

Bullock, C.H., Macmillan, D.C. and Crabtree, J.R. (1994) New perspectives on agro-forestry in lowland Britain, *Land Use Policy*, 11, pp. 222–32.

Bunce, R.G.H., Barr, C.J. and Whittaker, H.A. (1981) *Land Classes in Britain: Preliminary Descriptions for Users of the Merlewood Method of Land Classification*, Merlewood R&D Paper no. 86, ITE, Merlewood, Cumbria.

Burgess, J., Harrison, C.M. and Filius, P. (1998) Environmental communication and the cultural politics of environmental citizenship, *Environment and Planning A*, 30, pp. 1445–60.

Burley, F.W. (1988) Monitoring biological diversity for setting priorities in conservation, in Wilson, E.O. and Peter, F.M. (Eds), *Biodiversity*, National Academy Press, Washington D.C., pp. 227–30.

Burt, A. and Bradshaw, A. (1985) *Transforming our Waste Land: The Way Forward* (Environmental Advisory Unit, University of Liverpool), HMSO, London.

CAG/Land Use Consultants (1997) *Environmental Capital: A New Approach*, CAG/LUC for Countryside Commission, English Heritage, English Nature and the Environment Agency, London.

Centre for the Study of Environmental Change (CSEC) (1998) *Citizens and Wetlands*, CSEC, Lancaster University.

Chambers, R. (1994) The origins and practice of participatory rural appraisal, *World Development*, 22, pp. 953–69.

Clark, B.D., Chapman, K., Bisset, R., Wathern, P. and Barrett, M. (eds) (1981) *A Manual for the Assessment of Major Development Proposals*, HMSO, London.

Clayton, K. (1994) The threat of global warming, in O'Riordan, T. (ed.), *op. cit.*, 110–30.

Clayton, K. and O'Riordan, T. (1994) Coastal processes and management, in O'Riordan, T. (ed.), *op. cit.*, 151–64.

Climate Change Impacts Review Group (1996) *Review of Potential Effects of Climate Change in the United Kingdom*, Department of Environment, London.

Cobbing, P. and Slee, B. (1993) A contingent valuation of the Mar Lodge Estate, Cairngorm Mountains, Scotland, *Journal of Environmental Planning and Management*, 36, pp. 65–72.

Colman, D. (1994) Comparative evaluation of environmental policies: Environmentally Sensitive Areas in a policy context, in Whitby, M. (ed.), *op. cit.*, pp. 219–52.

Commission of the European Communities (1975) *Framework Directive (Waste) (75/442/EEC)*, CEC, Brussels.

Commission of the European Communities (1990) *Green Paper on the Urban Environment*, CEC, Brusssels.

Commission of the European Communities (CEC) (1992) *Towards Sustainability: A European Community Programme of Policy and Action in Relation to the Environment and Sustainable Development.* (COM (92) 23), CEC, Brussels.

Commission of the European Communities (1997a) *Agenda 2000: For a Stronger and Wider Union* (COM (97) 2000 final), Office for Official Publications of the European Communities, Luxembourg.

Commission of the European Communities (1997b) Council Directive 97/11/EC of 3rd March 1997 amending directive 85/337/EEC on the assessment of the effects of certain public and private projects on the environment, *Official Journal of the European Communities*, L73, pp. 5–15.

Commission of the European Communities (1998) *A European Community Biodiversity Strategy* (COM (1998) 42), CEC, Brussels.

Cooper, A. and Murray, R. (1992) A structured method of landscape assessment and countryside management, *Applied Geography*, 12, pp. 319–38.

Coopers & Lybrand (1993) *A Survey of English Local Authorities' Recycling Plans*, Coopers & Lybrand for DoE, London.

Cosgriff Dunn, B. and Steinemann, A. (1998) Industrial ecology for sustainable communities, *Journal of Environmental Planning and Management*, 41, pp. 661–72.

Council of Europe (1995) *European Landscape Convention*, Congress of Local and Regional Authorities of Europe, Strasbourg.

Council for the Protection of Rural England (CPRE) (1997) *Making Sense of Environmental Capacity*, CPRE, London.

Countryside Commission (1979) *Interpretive Planning*, The Commission, Cheltenham.

Countryside Commission (with the Forestry Commission) (1989) *Forests for the Community*, CC Paper 20, The Commission, Cheltenham.

Countryside Commission (1991) *Fit for the Future: Report of the National Parks Review Panel* (Chairman: R. Edwards), The Commission, Cheltenham.

Countryside Commission (1993) *Landscape Assessment Guidance*, CCP 423, The Commission, Cheltenham.

Countryside Commission (1998a) *Planning for Countryside Quality*, The Commission, Cheltenham.

Countryside Commission (1998b) *Protecting our Finest Countryside*, The Commission, Cheltenham.

Countryside Commission and English Nature (1996) *The Character of England: Landscape, Wildlife and Natural Features*, CC and EN, Cheltenham.

Countryside Commission/English Nature (undated) *North Northumberland Coastal Plain* (leaflet), CC, Cheltenham and EN, Peterborough.

Countryside Commission for Scotland (1986) *Lochshore Management*, CCS, Perth.

Countryside Council for Wales (CCW) (1995) *Guidelines for the Production of Countryside Strategies and Integrated Action Programmes*, CCW, Bangor.

Cox, G., Lowe, P. and Winter, M. (1990) *The Voluntary Principle in Conservation*, Packard, Chichester.

Crumley, C.L. (ed) (1993) *Historical Ecology: Cultural Knowledge and Changing Landscapes*, School of American Research (SAR) Press, Santa Fe, NM.

Curran, M.A. (1996) *Environmental Life Cycle Assessment*, McGraw-Hill, New York.

Dalal-Clayton, B. (1996) *Getting to Grips with Green Plans – National Level Experience in Industrial Countries*, Earthscan, London.

Dawson, D.G. (1994) *Are Habitat Corridors Conduits for Animals and Plants in a Fragmented Landscape?* English Nature Research Report 94, EN, Peterborough.

Department of Environment (1991) *Circular 16/91, Planning and Compensation Act 1991: Planning Obligations*, HMSO, London.

Department of Environment (1992a) *The Use of Planning Agreements. Research Report to the Department of Environment* (Grimley, J.R. Eve, with Vigers, Thames Polytechnic and Alsop Wilkinson), HMSO, London.

Department of Environment/Welsh Office (1992b) *Indicative Forestry Strategies, DoE Circular 29/1991, WO Circular 61/92*, HMSO, London.

Department of Environment (1992c) *Planning Policy Guidance Note 20: Coastal Planning*, HMSO, London.

Department of Environment (1992d) *Planning Policy Guidance Note 12: Development Plans and Regional Planning Guidance*, HMSO, London.

Department of Environment (1993a) *Alternative Development Patterns: New Settlements*, HMSO, London.

Department of Environment (1993b) *Making Markets Work for the Environment*, HMSO, London.

Department of Environment (1993c) *Planning Policy Guidance Note 22: Renewable Energy*, HMSO, London.

Department of Environment (1994a) *Planning Policy Guidance Note 24: Planning and Noise*, HMSO, London.

Department of Environment (1994b) *Planning Policy Guidance Note 9: Nature Conservation*, HMSO, London.

Department of Environment (1994c) *Planning Policy Guidance Note 23: Planning and Pollution Control*, HMSO, London.

Department of Environment (1994d) *Evaluation of Environmental Information for Planning Projects*, HMSO, London.

Department of Environment (1994e) *Planning Policy Guidance Note 22: Renewable Energy (Annexes)*, HMSO, London.

Department of Environment (1995a) *Guide on Preparing Environmental Statement for Planning Projects*, HMSO, London.

Department of Environment (1995b) *Planning Policy Guidance Note 2: Green Belts*, HMSO, London.

Department of Environment (1995c) *A Guide to Risk Assessment and Risk Management for Environmental Protection*, HMSO, London.

Department of the Environment (1995d) *Policy Guidelines for the Coast*, HMSO, London.

Department of Environment (1996a) *Indicators of Sustainable Development for the United Kingdom*, HMSO, London.

Department of Environment (1996b) *Planning Policy Guidance Note 14: Development on Unstable Land: Landslides and Planning*, HMSO, London.

Department of Environment (1996c) *Planning Policy Guidance Note 7: The Countryside – Environmental Quality and Economic and Social Development*, HMSO, London.

Department of the Environment (1996d) *Mineral Planning Guidance Note 1: General Considerations and the Development Plan System*, HMSO, London.

Department of Environment (1997) *Planning Policy Guidance Note 1: General Policy and Principles*, HMSO, London.

Department of Environment, Transport and the Regions (DETR) (1997) *The Application of Environmental Capacity to Land Use Planning*, DETR, London.

Department of Environment, Transport and the Regions (DETR) (1999) *A Better Quality of Life*, HMSO, London.

Department of Environment, Transport and the Regions (1998a) *Sustainability Counts*, DETR, London.

Department of Environment, Transport and the Regions (1998b) *A New Deal for Transport: Better for Everyone. The Government's White Paper on the Future of Transport*, HMSO, London.

Department of Transport (DoT) (1994) *Design Manual for Roads and Bridges, Volume 11, Environmental Assessment*, DoT, Scottish Office Industry Department, Welsh Office, London.

Dorney, R.S. (1989) *The Professional Practice of Environmental Management*, Springer, New York.

Dower, J. (1945) *National Parks in England and Wales* (Cmnd 6378), HMSO, London.

Eder, K. (1996) *The Social Construction of Nature*, Sage, London.

Edwards-Jones, E.S. (1997) The River Valleys Project: a participatory approach to integrated catchment planning and management in Scotland, *Journal of Environmental Planning and Management*, 40, pp. 125–42.

Ekins, P. (1992) *Wealth Beyond Measure: An Atlas of New Economics*, Gaia Press, London.

English Nature (1996) *A Space for Nature: Nature is Good for You*, leaflet, EN, Peterborough.

English Nature (EN) (1998) *Natural Areas* (CD-rom), EN, Peterborough.

Environment Agency (undated) *Rivers and Wetlands: Best Practice Guidelines*, EA, Bristol.

Environment Agency (1998) *Policy and Practice for the Protection of Floodplains*, EA, Bristol.

Environment Canada (1991) *The State of Canada's Environment*, EC, Ottawa.

Environment Canada (1994) *State of Environment Bulletin No. 94–6, Fall 1994, Stratospheric Ozone Depletion*, Government of Canada, Ottawa.

European Foundation for the Improvement of Living and Working Conditions (EFILWC) (1996) *What Future for Urban Environments in Europe?* Office for Official Publications of the European Communities, Luxembourg.

European Sustainable Cities and Towns Campaign (1994) *Charter of European Cities and Towns: Towards Sustainability*, ESCTC, Aalborg.

European Union Expert Group on the Urban Environment (EUEGUE) (1996) *European Sustainable Cities*, CEC, DG XI, Brussels.

Evans, J. (1984) *Silviculture of Broadleaved Woodland*, Forestry Commission Bulletin 62, HMSO, London.

Farina, A. (1998) *Principles and Methods in Landscape Ecology*, Chapman & Hall, London.

Farmer, A. (1997) *Managing Environmental Pollution*, Routledge, London.

Farming and Wildlife Advisory Group (FWAG) (1982) *Wildlife on the Farm* (pamphlet).

Feist, M. (1979) Management agreements: a valuable tool of rural planning, *Planner*, 65, pp.3–5.

Field, B.C. (1994) *Environmental Economics: An Introduction*, McGraw-Hill, New York.

Forestry Commission (1998a) *The UK Forestry Standard*, The Commission, Edinburgh.

Forestry Commission (1998b) *A New Focus for England's Woodlands*, The Commission, Edinburgh.

Forman, R. and Godron, M. (1986) *Landscape Ecology*, Wiley, New York.

French, P.W. (1997) *Coastal and Estuarine Management*, Routledge, London.

Gemmell, R.P. and Connell, R.K. (1984) Conservation and creation of wildlife habitats on industrial land in Greater Manchester, *Landscape Planning*, 11, pp. 175–86.

Gibbs, D. and Healey, M. (1997) Industrial geography and the environment, *Applied Geography*, 17, pp. 193–201.

Gibbs, D.C., Longhurst, J. and Braithwaite, C. (1998) 'Struggling with sustainability': weak and strong interpretations of sustainable development within local authority policy, *Environment and Planning A*, 30, pp. 1351–65.

Gilg, A. (1996) *Countryside Planning* (2nd edn.), Routledge, London.

Glasson, J. (1995) Socio-economic impacts: overview and economic impacts, in Morris, P. and Therivel, R. (eds), *op. cit.*, 9–28.

Goldsmith, F. (1983) Ecological effects of visitors and the restoration of damaged areas, in Warren, A. and Goldsmith, F. (eds) *Conservation in Perspective*, Wiley, London, 201–14.

Goodey, B. (1993) Urban design in central areas and beyond, in Hayward, R. and McGlynn, S. (eds) *Making Better Places: Urban Design Now*, Butterworth Architecture, London, pp. 53–8.

Grant, W. (ed.) (1985) *The Political Economy of Corporatism*, Macmillan, London.

Green, B. (1996) *Countryside Conservation* (3rd edn.) Spon, London.

Grime, J.P., Hodson, J.G. and Hunt, R. (1988) *Comparative Plant Ecology*, Unwin Hyman, London.

Grubb, M., Koch, M., Munson, A., Sullivan, F. and Thomson, K. (1993) *The Earth Summit Agreements: A Guide and Assessment*, Earthscan, London.

Guy, S. and Marvin, S. (1996) *Demand Side Management and Urban Infrastructure Provision* (ESRC GEC briefing), ESRC, Swindon.

Haines, M. and Davies, R. (1988) *Diversifying the Farm Business*, Blackwell, Oxford.

Haines-Young, R.H., Bunce, R.G.H. and Parr, T.W. (1994) Countryside Information System: an information system for environmental policy development and appraisal, *Geographical Systems*, 1, pp. 329–45.

Haines-Young, R.H., Watkins, C., Bunce, R.G.H. and Hallam, C.J. (1996) *Countryside Survey 1990: Environmental Assessment for Land Cover. Report to the Department of Environment*, DoE, London.

Halhead, A.V. (1987) *Farm Woodlands in Central Scotland. The Experience of the Central Scotland Woodlands Project*, Countryside Commission for Scotland, Perth.

Ham, C. and Hill, M. (1993) *The Policy Process in the Modern Capitalist State* (2nd edn.), Harvester Wheatsheaf, London.

Hammitt, W.E. and Cole, D.N. (1987) *Wildland Recreation: Ecology and Management*, Wiley, New York.

Handley, J., Wood, R. and Kidd, S. (1998) Defining coherence for landscape planning and management: a regional landscape strategy for North West England, *Landscape Research*, 23, pp. 133–58.

Harrison, C.M., Burgess, J. and Filius, P. (1996) Rationalizing environmental responsibilities: a comparison of lay publics in the UK and The Netherlands, *Global Environmental Change*, 6, pp. 215–34.

Harte, G. and Owen, D. (1991) Environmental disclosure in the annual reports of British companies: a research note, *Accounting, Auditing and Accountability Journal*, 4, pp. 51–61.

Haughton, G. and Hunter, C. (1994) *Sustainable Cities*, Regional Policy and Development Series no. 7, RSA, London.

Healey, P. (1997) *Collaborative Planning: Shaping Places in Fragmented Societies*, Macmillan, London.

Highlands and Islands Development Board (1987) *Cairngorm Estate Management Plan. A Report by ASH Consultants*, HIDB, Inverness.

Hill, D. (1994) *Citizens and Cities: Urban Policy in the 1990s*, Harvester Wheatsheaf, Hemel Hempstead.

HMSO (1990) *This Common Inheritance*, HMSO, London.

HMSO (1993) *Aspects of Britain: Conservation*, HMSO, London.

Hobhouse, Sir A. (Chairman) (1947) *Report of the National Parks Committee (England and Wales)* (Cmnd 6628), HMSO, London.

Hogwood, B.W. and Gunn, L.A. (1984) *Policy Analysis for the Real World*, Oxford University Press, Oxford.

Holden, R. (1989) British garden cities: the first eight years, *Landscape and Urban Planning*, 18, pp. 17–35.

House of Commons Select Committee on Agriculture (1990) *Agriculture Committee, Second Report, Land Use and Forestry, Volume I, Report and Proceedings of the Committee*, HMSO, London.

Hughes, G. (1989) Tourism in Scotland: the place of farm tourism. Paper Rural Resource Planning Conference, University of Dundee, 29 March.

Huxley, Sir J. (Chairman) (1947) *Report of the Committee on the Conservation of Nature in England and Wales* (Cmnd 7122), HMSO, London.

Ilbery, B. (1991) Farm diversification as an adjustment strategy on the urban fringe of the West Midlands, *Journal of Rural Studies*, 7, pp. 207–42.

Ilbery, B. and Bowler, I.R. (1998) From agricultural productivism to post-productivism, in Ilbery, B. (ed.) *The Geography of Rural Change*, Longman, London.

Independent Commission on International Development Issues (ICIDI) (1980) *North-South: A Programme for Survival*, Pion, London.

International Council for Local Environmental Initiatives (ICLEI) (1997) *Local Government Implementation of Climate Protection*, ICLEI, Toronto.

International Union for the Conservation of Nature and Natural Resources (IUCN) (1980) *World Conservation Strategy*, IUCN, Gland.

International Union for the Conservation of Nature and Natural Resources (1991) *Caring for the Earth*, IUCN, Gland.

International Union for the Conservation of Nature and Natural Resources (1994) *Guidelines for Protected Area Management Categories*, IUCN, Gland and Cambridge.

IPCC (1992) *Global Climate Change and the Rising Challenge of the Sea*, IPCC, Geneva.

IPCC (1996) *Second Assessment Synthesis of Scientific-Technical Information Relevant to Integrating Article 2 of the UN Framework Convention on Climate Change 1995*, WMO/ UNEP, Geneva.

Jones, C., Wood, C. and Dipper, B. (1998) Environmental assessment in the UK planning process, *Town Planning Review*, 69, pp. 315–39.

Jones, P. (1989) The developing programme of Groundwork projects, *Journal of Environmental Management*, 29, pp. 409–14.

Kidd, S. (1995) Planning for estuary resources: the Mersey Estuary Management Plan, *Journal of Environmental Planning and Management*, 38, pp. 435–42.

Kirby, K. (1986) Woodland and forest evaluation, in Usher, M. (ed.) *op. cit.*, 201–22.

Lancashire County Council (1997) *Lancashire's Green Audit 2*, LCC, Preston.

Leopold, L. (ed.) (1971) *A Procedure for Estimating Environmental Impact*, US Geological Survey, Washington, DC.

Lewis, G. and Williams, G. (1984) *Rivers and Wildlife Handbook: A Guide to Practices which Further the Conservation of Wildlife on Rivers*, Royal Society for the Protection of Birds, Sandy.

Liddle, M. (1996) *Recreation Ecology: The Ecological Impact of Outdoor Recreation and Ecotourism*, Kluwer, Amsterdam.

Lovejoy, T.E. (1997) Biodiversity: what is it? In Reaka-Kudla, M.L. *et al.* (eds), *op. cit.*, 7–14.

Lucas, R. (1985) The management of recreational visitors in wilderness areas in the United States, in Bayfield, N.G. and Barrow, G.C. (eds) *The Ecological Impacts of Outdoor Recreation on Mountain Areas in Europe and North America*, Recreation Ecology Research Group, 122–36

MacArthur, R.H. and Wilson, E.O. (1967) *The Theory of Island Biogeography*, Princeton University Press, Princeton, NJ.

Macfarlane, R. (1998) Implementing agri-environment policy: a landscape ecology perspective, *Journal of Environmental Planning and Management*, 41, pp. 576–96.

Malczewski, J., Moreno-Sanchez, R., Bojórquez-Tapia, L.A. and Ongay-Delhumeau, E. (1997) Multicriteria group decision-making model for environmental conflict analysis in the Cape Region, Mexico, *Journal of Environmental Planning and Management*, 40, pp. 349–74.

Marsden, T.K., Munton, R.J.C., Whatmore, S. and Little, J.K. (1989) Strategies for coping in capitalist agriculture: an evaluation of the responses of farm families in British agriculture, *Geoforum*, 20, pp. 201–14.

Marsden, T., Murdoch, J., Lowe, P., Munton, R. and Flynn, A. (1993) *Constructing the Countryside*, UCL Press, London.

Marsden, T.K., Whatmore, S, Munton, R.J.C. and Little, J.K. (1986) The restructuring process and economic centrality in capitalist agriculture, *Journal of Rural Studies*, 2, pp. 271–80.

Mason, R.J. (1994) The greenlining of America: managing private lands for public purposes, *Land Use Policy*, 11, pp. 208–221.

Mather, A.S. (1989) Patterns and modes of afforestation in Scotland. Paper delivered to the Rural Resource Planning Conference, University of Dundee, 29 March.

Mather, A.S. (ed.) (1993a) *Afforestation: Policies, Planning and Progress*, Belhaven, London.

Mather, A.S. (1993b) Afforestation in Britain, in Mather, A.S. (ed.), *op. cit.*, 13–33.

Meadows, D., Meadows, D., Randers, J. and Behrens III, W. (1974) *The Limits to Growth*, Pan, London.

Middleton, N. (1995) *The Global Casino*, Edward Arnold, London.

Miles, I. (1997) Review of Briggs-Erickson and Murphy, *Land Use Policy*, 15, pp. 171–3.

Morris, P. and Biggs, J. (1995) Water, in Morris, P. and Therivel, R. (eds), *op. cit.*, 161–96.

Morris, P. and Therivel, R. (eds) (1995) *Methods of Environmental Impact Assessment*, UCL Press, London.

Morris, P., Thurling, D. and Shreave, T. (1995) Terrestrial ecology, in Morris, P. and Therivel, R. (eds), *op. cit.*, 227–254.

Morrison, M.L., Marcot, B.G. and Mannon, R.W. (1992) *Wildlife – Habitat Relationships: Concepts and Applications*, University of Wisconsin Press, Madison, W.I.

Mowle, A. (1988) Integration: holy grail or sacred cow? In Selman, P.H. (ed.) *Countryside Planning in Practice: The Scottish Experience*, Stirling University Press, Stirling, 247–64.

Munton, R.J.C. (1990) Farm families in upland Britain: options, strategies and futures. Paper given to the Association of American Geographers, Toronto.

Murdoch, J. and Marsden, T. (1994) *Reconstituting Rurality: Changing Countryside in an Urban Context*, UCL Press, London.

Myerson, G. and Rydin, Y. (1996) *The Language of Environment: A New Rhetoric*, UCL Press, London.

National Parks Policies Review Panel (Chair, Edwards, R.) (1991) *Fit for the Future, Countryside Commission Paper* 334, The Commission, Cheltenham.

National Rivers Authority (NRA) (1992) *River Landscape Assessment*, NRA, Bristol.

National Rivers Authority (1993) *NRA Conservation Strategy*, NRA, Bristol.

Nature Conservancy Council (1988) *Site Management Plans for Nature Conservation – a Working Guide*, NCC, Peterborough.

Newbold, C., Honnor, J. and Buckley, K. (1989) *River Landscape Assessment*, Nature Conservancy Council, Peterborough.

Newson, M.D. (1991) Catchment control and planning: emerging patterns of definition, policy and legislation in UK water management, *Land Use Policy*, 8, pp. 9–16.

Nicholas Pearson Associates (1996) *Coastal Zone Management: Towards Best Practice*, DoE, London.

Northumberland National Park Authority (1995) *Northumberland National Park Minerals Local Plan: Consultation Draft*, NNPA, Hexham.

O'Callaghan, J.R. (1995) NELUP: an introduction, *Journal of Environmental Planning and Management*, 38, pp. 5–20.

Organisation for Economic Co-operation and Development (1995) *Urban Travel and Sustainable Development*, OECD, Paris.

O'Riordan, T. (1994a) Environmental science on the move, in O'Riordan, T. (ed.), *op. cit.*, 1–11.

O'Riordan, T. (ed.) (1994b) *Environmental Science for Environmental Management*, Longman.

O'Riordan, T. (1997) Democracy and the sustainability transition, in Lafferty, W.M. and Meadowcroft, J. (eds) *Democracy and the Environment: Problems and Prospects*, Edward Elgar, Cheltenham, 140–56.

O'Riordan, T. and Ward, R. (1997) Building trust in shoreline management: creating participatory consultation in shoreline management plans, *Land Use Policy*, 14, pp. 257–76.

Owens, P. and Owens, S. (1991) *The Environment, Resources and Conservation*, Cambridge University Press, Cambridge.

Owens, S. (1992) Energy, environment and sustainability in land-use planning, in Breheny, M. (ed.) *Sustainable Development and Urban Form. European Research in Regional Science No. 2*, Pion, London, pp. 79–105.

Parker, G. (1996) ELMs disease: stewardship, corporatism and citizenship in the English countryside, *Journal of Rural Studies*, 12, pp. 399–411.

Parker, K. (1985) The Peak District experiment, *Landscape Research*, 10, pp. 16–20.

Parry, M.L., Hossell, J.E. and Wright, L.J. (1992) Land use in the UK, in Whitby, M. (ed.) *Land Use Changes: the Causes and Consequences. ITE Symposium 27*, HMSO, London.

Peccol, E., Bird, C.A. and Brewer, T.R. (1996) GIS as a tool for assessing the influence of countryside designations and planning policies on landscape change, *Journal of Environmental Management*, 47, pp. 355–67.

Peterken, G. (1981) *Woodland Conservation and Management*, Chapman & Hall, London.

Peterken, G.F., Ausherman, D., Buchenau, M. and Forman, R.T.T. (1992) Old-growth conservation within British upland conifer plantations, *Forestry*, 65, pp. 127–44.

Phillips, A. (1998) The nature of cultural landscapes – a nature conservation perspective, *Landscape Research*, 23, pp. 21–38.

Pickering, K.T. and Owen, L.A. (1997) *An Introduction to Global Environmental Issues* (2nd edn), Routledge, London.

Potter, C. (1986) Processes of countryside change in lowland England, *Journal of Rural Studies*, 2, pp. 87–195.

Potter, C., Burnham, P., Edwards, A., Gasson, R. and Green, B. (1991) *The Diversion of Land: Conservation in a Period of Farming Contraction*, Routledge, London.

Potter, C. and Lobley, M. (1998) Landscapes and livelihoods: environmental protection and agricultural support in the wake of Agenda 2000, *Landscape Research*, 23, pp. 223–36.

Purseglove, J. (1988) *Taming the Flood*, Oxford University Press, Oxford.

Quinn, A.C.M. (1970) *Sand Dunes – Formation, Erosion and Management, with Particular Reference to Brittas Bay, Co. Wicklow*, National Institute for Physical Planning and Construction Research, Dublin.

Rackham, O. (1976) *Trees and Woodlands in the British Landscape*, Dent, London.

Ramsay, Sir J.D. (1945) *National Parks: A Scottish Survey* (Cmnd 6631), HMSO, Edinburgh.

Ramsay, Sir J.D. (1947) *National Parks and the Conservation of Nature in Scotland* (Cmnd 7235), HMSO, Edinburgh.

Ranwell, D.S. and Boar, R. (1986) *Coast Dune Management Guide*, Institute of Terrestrial Ecology, Huntingdon.

Ratcliffe, D. (ed.) (1977) *A Nature Conservation Review*, Cambridge University Press, Cambridge.

Ratcliffe, D. (1986) Selection of important areas for nature conservation in Great Britain: the NCC's approach, in Usher, M. (ed.), *op. cit.*, 135–60.

Reaka-Kudla, M.L., Wilson, D.E. and Wilson, E.O. (eds) (1997) *Biodiversity II: Understanding and Protecting our Biological Resources*, National Academy Press, Washington, DC.

Rhodes, R.A.W. and Marsh, D. (1992) Policy networks in British politics: a critique of existing approaches, in Marsh, D. and Rhodes, R. (eds) *Policy Networks in British Government*, Oxford University Press, Oxford, 1–26.

Ridd, M.K. (1965) *Area-Oriented Multiple Use Strategies, USDA Forest Service, Research Paper* INT-21, USDA, Washington, DC.

Rodwell, J.S. (ed.) (1991) *British Plant Communities, 1–2,* Cambridge University Press, Cambridge.

Rodwell, J.S. (ed.) (1993) *British Plant Communities, 3–5,* Cambridge University Press, Cambridge.

Rogerson, R. (1997) *Quality of Life in Britain,* Department of Geography, University of Strathclyde, Glasgow.

Rookwood, P. (1995) Landscape planning for biodiversity, *Landscape and Urban Planning,* 31, pp. 379–85.

Royal Commission on Environmental Pollution (RCEP) (1988) *Best Practicable Environmental Option, 12th Report (Cmd 310),* HMSO, London.

Royal Society for the Protection of Birds (RSPB) (1994) *The New Rivers and Wildlife Handbook.* RSPB, with National Rivers Authority and Royal Society for Nature Conservation. Sandy, Beds.

Royal Town Planning Institute (1995) *Environmental Assessment. Practice Advice Note 13,* RTPI, London.

Rural Strategy Advisory Group (1992) *Rural Strategy for Gloucestershire,* Gloucestershire County Council.

Rydin, Y. (1998) *Urban and Environmental Planning in the UK,* Macmillan, Basingstoke.

Sanderson, R.A., Rushton, S.P., Pickering, A.T. and Byrne J.P. (1995) A preliminary method of predicting plant species distributions using the British National Vegetation Classification, *Journal of Environmental Management,* 43, pp. 265–88.

Sandford, Lord (Chairman) (1974) *Report of the National Parks Policies Review Committee,* HMSO, London.

Scott, J.M. and Csuti, B. (1997) Gap analysis for biodiversity survey and maintenance, in Reaka-Kudla, M.L. *et al.* (eds), *op. cit.,* 321–40.

Scottish Development Department (SDD) (1990) *Indicative Forestry Strategies,* HMSO, Edinburgh.

Secter, J.P. and Wiebe, J.D. (1985) Environmental impact and monitoring: the Roberts Bank coal port experience, in Centre for Environmental Management and Planning (ed.) *Environmentally Sound Development in the Energy and Mining Industries,* CEMP, University of Aberdeen.

Selman, P. (ed.) (1988) *Countryside Planning in Practice: The Scottish Experience,* Stirling University Press, Stirling.

Selman, P. (1993) Landscape ecology and countryside planning: vision, theory and practice, *Journal of Rural Studies,* 9, pp. 1–21.

Selman, P. (1994) Systematic environmental reporting and planning: some lessons from Canada, *Journal of Environmental Planning and Management,* 37, pp. 461–76.

Selman, P. (1996) *Local Sustainability: Managing and Planning Ecologically Sound Places,* Paul Chapman, London.

Selman, P. (1997) The role of forestry in meeting planning objectives, *Land Use Policy,* 14, pp. 55–74.

Selman, P. (2000) Networks of knowledge and influence: connecting the planners and the planned, *Town Planning Review,* 71.

Shopley, J., Sowman, M. and Fuggle, R. (1990) Extending the capability of the component interaction matrix as a technique for addressing secondary impacts in environmental assessment, *Journal of Environmental Management,* 31, pp. 197–213.

Skinner, J.A., Lewis, K.A., Bardon, K.S., Tucker, P., Catt, J.A. and Chambers, B.J. (1997) An overview of the environmental impact of agriculture in the UK, *Journal of Environmental Management,* 50, pp. 111–28.

Slee, B. (1989) *Alternative Farm Enterprises,* Farming Press, Ipswich.

Smith, R.S. and Budd, R.E. (1982) *Land Use in Upland Cumbria: A Model for Farming/Forestry Strategies in the Sedbergh Area,* Technological Economics Research Unit, University of Stirling, Stirling.

Socolow, R., Andrews, C., Berkhout, F. and Thomas, V. (eds) (1994) *Industrial Ecology and Global Change*, Cambridge University Press, Cambridge.

Sowman, M.R. (1987) A procedure for assessing recreational carrying capacity of coastal resort areas, *Landscape and Urban Planning*, 14, pp. 331–44.

Sphere Environmental Consultants (1974) *Loch Carron Area: Comparative Analysis of Platform Construction Sites*, Sphere, London.

Spirn, A. (1984) *The Granite Garden*, Basic Books, New York.

Stevens, Sir R. (Chairman) (1975) *Planning Control over Mineral Workings*. HMSO, London.

Stocking, (1994) Soil erosion and land degradation, in O'Riordan T. (ed.), *op. cit.*, 223–42.

Tait, J., Lane, A. and Carr, S. (1988) *Practical Conservation: Site Assessment and Management Planning*, Hodder & Stoughton, London.

Tallis, J.H. and Yalden, D.W. (1983) *Peak District Moorland Restoration Project. Phase 2 Report. Regeneration Trials*, Peak Board Joint Planning Board, Bakewell.

Thompson, A., Hine, P.D., Poole, J.S. and Greig, J.R. (1998) *Environmental Geology in Land Use Planning: Advice for Planners and Developers*, Symonds Travers Morgan for DETR, East Grinstead.

Thompson, M.A. (1990) Determining impact significance in EIA: a review of 24 methodologies, *Journal of Environmental Management*, 31, pp. 235–50.

Thompson, S., Treweek, J.R. and Thurling, D.J. (1997) The ecological component of environmental impact assessment: a critical review of British environmental statements, *Journal of Environmental Planning and Management*, 40, pp. 157–72.

Trudgill, S. (1990) *Barriers to a Better Environment: What Stops us Solving Environmental Problems?* Belhaven, London.

Turner, R. (1994) Environmental economics and management, in O'Riordan, T. (ed.), *op. cit.*, 30–44.

UK Biodiversity Steering Group (1995) *Meeting the Rio Challenge (Volume 1)* and *Action Plans (Volume II)*, Department of the Environment, London.

UK Local Issues Advisory Group (UK LIAG) (1996) *Guidance for Local Biodiversity Action Plans: Guidance Note 1*. LGMB and UK Biodiversity Group, DETR, London.

United Nations Conference on Environment and Development (UNCED) (1992) *Agenda 21 – Action Plan for the Next Century*, UNCED, Rio de Janiero.

United Nations Environment Program (UNEP) (1980) *Guidelines for Assessing Industrial Environmental Impacts and Environmental Criteria for the Siting of Industry,* UNEP, Geneva.

United Nations Environment Program (UNEP) (1988) *Environmental Impact Assessment: Basic Procedure for Developing Countries*, UNEP, Bangkok.

University of West of England/Local Government Management Board (UWE/LGMB) (1995) *Sustainable Settlements: A Guide for Planners, Designers and Developers*, LGMB, Luton.

Usher, M. (ed.) (1986) *Wildlife Conservation and Evaluation*, Chapman & Hall, London.

van Huylenbrock, G. and Coppens, A. (1995) Multicriteria analysis of the conflicts between rural development scenarios in the Gordon District, Scotland, *Journal of Environmental Planning and Management*, 38, pp. 393–408.

Verney, Sir R. (Chairman) (1976) *The Way Ahead. Report of the Advisory Committee on Aggregates*, HMSO, London.

Wackernagel, M. (1998) The ecological footprint of Santiago de Chile, *Local Environment*, 3, pp. 7–25.

Wackernagel, M. and Rees, W. (1996) *Our Ecological Footprint: Reducing Human Impact on the Earth*, New Society Publishing, Gabriola Island, BC.

Walford, R. (ed.) (1997) *Land Use – UK: A Survey for the 21st Century*, The Geographical Association, Sheffield.

Ward, N., Marsden, T. and Munton, R. (1990) Farm landscape change: trends in upland and lowland England, *Land Use Policy*, 7, pp. 291–302.

Wathern, P. (ed.) (1988) *Environmental Impact Assessment: Theory and Practice*, Unwin Hyman, London.

West Sussex County Council (WSCC) (1996) *The Environmental Capacity of West Sussex*, The Council, Chichester.

Weston, J. (ed.) (1997) *Planning and Environmental Impact Assessment in Practice*, Addison Wesley Longman, Harlow.

Wettestad, J. (1997) Acid lessons? LRTAP implementation and effectiveness, *Land Use Policy*, 7, pp. 235–49.

Whitby, M. (ed.) (1994) *Incentives for Countryside Management: The Case of ESAs*, CAB International, Wallingford.

White, I.D., Mottershead, D.N. and Harrison, S.J. (1984) *Environmental Systems: An Introductory Study*, Allen & Unwin, London.

Wilcove, D.S., McLellan, C.H. and Dobson, A.P. (1986) Habitat fragmentation in the temperate zone, in Soulé, M.E. (ed) *Conservation Biology: The Science of Scarcity and Diversity*, Sinauer, Sunderland, MA, pp. 237–56.

Williams, A., Essex, S. and Pollard, A. (1998) Ecological and landscape effects of afforestation at a second rotation plantation: a case study of Fernworthy, Dartmoor, *Landscape Research*, 23, pp. 101–18.

Willis, K., Garrod, G.D., Benson, J.F. and Carter, M. (1996) Benefits and costs of the Wildlife Enhancement Scheme: a case study of the Pevensey Levels, *Journal of Environmental Planning and Management*, 39, pp. 387–402.

Wilson, E.O. (1997) Introduction, in Reaka-Kudla, M.L. *et al.* (eds) *op. cit.*, pp. 1–3.

Wilson, G.A. (1997) Factors influencing farmer participation in the Environmentally Sensitive Areas Scheme, *Journal of Environmental Management*, 50, pp. 67–93.

Winter, M. (1996) *Rural Politics: Policies for Agriculture and the Environment*, Routledge, London.

Wood, C.M. (1995) *Environmental Impact Assessment: A Comparative Review*, Longman, Harlow.

Wood, T.F. (1987) Methods of assessing relative risk of damage to soils and vegetation arising from winter sports development in the Scottish Highlands, *Journal of Environmental Management*, 25, pp. 253–70.

World Commission on Environment and Development (WCED) (1987) *Our Common Future*, Oxford University Press, Oxford.

World Resources Institute/International Institute for Environment and Development (1988 onwards) *World Resources*, Basic Books, New York.

World Wildlife Fund (WWF) UK and others (1983) *A Conservation and Development Programme for the UK*, Kogan Page, London.

Wu, J. and Levin, S.A. (1994) A spatial patch dynamic modeling approach to pattern and process in an annual grassland, *Ecological Monographs*, 64, pp. 447–64.

Index